Lead-Free Piezoelectric Materials

Lead-Free Piezoelectric Materials

Jing-Feng Li

WILEY-VCH

Author

Jing-Feng Li
Tsinghua University
Materials Science and Engineering
No. 30 Shuangqing Road
Haidian District
100084 Beijing
China

Cover Image:
© Coffeemill/Shutterstock

Library of Congress Card No.:
applied for

British Library Cataloguing-in-Publication Data
A catalogue record for this book is available from the British Library.

Bibliographic information published by the Deutsche Nationalbibliothek
The Deutsche Nationalbibliothek lists this publication in the Deutsche Nationalbibliografie; detailed bibliographic data are available on the Internet at <http://dnb.d-nb.de>.

Print ISBN: 978-3-527-34512-0
ePDF ISBN: 978-3-527-81707-8
ePub ISBN: 978-3-527-81705-4
oBook ISBN: 978-3-527-81704-7

Typesetting SPi Global, Chennai, India
Printing and Binding
CPI Group (UK) Ltd, Croydon CR0 4YY
Printed on acid-free paper

10 9 8 7 6 5 4 3 2 1

Contents

About the Author

Jing-Feng Li is Changjiang Scholar Distinguished Professor of the School of Materials Science and Engineering at Tsinghua University, Beijing, China, and also serves as deputy director of Tsinghua University-Toyota Research Center. Dr. Li has received master and doctor degrees in materials science and engineering both from Tohoku University (Japan) in 1988 and 1991 and a bachelor of engineering from Huazhong University of Science and Technology, China, in 1984. After temporarily working at Nihon Ceratec Co. Ltd., he was appointed to assistant professor in Tohoku University in 1992 and promoted to associate professor in 1997, then joined Tsinghua University as full professor in 2002. He also served as vice department chair (2005–2012) and vice school dean (2013–2016) of Materials Science and Engineering Department/School at Tsinghua University.

Dr. Li has conducted a wide range of research. His early work in Tohoku University focused on phase transition of zirconia ceramics, ceramic processing, and mechanical properties of silicon carbide and aluminum nitride ceramics, and ceramic microfabrication processing. From 2002 at Tsinghua, Li has been leading a research group with research interests on lead-free ferroelectric ceramics for piezoelectric and energy storage applications, piezoelectric films for micro-electromechanical system (MEMS) applications, thermoelectric materials, and microdevices for energy conversion, MEMS-based microfabrication processing. He has published >501 papers (H-index = 70) in prestigious journals and received several awards including distinguished young researcher award from the Japan Institute of Metals and the first-class science and technology award from the Chinese Ceramic Society, distinguished young scholars fund from National Natural Science Foundation of China, and Journal of the American Ceramic Society Loyalty Award.

He serves as Editor-in-Chief of Journal of Materiomics, associate Editor-in-Chief for Journal of the Chinese Ceramic Society, Reginal/Subject Editor for Journal of Materials Processing Technology (2005–2013), and editorial board member for NPG Asia Materials, Journal of Advanced Ceramics, Journal of Asian Ceramic Societies, Science China Technological Science, Energy Harvesting and Systems, Journal of Inorganic Materials, Electronic Components and Materials (Chinese), Powder Metallurgy Technology (Chinese). He has been elected to the President of Thermoelectric Materials and Applications Subsociety of Chinese Materials

Research Society and Vice President of Micro/Nano Technology Subsociety of Chinese Ceramic Society in 2018. He is a fellow of the American Ceramic Society, board member of International Thermoelectric Society, full member of International Institute for the Science of Sintering, and ferroelectric committee member of IEEE Ultrasonics, Ferroelectrics, and Frequency Control Society.

Foreword by Professor Longtu Li

Piezoelectric materials possess the functionality of interconversion between electrical energy and mechanical energy, which have been widely used in electronic information, communication, automobile, military defense, aeronautics and space technology, medical diagnostics, energy storage and harvesting, artificial intelligence, and advanced manufacturing, among other applications. The market-dominating piezoelectric ceramics are based on the $Pb(Zr,Ti)O_3$ (abbreviated as PZT) system, which is a big family with diverse chemical modifications designed for specific applications. PZT-based ceramics possess comprehensive excellent properties, but its environmental incompatibility has generated increasing concerns since the beginning of the twenty-first century. In addition, the European Union's legislation about Restriction of the use of certain Hazardous Substances (RoHS) in electrical and electronic equipment has accelerated the research and development of high-performance lead-free piezoelectric ceramics.

Piezoelectric ceramics and device applications have been one of the key research directions in the State-Key Laboratory of New Ceramics and Fine Processing, Tsinghua University. At Tsinghua University, our research groups mainly focused on lead-based piezoelectric ceramics and their applications before the twenty-first century. Since the new century I became deeply interested in lead-free piezoelectrics and have been heavily involved in the related research activities with several research groups within our state-key laboratory. Among them, Prof. Jing-Feng Li is the leader of the biggest research group working on lead-free piezoelectric ceramics since 2002 when he moved to Tsinghua University from Tohoku University. With the financial support from the National Natural Science Foundation and the Ministry of Science and Technology of China, Tsinghua University and Toyota Motor Corporation, Prof. Jing-Feng Li has been focusing on the $(K,Na)NbO_3$ lead-free piezoelectric system and made substantial progress in numerous research areas, including sintering process, characterization and manipulation of hierarchical structure, chemical modification, and piezoelectricity enhancement mechanism, etc. Because of these impressive achievements, he has received the first-class science and technology award from the Chinese Ceramic Society. Meanwhile, a start-up company was also founded by his former students quite recently to promote the mass production and industrial applications of the high-performance $(K,Na)NbO_3$-based piezoceramics.

Writing a book especially in foreign language is not easy, which requires not only extensive knowledge about the field but also hard work and perseverance. Prof. Jing-Feng Li has spent more than two years to complete this monograph about lead-free piezoelectric materials, which adds his contributions to the field of ferroelectrics and piezoelectrics. This book has systematically compiled the research progress of several promising lead-free piezoelectric systems, which could synchronize the perspectives of researchers from both academia and industry over the development of lead-free piezoelectric systems, and will undoubtedly inspire future research endeavors in both fundamental and application-oriented studies.

Owing to the increasing emphasis and requirement of green manufacturing and environmental friendly materials, study on the lead-free piezoelectric ceramics has become more popular around the world. Many significant results and findings of theoretical and experimental studies related to lead-free piezoelectric ceramics have been published. This book has captured some of the latest developments and most recent advancements in this exciting field. I hope and believe that readers will find interesting and useful information from this book.

Longtu Li
Professor, Tsinghua University
Member of Chinese Academy of Engineering

Foreword by Professor Jürgen Rödel

Piezoelectricity allows the generation of electrical charges on the surface of certain insulating materials in response to applied mechanical stress or, conversely, allows the generation of mechanical strain when subjected to an applied electric field. With such functionality, piezoelectric materials have been increasingly used nowadays in a wide range of technologies, as an intermediary between mechanical and electric energies. Typical examples include piezoelectric sensors and actuators in smartphones, piezoelectric probes in ultrasonic diagnostic machines, piezoelectric buzzers in industrial machines, and consumer electronics including toys, high-precision positioning units in scientific research equipment, ignitors for lighters, pressure sensors, and so on. $PbZrO_3$–$PbTiO_3$ solid solutions have formed a major commercial system, abbreviated as PZT. The PZT-based ceramics have dominated the market for about 70 years because of their excellent piezoelectric properties and a mature understanding of the processing technique. However, this "king" of piezoceramics has been challenged by the awareness of environmental protection, as European Union (EU) issued legislation on the restriction of hazardous substances (RoHS) in electrical and electronic equipment at the beginning of this century. RoHs and many other similar legislations throughout the world have stimulated the research searching for high-performance lead-free piezoceramics.

Significant progress has been witnessed in the past two decades. For select applications, replacements for PZT are being made available and some new materials even have advantages over PZT. For example, some of them feature higher fracture toughness, higher thermal conductivity, or simply lower density. However, transition into application demands from companies to run two production lines, one for PZT and one for the new material and customers need to be convinced. At present, there is no single system that can fully replace PZT. Nevertheless, there are competitive lead-free substitutions, such as $(K,Na)NbO_3$ (KNN), $(Bi_{0.5}Na_{0.5})TiO_3$ (BNT), and $BaTiO_3$-based (BT) systems. Prof. Jing-Feng Li has been working mainly on the KNN system, while my group is working on the BNT system. Interestingly, both systems have complementary piezoelectric properties: KNN-based ceramics possess high quasi-static or small-signal piezoelectric coefficients, whereas large piezoelectric strains can be achieved in BNT-based ceramics. For many years, we had intimate collaborations on research projects and academic exchange. I would like to mention how amazing our collaborations are, especially once we both worked as guest co-editors for a special issue about lead-free piezoelectrics in MRS Bulletin in August 2018.

I am very pleased to know that Prof. Jing-Feng Li has completed the writing of a specialized book on lead-free piezoelectric materials, which is truly a tough yet important task for our community. This comes at exactly the right time where first high-volume applications are expected in the next three years. I noticed that this book has two introductory chapters about the fundamentals of piezoelectricity and lead-free piezoelectrics. I particularly like the detailed descriptions on measurement technology, which is so important to newcomers into this field. The second chapter gives an overview of the lead-free piezoelectrics, including the background, the general classification, and the research progress of the lead-free piezoelectrics. These two chapters should be very helpful for young researchers who have just started working on research topics related to ferroelectrics and piezoelectrics. Four representative lead-free piezoelectric systems are introduced in the later chapters, which are then followed by a final chapter about lead-free piezoelectric applications. Such an arrangement is excellent. Explanations are clear, diagram and sketches are simple but insightful, and references are well-chosen and complete.

This book is not meant as a stand-alone. Scientists and engineers need to consider a basic book on ferroelectrics for all the fundamental physics and possibly a book on applications. However, this book by Jing-Feng Li is a definite must-read for graduate students and young researchers working on the research topics related to lead-free ferroelectric and piezoelectric materials. It suffices both as an introduction to the field as well as the first application-oriented book on piezoceramics. This work should also be a valuable reference for the industry who wishes to obtain a comprehensive picture of the development status of lead-free piezoceramics.

Jürgen Rödel
Professor, Department of Materials and Earth Sciences
Technical University of Darmstadt, Germany
Guest Professor at Tsinghua University and Tokyo Institute of Technology
Member of German National Academy of Science and Engineering

Preface

Ferroelectric/piezoelectric materials have been a hot research topic for a diversity of fields not only due to their wide applications but also because of their fascinating physical and chemical nature. The market has been dominated by lead zirconate titanate (PZT) ceramics over decades, due to their excellent properties and flexibility in terms of compositional modifications. Since 2006, the European Union has already adopted some well-known directives, e.g. the Waste Electrical and Electronic Equipment (WEEE) directive and the Restriction of the use of certain Hazardous Substances (RoHS) in electrical and electronic equipment directive, to protect human health and the environment by prohibiting the uses of hazardous substances in electrical and electronic devices. Therefore, lead-free piezoelectrics have been attracting more and more interest since the beginning of the twenty-first century. Encouraging progress in the development of high-performance lead-free piezoelectrics has been witnessed over the last decade. Despite a very limited scale, lead-free piezoelectric ceramics have successfully entered the market, which is estimated to be US$ 172 million in 2019. Given the significant R&D progress of lead-free piezoelectrics, a renewal of the RoHS directive to terminate the exemption of lead-containing piezoelectric materials should be approaching soon, at least for some applications that do not require a comprehensively excellent performance of high piezoelectricity and good temperature stability, equivalent to those of the PZT-based materials.

I believe that now is the perfect time for reviewing the research of the lead-free piezoelectrics, which is not only important for the young generation of researchers who are looking for a good starting point but also valuable for those who are already in the fields to think and innovate new strategies for enhancing the piezoelectric performances. Certainly, there have been many excellent review articles but most of them rather focused on some specific topics in individual material systems. There are four promising lead-free piezoelectric systems, which are $BaTiO_3$, $(K,Na)NbO_3$, $(Bi_{0.5},Na_{0.5})TiO_3$, and $BiFeO_3$. These systems share similarities but they are not cut from the same cloth. To have a comprehensive picture of the world of the lead-free piezoelectrics, we need a book that covers the fundamental knowledge about piezoelectrics and also the cutting-edge research progress about lead-free piezoelectrics. With such motivation, I wrote this book.

This book consists of seven chapters, starting with two introductory ones about the fundamentals of piezoelectricity and lead-free piezoelectrics. These two chapters should be beneficial for the readers, especially for those who wish to reinforce their basic knowledge. The first chapter will provide the fundamental background for piezoelectricity, emphasizing on the piezoelectric effects, the relationship between ferroelectrics and piezoelectrics, the meanings of the piezoelectric parameters, and common characterization techniques. In the second chapter, an overview of the lead-free piezoelectrics will be given, including the background, the general classification, and the research progress of the lead-free piezoelectrics. Next, four representative lead-free piezoelectric systems will be introduced in the following chapters. Each chapter consists of a brief introduction of the history and crystal structure as well as the characteristics of the focused lead-free piezoelectric system. Finally, this book will overview some applications based on the lead-free piezoelectrics. By comprehensively covering the aforementioned contents, I expect this book to be informative to the readers. Other lead-free piezoelectric materials such as AlN and ZnO are not involved in this book, since they are non-perovskite-structured and largely differ from the focused material systems in properties and applications.

I would like to thank my group members for their contributions to this book. Special thanks go to Dr. Yichi Zhang (postdoc), Dr. Zhen Zhou (graduated Ph.D. student), Dr. Lei Zhao (postdoc), Dr. Lisha Liu (postdoc), and Dr. Ke Wang (associate professor) for their deep involvement in the writing of Chapters 3–7, respectively. Also, I deeply thank Mr. Hao-Cheng Thong (Ph.D. student), Dr. Qian Li (assistant professor), Mr. Yi-Xuan Liu (Ph.D. student), Dr. Qing Liu (graduated Ph.D. student), Dr. Fang-Zhou Yao (graduated Ph.D. student), Mr. Hua-Lu Zhuang (Ph.D. student) and Miss Jing Gao (Ph.D. student) for their great help in editing the manuscripts.

Finally, I also want to express my thanks to my family and colleagues for their continuous encouragement and extensive support. I sincerely wish my book will be helpful to the readers who are interested in ferroelectrics and piezoelectrics.

30 April 2020 *Jing-Feng Li*
Beijing, China

1
Fundamentals of Piezoelectricity

1.1 Introduction

In 1880, Pierre Curie and Jacques Curie discovered the (direct) piezoelectric effect in quartz (SiO_2) and other single crystals, which generates an electric charge proportional to a mechanical stress. The converse piezoelectric effect, a geometric strain proportional to an applied voltage, was also soon realized. Since then, quartz has been one of the most well-known and widely used piezoelectric materials. Many decades later, polycrystalline piezoelectric ceramics (oxides) have been discovered. The first one is $BaTiO_3$ that was discovered during the World War II, which was used as dielectric materials for solid condensers at first [1]. In 1947, Roberts found that $BaTiO_3$ ceramics (polycrystals) showed good piezoelectricity, about 100 times higher than that of quartz, after they were poled under a high voltage [2]. Since then, $BaTiO_3$ ceramics have been widely applied to transducers, sensors, and filters, particularly in Japan. In 1952, Shirane et al., reported that solid solutions can be formed between $PbTiO_3$ and $PbZrO_3$ [3, 4]. One year later, ferroelectricity and antiferroelectricity were found in the solid solutions [5]. In 1954, Jaffe et al. studied the piezoelectric properties of $PbTiO_3$–$PbZrO_3$ solid solution ceramics, and found that its piezoelectric constants were twice as high as that of $BaTiO_3$, and its Curie temperature (above which the piezoelectricity disappears) was over 300 °C [6]. Now, the $PbTiO_3$–$PbZrO_3$ solid solutions, abbreviated as PZT, are the most widely used piezoelectric ceramics [7–10]. The PZT ceramics show greatly enhanced piezoelectric and dielectric properties when the Zr/Ti ratio is close to 52/48, where exists a morphotropic phase boundary (MPB) separating the rhombohedral and tetragonal regions [7]. It is generally understood that the piezoelectricity enhancement stems from the effect of phase coexistence enabled by the existence of MPB.

Despite the facts that $BaTiO_3$ is lead-free and was also discovered before PZT, the markets of piezoceramic applications have been dominated by PZT-based ceramics mainly because of its following advantages compared with $BaTiO_3$: (i) excellent and adjustable piezoelectric properties, (ii) relatively high Curie temperature, and (iii) relatively low sintering temperature. Recently, environmental protection has become a major global concern, and environmental-friendly materials and technology are one of the main tasks to be resolved in this new century. The manufacturing,

Lead-Free Piezoelectric Materials, First Edition. Jing-Feng Li.

handling, and disposal of PZT ceramics, which contain >60 wt% of lead, pose harmful influences on the workers' safety and soil environment as well as water supply. That is why many countries have incentivized the development of lead-free piezoelectric materials [11–18].

For the R&D of lead-free piezoelectric materials, it is very important to get a full understanding of piezoelectric principles and the piezoelectric mechanisms of existing piezoelectric materials, especially PZT ceramics. However, because PZT ceramics have many important applications, and in some sense, its application research has moved faster compared with the fundamental research on its piezoelectric mechanism, there are still a lot of things remaining very unclear. For example, the phase diagram of PZT around the MPB has been renewed even after half a century passed since the discovery of PZT [19–21], and rigorous descriptions still lack for unambiguous understanding of the MPB's contribution to piezoelectricity. The fundamental structure–property mechanisms revealed in lead-containing piezoelectric materials can be also operational in lead-free systems and at a minimum, should be considered as starting guidelines for the development of lead-free piezoelectrics from the aspects of composition modification, microstructure tailoring, property characterizations, device applications, etc.

1.2 Piezoelectric Effects and Related Equations

The piezoelectric effect or piezoelectricity is the generation of electric charges on the surface of certain non-conducting materials in response to applied mechanical stress, or conversely, the generation of a mechanical strain in such materials when they are subjected to an electric field, as schematically shown in Figure 1.1 [17]. The piezoelectric effect is a reversible process, so the materials exhibiting the direct

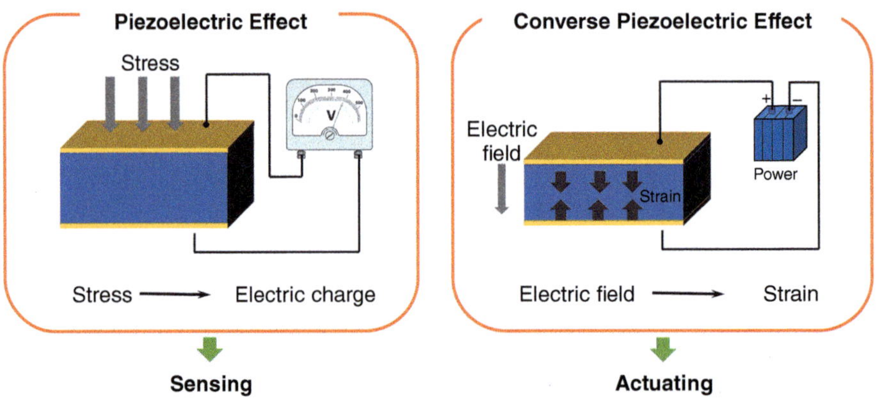

Figure 1.1 (a) The direct piezoelectric effect provides an electric charge upon application of a mechanical stress, whereas (b) the converse piezoelectric effect describes the situation where strain develops under an applied electric field. Source: Reproduced with permission from Roedel and Li [17]. Copyright 2018, Cambridge University Press.

piezoelectric effect also exhibit the converse piezoelectric effect. As such, piezoelectricity is referred to as both direct and converse effects, even though the word "piezoelectricity" often leads us to the meaning of the direct piezoelectric effect of the internal generation of electrical charges resulting from an applied mechanical force.

In a narrow sense, piezoelectricity can be understood as a linear electromechanical interaction between the mechanical and the electrical states. The constant for such a linearly proportional relation is defined as the piezoelectric coefficient d, which is a third-rank tensor coupling the first-rank tensor or vector (electric displacement or field) and the second-rank tensor (stress or strain). Hence, the piezoelectric equations may be written in the following form ($i, j, k = 1, 2, 3$) [22]

$$D_k = d_{kij}T_{ij} \tag{1.1}$$

$$S_{ij} = d_{kij}{}^{*}E_k \tag{1.2}$$

where D_i is electric displacement (C/m^2), E_i is electric field component (V/m), S_{ij} is strain component, T_{ij} is stress component (N/m^2), and d_{kij} or $d_{kij}{}^{*}$ is component of the piezoelectric charge or strain constant. It should be noted that the subscripts of piezoelectric constant are commonly expressed using the reduced Voigt matrix notation d_{km}, where k denotes the component of electric displacement D or field E in the Cartesian reference frame (x_1, x_2, x_3), and the index $m = 1, \ldots, 6$ is used to define the mechanical stress or strain. In this case, $m = 1$, 2, and 3 correspond to the normal stresses along the x_1, x_2, and x_3 axes, respectively, whereas $m = 4$, 5, and 6 stand for the shear stresses T_{23}, T_{13}, and T_{12}, respectively. Both d and d^* are called the piezoelectric constant or coefficient, but they have different units, which are pC/N and pm/V (here, p stands for 10^{-9}), respectively. It follows from thermodynamic considerations that $d_{km} = d_{km}{}^{*}$, namely, the coefficients that connect the field and strain are equal to those connecting the stress and the polarization.

In addition to the piezoelectric charge or strain constant, other forms of piezoelectric constants are also used in specialized design cases. Totally, there are four piezoelectric constants including the abovementioned piezoelectric charge or strain coefficient d, which are listed in Table 1.1 with their names and definitions [22]. These piezoelectric constants are defined as partial derivatives evaluated at constant stress (subscript T), constant electrical field (subscript E), constant electrical displacement (subscript D), or constant strain (subscript S). These conditions can be regarded as "mechanically free," "short circuit," "open circuit," and "mechanically clamped," respectively.

1.3 Ferroelectric Properties and Its Contribution to Piezoelectricity

Since most high-performance piezoelectric materials are also ferroelectric materials, it is necessary to review ferroelectric properties and their contribution to piezoelectricity [23–28]. Ferroelectricity is a character of certain materials that have a spontaneous electric polarization that can be reversed by the application of an external electric field. As illustrated in Figure 1.2, dielectrics are the big family

Table 1.1 Piezoelectric constants.

Symbol	Name	Definition
d	Piezoelectric charge coefficient or piezoelectric strain coefficient	$d_{ij} = \left(\frac{\partial D_i}{\partial T_j}\right)_E = \left(\frac{\partial S_j}{\partial E_i}\right)_T$
g	Piezoelectric voltage coefficient (voltage output constant)	$g_{ij} = -\left(\frac{\partial E_i}{\partial T_j}\right)_D = \left(\frac{\partial S_j}{\partial D_i}\right)_T$
e	Piezoelectric stress coefficient	$e_{ij} = -\left(\frac{\partial T_j}{\partial E_i}\right)_S = \left(\frac{\partial D_i}{\partial S_j}\right)_E$
h	Piezoelectric stiffness coefficient	$h_{ij} = -\left(\frac{\partial E_i}{\partial S_j}\right)_D - \left(\frac{\partial T_j}{\partial D_i}\right)_S$

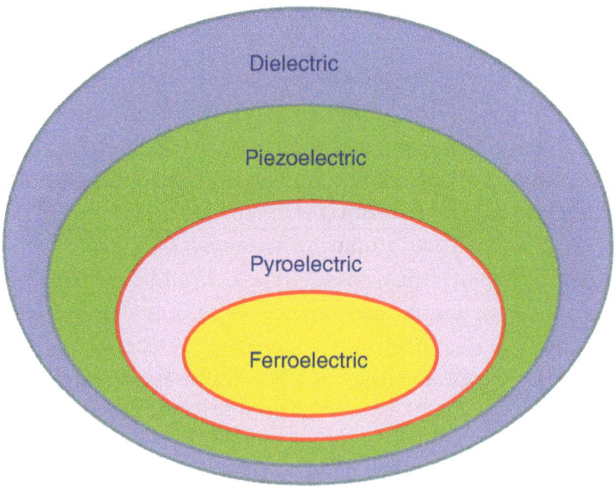

Figure 1.2 The relationship among dielectric, piezoelectric, pyroelectric, and ferroelectric materials.

with the core subset being ferroelectrics. Dielectric materials are basically electrical insulators, which become polarized by the peripheral application of electrical field when placed across the plates of a capacitor. Piezoelectric materials belong to the dielectric group, but a stress can create a net separation of positive and negative charges in a piezoelectric crystal that has a non-centrosymmetric crystal structure. Pyroelectrics are those materials with the ability to generate a temporary voltage when they are heated or cooled, since the polarization magnitude in a pyroelectric crystal can be thermally changed by the temperature change. By comparison, for a piezoelectric crystal, it is the mechanical stimuli resulting in the polarization change and as a consequence, charges build up at its surfaces. Ferroelectrics are an experimental subset of pyroelectric materials. All ferroelectric materials are pyroelectrics, and all pyroelectrics are piezoelectric; however, not all piezoelectric materials are pyroelectric and not all pyroelectrics are ferroelectric. It is known that crystal symmetry governs the aforementioned categorization. All crystalline substances belong to one of the 32 crystallographic point groups. There are 20 piezoelectric point groups and 10 ferroelectric point groups.

The direction of electric dipoles in both piezoelectric and pyroelectric (but not ferroelectric) materials cannot be changed, whereas it can be reversed by an electric field for ferroelectric materials. Therefore, the distinguishing feature of ferroelectrics is that the spontaneous polarization can be reversed by a sufficiently high applied electric field along the opposite direction. Furthermore, the polarization is dependent not only on the electric field but also on its history that the material has experienced, thereby yielding a hysteresis *P*–*E* (polarization–electric field) loop, as shown in Figure 1.3. Starting from point A, the polarization initially increases slowly with E-field, but turns to a sharp rise when the applied field is sufficiently high. Then, after a long and slow stage, the polarization reaches a saturation level (saturation polarization, P_s). The P_s is normally estimated by intersecting the polarization axis with the saturated linear part. The polarization does not go back to the starting point after the removal of E-field but instead results into non-zero values, which is defined as the remnant polarization, P_r. In order to reach a zero polarization state, an E-field applied along the opposite direction is required. This E-field is named as the coercive field, E_C, which stands for the magnitude of the applied electric field to reverse the direction of ferroelectric polarization.

The appearance of such a *P*–*E* loop is an important criterion to distinguish whether a material is ferroelectric or not. Ferroelectric materials display such a hysteretic behavior as a result of the response of electric domains to electric field, analogous to that of magnetic domains of a ferromagnetic material against a magnetic field. It should be emphasized that a polar material may be piezo-/pyro-electric but not ferroelectric if the direction of its dipoles is not switchable even under exceedingly high external electrical fields. For example, single crystalline quartz is a conventional piezoelectric material, but has no ferroelectric properties. Similarly, ZnO is a piezoelectric but non-ferroelectric material in general.

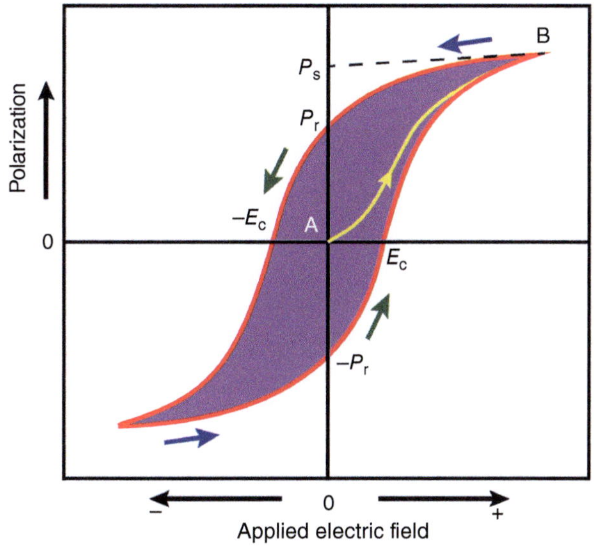

Figure 1.3 Polarization vs. electric field hysteresis loop in ferroelectric materials.

Once a ferroelectric crystal is cooled across the Curie temperature, a polarization develops. The ferroelectric phase transition is a structural phase transition, during which the displacements of ions produce lattice distortions and change the symmetry of the crystal. The magnitude of the ion displacements along certain crystallographic directions in the materials is specific to a given crystal structure and composition. If the polarization develops uniformly throughout the whole crystal, a depolarizing electric field will be produced. To minimize the electrostatic energy associated with this field, the crystal often splits into regions, called *domains*; a region in which the polarization is uniform is called a domain. The regions between two adjacent domains are called domain walls. Their thickness is typically of the order of 10–100 Å. Domain represents a region within a ferroelectric material in which the direction of polarization is uniform. The saturation polarization, P_s, corresponds to the total polarization at an extreme state where (almost) all domains are aligned along the direction of applied electric field. Some of these domains stay at the same direction even after the removal of electric field, resulting in the remnant polarization. It can be readily envisaged that a ferroelectric material at a state with remnant polarization can be used a piezoelectric material, since it can generate electric charges when subjected to mechanical stress. In other words, if a ferroelectric material, at least polycrystalline bulk materials should show no piezoelectric response if it has not been subjected to an electric field. This is because the charges will be canceled collectively if the domains are randomly distributed along different directions, resulting in zero change when the whole material receives mechanical deformation. As such, piezoelectricity can be regarded as one of the functionalities of ferroelectric materials, and in general, ferroelectric materials need to be poled before they can be used as piezoelectric materials. Therefore, electrical poling is an indispensable process for ferroelectric piezoelectric materials. During poling, a strong electric field is applied across ferroelectric materials and consequently, a majority of the domains switch their pristine polarization and become aligned along the electric field direction. Figure 1.4 schematically shows the poling process. The virgin materials are subjected to an electric field, which should be sufficiently higher

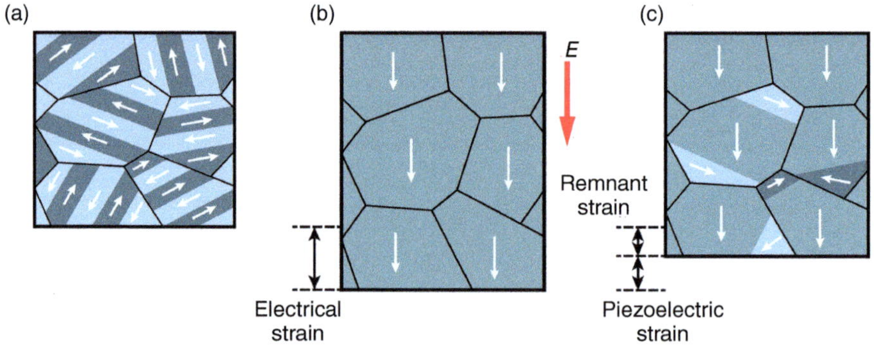

Figure 1.4 The schematic illustrations showing the alignment of ferroelectric domain and macroscopic strains when a ferroelectric material is subjected to a poling treatment under an electric field. (a) Virgin state. (b) Saturation state. (c) Remnant state.

than the coercive field (E_C), so that the domains can be re-orientated almost along the same direction. As shown in Figure 1.4b, the poling process is accompanied with an expansion of the poled materials or tensile strains, which is basically consistent with the converse piezoelectric phenomenon. As shown in Figure 1.4c, although most domains are kept along the poling direction, part of them revert back or change their orientations after the removal of the poling electrical field in order to reduce the mechanical strains. After the poling treatment, the material possesses a macroscopic polarization, which is equal to the remnant polarization (P_r) in the P–E loop shown earlier. Therefore, the poling process is very important for piezoelectric materials. Even for the same materials, if not completely poled, the resultant piezoelectric properties, especially piezoelectric charge coefficient (d_{33}), will be very low. Also, it is clear that the poling process is not applicable to non-ferroelectric materials. That is why high-performance piezoelectric materials must be ferroelectric in the first place.

1.4 Piezoelectric Parameters

1.4.1 Piezoelectric Constants

1.4.1.1 Piezoelectric Charge (Strain) Constant

The piezoelectric charge coefficient relates the electric charge generated per unit area with an applied mechanical force and is expressed in the unit of Coulomb/Newton (C/N) [7, 22]. This constant is most frequently used to evaluate the goodness of a piezoelectric material.

$$d = \frac{\text{Strain developed}}{\text{Applied field}} = \frac{\text{Charge density (open circuit)}}{\text{Applied stress}} \tag{1.3}$$

The d constant is associated with three important materials properties through the following the equation:

$$d = k\sqrt{\varepsilon_0 k^{\mathrm{T}} s^{\mathrm{E}}} \quad (\mathrm{C/N}) \tag{1.4}$$

where k is electro-mechanical coupling coefficient, k^{T} denotes relative dielectric constant at a constant stress, and s^{E} is elastic compliance (10 m/N) at a constant electrical field.

There are two important d constants:

$$d_{31} = k_{31}\sqrt{\varepsilon_0 k_3^{\mathrm{T}} s_{11}^{\mathrm{E}}} \quad (\mathrm{C/N}) \tag{1.5}$$

$$d_{33} = k_{33}\sqrt{\varepsilon_0 k_3^{\mathrm{T}} s_{33}^{\mathrm{E}}} \quad (\mathrm{C/N}) \tag{1.6}$$

It is useful to remember that large d constants relate to large mechanical displacements, which are usually sought in motional transducer devices. Conversely, the coefficient may be viewed as relating the charge collected on the electrodes, to the applied mechanical stress. d_{33} applies when the force is along the three direction (parallel with the polarization axis) and is impressed on the same surface from which the charge is collected. d_{31} applies when the charge is collected from the same surface as with d_{33}, but the force is applied at right angles to the polarization axis. It is commonly known that they have the following empirical relation.

$$d_{33} \approx -2.5 \cdot d_{31} \tag{1.7}$$

1.4.1.2 Piezoelectric Voltage Coefficient (G-constant)

The piezoelectric voltage coefficient is also called voltage output constant, which is defined as the ratio of the electric field produced to the mechanical stress applied and is expressed in the unit of voltage-meter/Newton (Vm/N).

$$g = \frac{\text{Strain developed}}{\text{Applied charge density}} = \frac{\text{Field developed}}{\text{Applied mechanical stress}} \tag{1.8}$$

The g-constants are calculated from the piezoelectric charge (strain) constant (d) and relative permittivity (ε) from the equation:

$$g = \frac{d}{\varepsilon}\ (\text{Vm/N}) \tag{1.9}$$

Depending on the type of relative directions, the g constant can be categorized as g_{33}, g_{31}, or g_{15}, corresponding to d_{33}, d_{31}, or d_{15}, respectively.

1.4.2 Piezoelectric Coupling Coefficient

The piezoelectric coupling coefficient (sometimes referred as the electromechanical coupling coefficient) is defined as the ratio of the mechanical energy accumulated in response to an electrical input or vice versa. It also corresponds to the fraction of electrical energy that can be converted into mechanical energy and vice versa. Thus, the piezoelectric coupling coefficient can be expressed by the following equation:

$$x = \sqrt{\frac{\text{Mechanical energy stored}}{\text{Electrical energy applied}}} = \sqrt{\frac{\text{Electrical energy stored}}{\text{Mechanical energy applied}}} \tag{1.10}$$

The coupling factor can be calculated based on the measured resonance and anti-resonance frequencies of a piezoelectric element, depending on the vibration mode at which the element is excited. The most used coupling factors are k_p and k_t for the vibration along the radial and thickness directions in a circle-shaped disk, respectively. In general, a useful parameter k_eff is frequently used to express the effective coupling coefficient of an resonator with an arbitrary shape, either at its fundamental resonance or at any overtone modes, and is expressed as follows:

$$k_\text{eff}^2 = 1 - \left(\frac{f_\text{r}}{f_\text{a}}\right)^2 \tag{1.11}$$

where f_r and f_a stand for resonating frequency and anti-resonating frequency, respectively. The coupling coefficients can be calculated for the various modes of vibration from the following equations:

$$\frac{k_\text{p}^2}{1-k_\text{p}^2} = \frac{(1-\sigma^\text{E})J_1\left[\eta_1\left(1+\frac{\Delta F}{F_\text{r}}\right)\right] - \eta_1\left(1+\frac{\Delta F}{F_\text{r}}\right)J_0\left[\eta_1\left(1+\frac{\Delta F}{F_\text{r}}\right)\right]}{(1+\sigma^\text{E})J_1\left[\eta_1\left(1+\frac{\Delta F}{F_\text{r}}\right)\right]} \tag{1.12}$$

where

J Bessel function of the first kind and zero order

J_1 Bessel function of the first kind and first order

σ^{E} Poisson's ratio

η_1 Lowest positive root of $(1+\sigma^{E})\cdot J_1\eta = \eta J_0(\eta)$

F_r Resonance frequency (Hz)

F_a Anti-resonance frequency (Hz)

ΔF $= F_a - F_r$ (Hz)

Assuming that $\sigma^{E} = 0.31$ for PZT ceramics and $\eta_1 = 2.05$, the following simplified equations holds:

$$k_{33}^2 = \frac{\frac{\pi}{2}}{1+\frac{\Delta F}{F_r}} \tan \frac{\frac{\pi}{2}\frac{\Delta F}{F_r}}{1+\frac{\Delta F}{F_r}} \tag{1.13}$$

$$\frac{k_{31}^2}{1-k_{31}^2} = \frac{\pi}{2}\left(1+\frac{\Delta F}{F_r}\right)\tan\frac{\pi}{2}\frac{\Delta F}{F_r} \tag{1.14}$$

1.4.3 Mechanical Quality Factor

The mechanical Q_m (also referred to as Q) is the ratio of the reactance to the resistance in the series equivalent circuit representing the piezoelectric resonator, which is related to the sharpness of the resonance frequency. The mechanical Q_m can be calculated using the equation:

$$Q_m = \frac{f_r}{f_2 - f_1} \tag{1.15}$$

where f_r is the resonance frequency, f_1 and f_2 are frequencies at -3 dB of the maximum admittance. The mechanical Q_m is also related to the electromechanical coupling factor k, following the equation:

$$Q_m = \frac{1}{2\pi F_r Z_m C_0}\left(\frac{F_a^2}{F_a^2 F_r^2}\right) \tag{1.16}$$

where

F_r Resonance frequency (Hz)

F_a Anti-resonance frequency (Hz)

Z_m Impedance at F_r (ohm)

C_0 Static capacitance (Farad)

1.5 Issues for Measuring Piezoelectric Properties

1.5.1 Measurement of Direct Piezoelectric Coefficient Using the Berlincourt Method

One of the most crucial figures of merits characterizing a piezoelectric material is the piezoelectric charge constant, also called the direct piezoelectric coefficient. It reflects the internal generation of electrical charges resulting from an applied mechanical force, as previously mentioned. Basically, the higher the piezoelectric charge constant, the more active a piezoelectric material is. A fast and accurate evaluation of the direct piezoelectric coefficient can be realized by the Berlincourt method associated with a quasi-static piezo d_{33}-meter [6, 28–31]. In this method, sample size or geometric shape becomes a factor that need not be strictly taken into account. Besides, the availability and convenient operation of a d_{33}-meter are obvious, which make it a predominantly used method in practice. However, the name of "Berlincourt method" is often mistaken by some people nowadays as being synonymous with the quasi-static measurements of the direct piezoelectric coefficient. The latter more broadly refers to the methods operating at low or quasi-static frequencies, and its basic principle of was proposed in "Piezoelectric Ceramics" by Jaffe et al. The name of "Berlincourt" actually derives from the researcher, Don Berlincourt, who devoted a lot of effort to the development of the initial commercial d_{33} apparatus based on the quasi-static measurement principle [28].

Here, we consider a simple case of measuring the d_{33} value of a ceramic sample poled along the three-direction (z) to elucidate the mathematical basis for the Berlincourt method. In the common case, the interaction between the mechanical and electrical behavior can be described by the equation $d_{33} = [\delta D_3/\delta T_3]_E$, where D_3 denotes electric displacement along the three-direction (z) and T_3 denotes applied stress also along the three-direction (z). For the practical measurement of d_{33}, this equation can be altered as $d_{33} = [(Q/A)\cdot\cdot\cdot(F/A)] = (Q/F)$, where F is applied force, A is the acting area, and Q is charge developed. It is obvious that d_{33} can be determined via measuring the charge induced by a certain force applied on the piezoelectric samples, while the measurement of the areas can be neglected as they cancel out. It should be noted that a constant electric field as fulfilled in the short-circuit condition is the prerequisite of this measurement. To achieve this condition, a large capacitor across the Device Under Test (DUT) or a virtual-ground amplifier is often embedded in the test system.

Based on the aforementioned consideration, one can easily realize that the d_{33} meter must include at least two parts: the force loading system for applying a small oscillating force, and the electronics for the circuit control and the charge measurement. As shown in Figure 1.5, the force loading system can further be divided to three parts, namely, contact probes, loading actuator, and reference sample [31]. The loading actuator is usually in the form of a loudspeaker type coil, which is cheap and easily controllable using electronic signals. The reference sample is used to monitor the applied force. It should be a piezoelectric material with a known piezoelectric coefficient and under the same force condition as the DUT. Thus, it is always put in

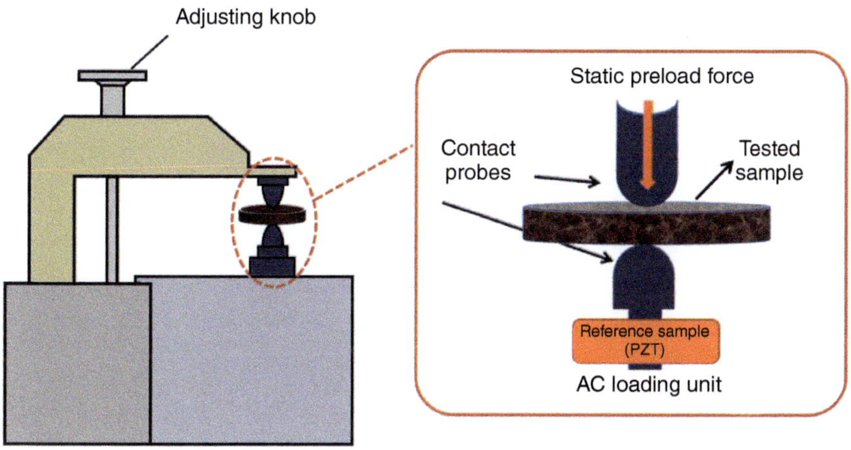

Figure 1.5 Schematic illustration of the components in the force loading system. Source: Modified from Cain [31].

the same loading line. PZT ceramics are usually served as the reference sample due to their high sensitivity. When the DUT is stimulated under the oscillating force with a certain frequency controlled by an amplified AC signal generated by the electronics, the corresponding charge is simultaneously measured by the electronics and the charge detected from the reference sample is converted to the actual force amplitude. Root Mean Squared (RMS) values of these signals are collected to determine the d_{33} value. It is worth nothing that some means of calibration must be conducted to give the correct value as the processing results from RMS signals are only proportional to the d_{33}. Thus, reference calibration sample with a certified value is usually used. Finally, the electronics display a digital readout of the calculated d_{33} value.

The main advantage of the Berlincourt method lies in its simplicity. However, due to its simplicity, anyone can design their own measurement systems and no uniform standard about this method were established. We have found various commercial systems with different measurement performance. Though it is still reliable to compare the results measured in the same apparatus, large variabilities exist in the test results from different measurement systems. There are several factors controlling the accuracy and reliability of the results associated with the Berlincourt method. This issue might be briefly introduced by taking, for example, the magnitude and frequency of the applied AC load force. The magnitude of the force does not make much difference as long as if the piezoelectric sample behaves linearly within the stress range. Nonetheless, the increased magnitude of the force is expected to generate a larger charge signal, which helps increase the signal to noise ratio. It is thus better to set the force to at high levels as long as it can still keep the piezoelectric operating in the linear regime. For a typical Berlincourt type instrument, the frequency of the stimulus force usually ranges from 10 Hz to 1 kHz. This range is governed by both the charge measurement system (for the lower limit) and the force loading system (for the upper limit). The frequency for the mechanical or electrical resonance should be avoided in case of the corresponding measurement anomalies. Thus, some specific frequencies are not used in some countries, such as 97 Hz in the United Kingdom and 110 Hz in the United States. The frequency response also varies in different materials, which can result in a frequency dependent gain issue. At the low frequency range, the measured d_{33} for "soft" piezoelectric materials usually show a pronounced downturn behavior with increasing frequency. This can be attributed to the inhibited domain movement induced by the increasing frequencies, which is depicted by the Rayleigh law. In contrast, the measured d_{33} for "hard" piezoelectric materials often appears to increase linearly with the frequency as the domain wall motion is not dominant in the low frequency range. This latter behavior is tentatively assumed to be influenced by the proximity to resonance peaks in the kHz region.

In summary, the quasi-static method is very simple and straightforward. If the relative magnitudes of the charge output and the applied small oscillating force can be measured, one can easily read d_{33} value by reference to a sample with a known and certificated piezoelectric coefficient.

1.5.2 Measurement of Converse Piezoelectric Coefficient by Laser Interferometer

The displacement of piezoelectric materials under an electric field is concerned as piezoelectric materials usually serve as actuators. However, the displacement is too small to be easily measured in a routine method. With very high resolution and no need for calibration on the length scale, optical interferometry provides the chance for the precise measurement of small displacements within units of nanometer [31–35]. Besides, optical interferometry can achieve the measurement without mechanical contact. Interferometry techniques for the detection of strains have been developed for nearly 50 years. However, at present, this method is still mainly used in research laboratories due to its high price and the need of vibration insulation system.

Piezoelectric coefficients can be measured using single beam, double beam, and heterodyne laser interferometers. For simplicity, the principle for the case based on a single beam laser interferometer is mainly discussed here.

Figure 1.6 depicts the schematic diagram of a Michelson interferometer. When a monochromatic light of wavelength λ interferes with a reference beam, the interference light intensity can be described as follows:

$$\begin{aligned} I &= |E_\mathrm{p}(t) + E_\mathrm{r}(t)|^2 = |E_{\mathrm{p}0}e^{i\{wt-2k(d+\Delta d)\}} + E_{\mathrm{r}0}e^{i(wt-2kd)}|^2 \\ &= E_{\mathrm{p}0}^2 + E_{\mathrm{r}0}^2 + 2E_{\mathrm{p}0}E_{\mathrm{r}0}\cos(2k\Delta d) = I_\mathrm{p} + I_\mathrm{r} + 2\sqrt{I_\mathrm{p}I_\mathrm{r}}\cos(4\pi\Delta d/\lambda) \end{aligned} \quad (1.17)$$

where I_p and I_r are the intensities of the probing and reference beams, respectively, Δd is the optical path-length difference between the two beams, and $k = 2\pi/\lambda$ is the wave number. Actually, the abovementioned relation can be converted into the formula with the parameters, the maximum and minimum interference light intensities $I_{\max}$ and $I_{\min}$, which can be measured by a photo-detector.

$$I = 1/2\,(I_{\max} + I_{\min}) + 1/2\,(I_{\max} - I_{\min})\cos(4\pi\Delta d/\lambda) \quad (1.18)$$

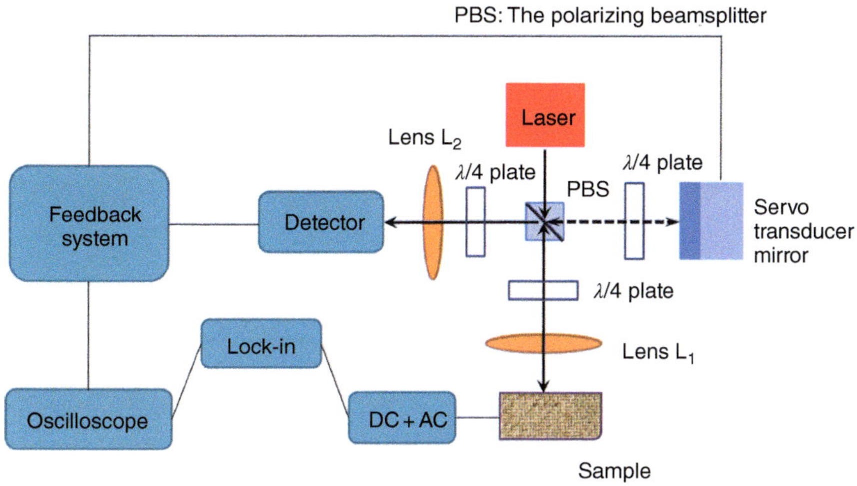

Figure 1.6 Schematic diagram of Michelson interferometer for the measurement of displacement.

In a photodetector, the corresponding photodiode output is determined by the light intensity related to the optical path-length Δd, which is directly related with the sample displacement. An amplified output voltage signal corresponding to the displacement can be obtained, which is usually monitored by an oscilloscope. A definite relationship exists between the voltage output and the displacement. According to the relationship, interferometer sensitivity can be actually set to a specific value, say, 10 nm/V. The changes of the sample dimension are induced by the connection and the disconnection of voltage to the sample. The dimension changes by the identical applied voltage are measured multiple times. The averaged measured value and the connected voltage are used to calculate the piezoelectric charge constant d_{33}, which is governed by the following equation:

$$d_{33} = \Delta l/U \tag{1.19}$$

where U is the applied voltage and Δl is the change of length determined by multiplying the voltage output of the interferometer and its sensitivity.

Single Beam Michelson interferometers are widely used by different research groups to measure piezoelectric and electrostrictive strains. It should be noted that there are various factors, including sample shape and optical alignment, which can affect the measurement validity and accuracy. To obtain the actual value of the sample dimension change parallel to the laser beam, it should be ensured that only the front surface of a sample can move while the back surface of the sample is fixed, and the sample dimension along the lateral direction can expand and contract freely. Due to these requirements, the specimen shape and dimensions are limited. The bulk sample is usually in the form of a cylinder or a thin plate with a suitable ratio between the lateral dimension and the thickness. If this ratio is very large, the clamping effect and the sample bending effect will become significant. One should not neglect such an impact because the resultant errors can even surpass the longitudinal piezoelectric displacement. When measuring the d_{33} of a thin film deposited on a very thick substrate, the clamping effect should be concerned as the substrate bonding is usually assumed to be infinitely rigid. The measured d_{33} is actually the effective converse longitudinal piezoelectric coefficient $d_{33,\text{eff}}$, which can be determined as follows:

$$d_{33,\text{eff}} = d_{33} - 2d_{31}\frac{s_{13}^{\text{E}}}{s_{11}^{\text{E}} + s_{12}^{\text{E}}} \tag{1.20}$$

where S_{ij} is the mechanical compliance of the piezoelectric film and d_{31} is the transverse piezoelectric coefficient.

Proper mounting of the sample is also very important to obtain the actual piezoelectric coefficients. It has been reported that the measured d_{33} of the same disk specimen of PC5H (Morgan Electro Ceramics) showed different values ranging from 750 to 1250 pC/N when simply changing the way the sample was mounted [34]. The accuracy is also strongly affected by small vibrations or abnormal conditions during the measurement. Thus, the measurements using the laser interferometry should be carried out with great caution and patience, though this type of measurement method is simply based on acquiring the displacement value of the sample surface induced by the corresponding applied voltage.

1.5.3 Resonance and Anti-resonance Method

Both the quasi-static method and the laser interferometry method are used for the direct measurements of d_{33}/d_{31}. In contrast to the two aforementioned methods, the resonance and anti-resonance method, or the frequency method can be applied to determine the complete tensor matrix of the material coefficients [21, 36, 37]. These coefficients can be derived by the raw set of measured parameters including resonant frequencies, density, and sample dimensions. The detailed procedure for the determination of these coefficients can be found in the European standard EN 50324-2:2002 and the world standard CEI/IEC 60483:1976.

For this method, an accurate impedance analyzer is a crucial requirement as use for the testing of the resonant frequencies f_{s} and antiresonant frequencies f_{a} as well as the free capacitance C^{T} at 1 kHz. The commercially available lineups of impedance analyzers differ in terms of frequency range and sensitivity. The most regularly used one are, for example, Agilent E4294A and Wayne Kerr 65120B, produced by Agilent and Wayne Kerr, respectively. The other thing to notice is that a set of samples in the forms of a disk, a plate, and a cylinder, which have different vibration modes, should be prepared before the measurements are carried out. Obtaining all the needed sample items from a single bulk material of appropriate size is preferred. The sample dimensions should conform to the world standard CEI/IEC 60483:1976.

Material constants such as piezoelectric coefficient d, dielectric coefficient ε, and elastic coefficient s are anisotropic in general. They are usually described by tensor components written in a simplified matrix form according to the point-group symmetry of the materials. For piezoelectric ceramics, the complete matrix of the coefficients of electromechanical properties can be written as follows:

$$\begin{bmatrix} S_1 \\ S_2 \\ S_3 \\ S_4 \\ S_5 \\ S_6 \\ D_1 \\ D_2 \\ D_3 \end{bmatrix} = \begin{bmatrix} s_{11}^{E} & s_{12}^{E} & s_{13}^{E} & & & & & & d_{31} \\ s_{12}^{E} & s_{11}^{E} & s_{13}^{E} & & & & & & d_{31} \\ s_{13}^{E} & s_{13}^{E} & s_{33}^{E} & & & & & & d_{33} \\ & & & s_{55}^{E} & & & & d_{15} & \\ & & & & s_{55}^{E} & & d_{15} & & \\ & & & & & 2\cdot(s_{11}^{E}-s_{12}^{E}) & & & \\ & & & & d_{15} & & & & \\ & & & d_{15} & & & & & \\ d_{31} & d_{31} & d_{33} & & & & & & \end{bmatrix} \cdot \begin{bmatrix} T_1 \\ T_2 \\ T_3 \\ T_4 \\ T_5 \\ T_6 \\ E_1 \\ E_2 \\ E_3 \end{bmatrix} \tag{1.21}$$

Based on the IEEE standards [22], the requirement of the sample dimensions and the complete matrix of piezoelectric constants can be established as illustrated in Figure 1.7 [37].

After acquiring the entry parameters, namely, resonant frequencies, antiresonant frequencies, capacitance, density, and sample dimensions, we can calculate the electromechanical coupling coefficients, according to which elastic coefficients s_{11}^{E},

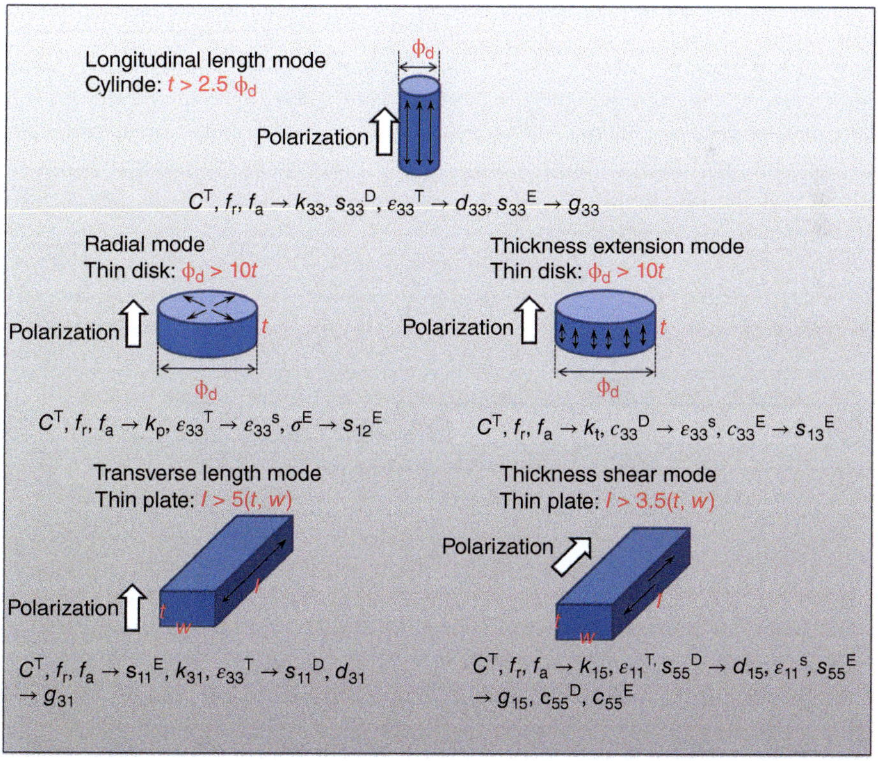

Figure 1.7 Establishing the complete set of material coefficients defined by IEEE standards [37]. Source: Reprinted with permission from Fialka and Benes [37]. Copyright 2013, IEEE.

s_{12}^{E}, s_{13}^{E}, s_{33}^{E}, s_{55}^{E}, and s_{66}^{E} can be derived. As an example, the related formulas for the calculation of the elastic coefficient s_{33}^{E}, the piezoelectric charge coefficient d_{33}, and the piezoelectric voltage coefficient g_{33} for the longitudinal length mode in the cylinder are deduced based on the IEEE standards as follows:

$$s_{33}^{E} = \frac{s_{33}^{D}}{1 - k_{33}^{2}} = \frac{\frac{1}{4 \cdot \rho \cdot f_{p}^{2} \cdot t^{2}}}{1 - \frac{\pi}{2} \cdot \frac{f_{r}}{f_{a}} \cdot \tan\left(\frac{\pi}{2} \cdot \frac{f_{a}-f_{r}}{f_{a}}\right)} \tag{1.22}$$

$$d_{33} = k_{33}(\varepsilon_{33}^{T} \cdot s_{33}^{E})^{\frac{1}{2}} = \left(\varepsilon_{33}^{T} \cdot s_{33}^{D} \cdot \frac{k_{33}^{2}}{1 - k_{33}^{2}}\right)^{\frac{1}{2}}$$

$$= \left(C^{T} \cdot \frac{t}{\frac{\pi \cdot d^{2}}{4}} \cdot \frac{1}{4 \cdot \rho \cdot f_{p}^{2} \cdot t^{2}} \cdot \frac{\frac{\pi}{2} \cdot \frac{f_{r}}{f_{a}} \cdot \tan\left(\frac{\pi}{2} \cdot \frac{f_{a}-f_{r}}{f_{a}}\right)}{1 - \frac{\pi}{2} \cdot \frac{f_{r}}{f_{a}} \cdot \tan\left(\frac{\pi}{2} \cdot \frac{f_{a}-f_{r}}{f_{a}}\right)}\right)^{\frac{1}{2}} \tag{1.23}$$

$$g_{33} = \frac{d_{33}}{\varepsilon_{33}^{T}} = \frac{d_{33}}{C^{T} \cdot \frac{t}{\frac{\pi \cdot d^{2}}{4}}} \tag{1.24}$$

where f_r is the resonance frequency, f_a is anti-resonance frequency, and f_p is parallel resonant frequency.

The frequency method is advantageous since it can determine the complete matrix of the material coefficients, though a complete set of samples needs to be manufactured. The accuracy of the calculated coefficients depends on the overall measurement accuracy of the initial parameters including resonant frequencies, anti-resonant frequencies, density, and sample dimensions. Besides, it is worth nothing that the measured samples should comply with a minimum aspect ratio stipulated by the IEEE standards.

References

1 Coursey, P.R. and Brand, K.G. (1946). Dielectric constants of some titanates. *Nature* 157: 297–298.

2 Roberts, S. (1947). Dielectric and piezoelectric properties of barium titanate. *Physical Review* 71 (12): 890.

3 Shirane, G. and Takeda, A. (1952). Phase transitions in solid solutions of $PbZrO_3$ and $PbTiO_3$: I. Small concentrations of $PbTiO_3$. *Journal of the Physical Society of Japan* 7: 5–11.

4 Shirane, G. and Suzuki, K. (1952). Crystal structure of $Pb(Zr–Ti)O_3$. *Journal of the Physical Society of Japan* 7: 333–333.

5 Sawaguchi, E. (1953). Ferroelectricity versus antiferroelectricity in the solid solutions of $PbZrO_3$ and $PbTiO_3$. *Journal of the Physical Society of Japan* 8: 615–629.

6 Jaffe, B., Roth, R.S., and Marzullo, S. (1954). Piezoelectric properties of lead zirconate–lead titanate solid-solution ceramics. *Journal of Applied Physics* 25: 809–810.

7 Jaffe, B., Cook, W.R., and Jaffe, H. (1971). *Piezoelectric Ceramics*. Academic Press.
8 Haertling, G.H. (1999). Ferroelectric ceramics: history and technology. *Journal of the American Ceramic Society* 82: 797–818.
9 Uchino, K. (1997). *Piezoelectric Actuators and Ultrasonic Motors*. Boston, MA: Kluwer Academic Publishers.
10 Trolier-McKinstry, S. and Randall, C.A. (2018). Movers, shakers, and storers of charge: the legacy of ferroelectricians L. Eric Cross and Robert E. Newnham. *Journal of the American Ceramic Society* 100: 3346–3359.
11 EU-Directive 2002/95/EC (2003). Restriction of the Use of Certain Hazardous Substances in Electrical and Electronic Equipment (RoHS). *Official Journal of the European Union* 46 (L37): 19.
12 Takenaka, T. and Nagata, H. (2005). Current status and prospects of lead-free piezoelectric ceramics. *Journal of the European Ceramic Society* 25: 2693–2700.
13 Saito, Y., Takao, H., Tani, T. et al. (2004). Lead-free piezoceramics. *Nature* 432: 84–87.
14 Shrout, T.R. and Zhang, S.J. (2007). Lead-free piezoelectric ceramics: alternatives for PZT? *Journal of Electroceramics* 19: 111–124.
15 Roedel, J., Jo, W., Seifert, K.T.P. et al. (2009). Perspective on the development of lead-free piezoceramics. *Journal of the American Ceramic Society* 92: 1153–1177.
16 Li, J.-F., Wang, K., Zhu, F.-Y. et al. (2013). (K,Na)NbO_3-based lead-free piezoceramics: fundamental aspects, processing technologies, and remaining challenges. *Journal of the American Ceramic Society* 82: 797–818.
17 Roedel, J. and Li, J.-F. (2018). Lead-free piezoceramics: status and perspectives. *MRS Bulletin* 43: 576–580.
18 Trolier-McKinstry, S., Zhang, S.J., Bell, A.J., and Tan, X.L. (2018). High-performance piezoelectric crystals, ceramics, and films. *Annual Review of Materials Research* 48: 191–217.
19 Guo, R., Cross, L.E., Park, S.E. et al. (2000). Origin of the high piezoelectric response in $PbZr_{1-x}Ti_xO_3$. *Physical Review Letters* 84: 5423–5426.
20 Noheda, B., Cox, D.E., Shirane, G. et al. (2001). Stability of the monoclinic phase in the ferroelectric perovskite $PbZr_{1-x}Ti_xO_3$. *Physical Review B* 63: 014103.
21 Noheda, B. (2002). Structure and high-piezoelectricity in lead oxide solid solutions. *Current Opinion in Solid State and Materials Science* 6: 27–34.
22 IEEE Standard on Piezoelectricity. (1987). ANSI/IEEE, Standard 176-1987.
23 Lines, M.E. and Glass, A.M. (1979). *Principles and Applications of Ferroelectrics and Related Materials*. Oxford: Clarendon.
24 Cady, W.G. (1946). *Piezoelectricity: An Introduction to the Theory and Applications of Electromechanical Phenomena in Crystals*. McGraw-Hill.
25 Sawyer, C.B. and Tower, C.H. (1930). Rochelle salt as a dielectric. *Physical Review* 35: 269–273.
26 Mitsui, T., Tatsuzaki, I., and Nakamura, E. (1976). *An Introduction to the Physics of Ferroelectrics*. London: Gordon and Breach.
27 Damjanovic, D. (1998). Ferroelectric, dielectric and piezoelectric properties of ferroelectric thin films and ceramics. *Reports on Progress in Physics* 61: 1267–1324.

28 Jin, L., Li, F., and Zhang, S.J. (2014). Decoding the fingerprint of ferroelectric loops: comprehension of the material properties and structures. *Journal of the American Ceramic Society* 97: 1–27.

29 Berlincourt, D. and Jaffe, H. (1958). Elastic and piezoelectric coefficients of single-crystal barium titanate. *Physical Review* 111: 143–148.

30 Jaffe, H. and Berlincourt, D.A. (1965). Piezoelectric transducer materials. *Proceedings of the IEEE* 53: 1372–1385.

31 Cain, M.G. (2014). *Characterisation of Ferroelectric Bulk Materials and Thin Films*. Springer.

32 Yamaguchi, T. and Hamano, K. (1979). Inteferometric method of measuring complex piezoelectric constants of crystals in a frequency range up to about 50 kHz. *Japanese Journal of Applied Physics* 18: 927–932.

33 Zhang, Q.M., Pan, W.Y., and Cross, L.E. (1988). Laser interferometer for the study of piezoelectric and electrostrictive strains. *Journal of Applied Physics* 63: 2492–2496.

34 Li, J.-F., Moses, P., and Viehland, D. (1995). Simple, high-resolution interferometer for the measurement of frequency-dependent complex piezoelectric responses in ferroelectric ceramics. *Review of Scientific Instruments* 66: 215–221.

35 Moilanen, H. and Leppävuori, S. (2001). Laser interferometric measurement of displacement-field characteristics of piezoelectric actuators and actuator materials. *Sensors and Actuators A* 92: 326–334.

36 Warner, A.W., Onoe, M., and Coquin, G.A. (1966). Determination of elastic and piezoelectric constant in class (3M). *Journal of the Acoustical Society of America* 42: 1223–1231.

37 Fialka, J. and Benes, P. (2013). Comparison of methods for the measurement of piezoelectric coefficients. *IEEE Transactions on Instrumentation and Measurement* 62: 1047–1057.

2

High-Performance Lead-Free Piezoelectrics

2.1 Introduction

Piezoelectric materials are used in a large variety of applications including actuators, sensors, electromechanical motors, and sonar systems. In the future, the applications will be extended to energy harvesting systems for distributed low-power systems. The global market for piezoelectric devices is estimated to be valued at US$ 25.1 billion in 2019 and projected to expand at a compounded annual growth rate of 6.2% through the next five years [1]. One of the most widely used piezoelectric ceramic materials is the solid solution of lead zirconate ($PbZrO_3$) and lead titanate ($PbTiO_3$), commonly known as PZT, with a general formula $PbZr_{1-x}Ti_xO_3$ [2]. As mentioned in Chapter 1, although $BaTiO_3$ was discovered before lead zirconate titanate (PZT), the latter soon dominated the market for piezoelectric applications because of its comprehensively excellent piezoelectric properties and high Curie temperature as well as the flexibility for compositional modification to adjust to specific applications.

The outstanding properties of PZT originate from the nature of end-number compounds and the phase structure evolution in their solid solutions. Both $PbZrO_3$ and $PbTiO_3$ have perovskite structure with Pb ions at the same A-site. $PbZrO_3$ is antiferroelectric, but $PbTiO_3$ is ferroelectric [3]. Solid solutions are formed between them in a full range, but there exists a steep boundary where a transition occurs between the rhombohedral Zr-rich phase and the tetragonal Ti-rich phase, as shown in Figure 2.1. It should be noted that the tetragonal $PbTiO_3$ has a high transition temperature to cubic phase, which endows high Curie temperature to PZT. The phase boundary separating two phases in a solid solution system is called morphotropic phase boundary (MPB), where piezoelectric coefficients, dielectric constant, and electromechanical coupling factor are greatly enhanced [4]. The "phase boundary" is also considered as a composition where the two phases are believed to coexist, which renders the materials more responsive to extra stimuli. It should be noticed that the MPB is nearly temperature independent, which is why the properties of PZT ceramics have excellent temperature stability. More in-depth studies had been conducted regarding the nature of the MPB using advanced structural characterization techniques, leading to several modified versions of the PZT phase diagram. One representative work suggested that the MPB was a

Lead-Free Piezoelectric Materials, First Edition. Jing-Feng Li.

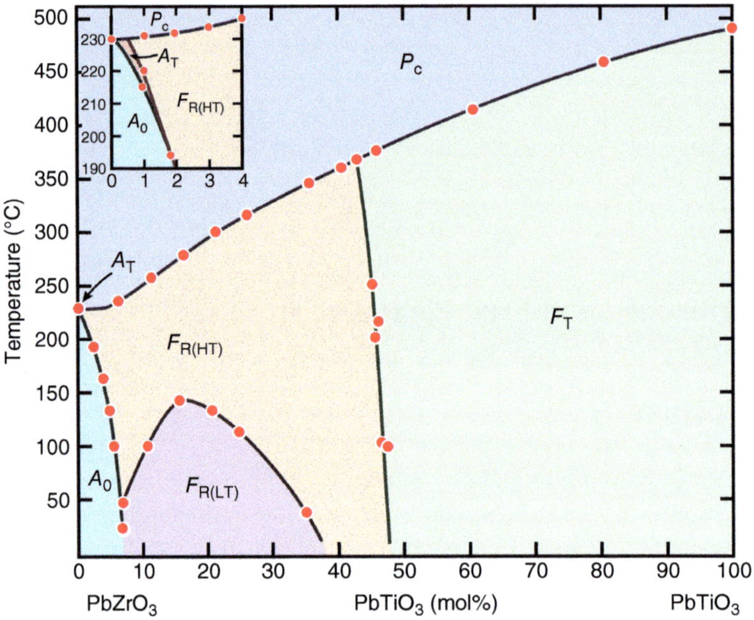

Figure 2.1 Phase diagram of the $PbZrO_3-PbTiO_3$ system. Note that there are revised phase diagrams showing an area instead of a line in the middle composition.

narrow region where monoclinic phases existed in between the tetragonal and rhombohedral phases [5, 6]. It seems that the hierarchical structures ubiquitously present in ferroelectric materials make the phase diagram studies more complicated than non-ferroelectric materials [7]. Nevertheless, a similar concept of MPB is useful and highly desirable for the development of lead-free piezoelectric materials.

In addition to the existence of MPB, the properties of PZT ceramics can be adjusted relatively easily by chemical doping or substitution, which makes PZT like "all-around player" to meet a wide range of applications. In fact, piezoelectric ceramic devices can be classified as actuators, sensors, and transducers, which have specific property requirements. For example, "hard" PZT ceramics for high-frequency transducers actuators need high mechanical quality factor (Q_m), but sensors usually use "soft" PZT with enhanced d_{33}. All kinds of PZT ceramics are commercially available. However, the large amount of lead contained in PZT materials has drawn much attention during the past decade, due to the environmental concerns as well as governmental regulations against hazardous substances. In 2006, the European Union has already adopted the well-known directives: Waste Electrical and Electronic Equipment (WEEE) and Restriction of the use of certain Hazardous Substances (RoHS) in electrical and electronic equipment, with the purpose of protecting human health as well as environment by exclusion or substitution of hazardous substances used in electrical and electronic devices [8]. Similar regulations have also been planned or established in North America, Japan, Korea, and China. Lead ranks foremost among the hazardous substances list due to its notoriety, with maximum allowed concentration of merely

0.1 wt%. Bearing the goal to reduce the production and waste disposal of toxic lead, the research on lead-free piezoceramics strives to make PZT and similar perovskite materials redundant. Substantial effort has been devoted to the development of high-performance lead-free piezoelectric ceramics based on the several perovskite compounds [9], including $BaTiO_3$, $(K,Na)NbO_3$, $(Bi_{0.5}Na_{0.5})TiO_3$, and $BiFeO_3$. In particular, great progress has been made since 2004 when Saito et al. reported $(K,Na)NbO_3$ based lead-free piezoelectrics with properties comparable with those found in PZT [10].

This chapter is set up to enable the readers to quickly learn about the lead-free piezoelectrics by introducing their brief history, crystal structure, and physical natures for $BaTiO_3$, $(K,Na)NbO_3$, $(Bi_{0.5}Na_{0.5})TiO_3$, and $BiFeO_3$ systems. We first introduce $BaTiO_3$ not only because it is the first lead-free ferroelectric compound, but also for the purpose to show the intrinsic role of Pb by comparing $BaTiO_3$ with $PbTiO_3$, which is a representative lead-containing ferroelectric. The detailed research progress about each system will be given in Chapters 3–7.

2.2 $BaTiO_3$

$BaTiO_3$ is a well-known ferroelectric material that was discovered before PZT in the 1940s during the World War II [11]. Before the emergence of PZT, $BaTiO_3$ had been the most important piezoelectric material, which has a higher piezoelectric coefficient d_{33} = 120–190 pC/N than that of preceding piezoelectric materials such as triglycine sulfate (TGS), potassium dihydrogen phosphate (KDP), and quartz. $BaTiO_3$ undergoes a series of phase transitions, from cubic to tetragonal approximately at 130 °C, then to orthorhombic at 0 °C, and further to a rhombohedral crystal structure at −90 °C, as shown in Figure 2.2 [12].

Although $BaTiO_3$ is the first lead-free piezoelectric compound, its Curie point (approximately 130 °C) is moderate, which limits its piezoelectric applications. Nevertheless, there are also many studies on lead-free piezoelectrics based on $BaTiO_3$ [13, 14]. High piezoelectricity has been reported in pure $BaTiO_3$ single crystals [15, 16] and fine-grained ceramics [17, 18]. In 2009, Liu and Ren reported a large piezoelectric response for the $BaTiO_3$-based ceramic system with a d_{33} coefficient >600 pC/N [19], which is comparable with soft PZT ceramics. The details will be given in Chapter 5.

$BaTiO_3$ is often used as a model ferroelectric material for fundamental studies because of its simple composition and rich phase structures. For example, early studies have been conducted to reveal the role of Pb from a comparison between $BaTiO_3$ and $PbTiO_3$ [20]. Both compounds have similar cohesive properties such as unit-cell volume (0.0632 and 0.0642 nm^3, respectively), yet with distinct ferroelectric behavior. $PbTiO_3$ has a high Curie temperature of 490 °C, and a large value of spontaneous polarization P_s (57 $\mu C/cm^2$), whereas those of $BaTiO_3$ are much lower ($P_s \sim 26\ \mu C/cm^2$). $PbTiO_3$ has no low-temperature phase transitions, whereas $BaTiO_3$ is known to have successive phase transitions, from the cubic structure to tetragonal, orthorhombic, and rhombohedral structure at low temperature.

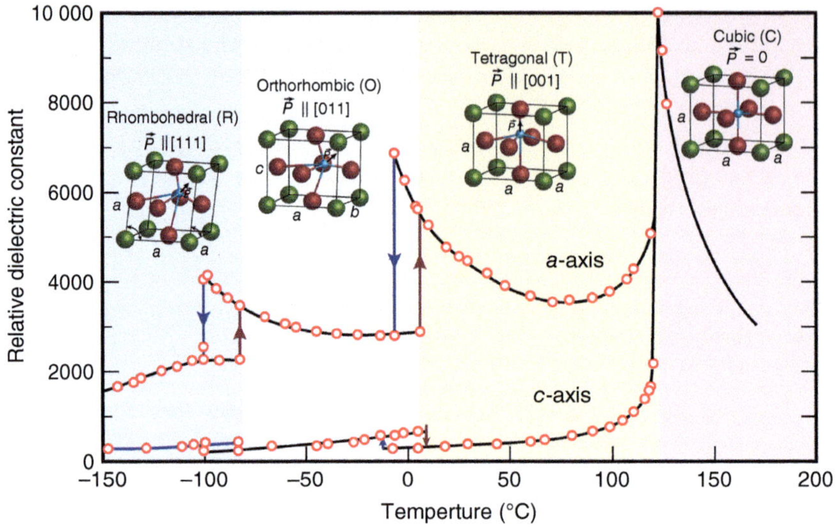

Figure 2.2 Change in relative dielectric constants of $BaTiO_3$ ceramics as a function of temperature. Inserted are the crystal structures at the corresponding temperatures. Source: Modified from Merz [12].

To explain why $BaTiO_3$ and $PbTiO_3$ show different ferroelectric behaviors, the studies have been focused on the difference in the chemical bonds of the two perovskite compounds. Kuroiwa et al. demonstrated through calculations that the Pb and O electronic states hybridize in tetragonal $PbTiO_3$, whereas the interaction between Ba and O is almost purely ionic in the tetragonal $BaTiO_3$, as shown by the charge-density distributions of cubic and tetragonal $PbTiO_3$ and $BaTiO_3$ in Figure 2.3 [20]. It clearly revealed that no apparent difference could be seen between the cubic phases of $PbTiO_3$ and $BaTiO_3$, but between their tetragonal phases. Within the latter phases, the distributions in the Pb–O and Ti–O planes of $PbTiO_3$ are significantly different from those in the Ba–O and Ti–O planes of $BaTiO_3$. The electron-density distribution around the Pb and O2 sites in tetragonal $PbTiO_3$ is quite anisotropic as shown in Figure 2.3c, i.e. extending toward the O2 and Pb, respectively. The electron densities of Pb and O2 are overlapped along the Pb–O2 direction in the tetragonal $PbTiO_3$. However, the electron densities around Ba and O are isotropic for tetragonal $BaTiO_3$, as shown in Figure 2.3a. These results indicate that the Pb—O bonds of tetragonal $PbTiO_3$ are covalent, although they are ionic in the cubic phase. By comparing Figure 2.3b,d, it is clear that the Ti in $PbTiO_3$ has a larger displacement than that in $BaTiO_3$, consistent with the fact that the spontaneous polarization of $PbTiO_3$ is larger than that of $BaTiO_3$. This provides clear evidence for the Pb–O hybridization in tetragonal $PbTiO_3$, which has been theoretically predicted as a key factor of much larger ferroelectricity of $PbTiO_3$ as a representative Pb-containing ferroelectric compound than that of $BaTiO_3$ as a lead-free counterpart.

The aforementioned comparison between $BaTiO_3$ and $PbTiO_3$ indicates that partial covalent bonding due to the Pb–O hybridization should be responsible for

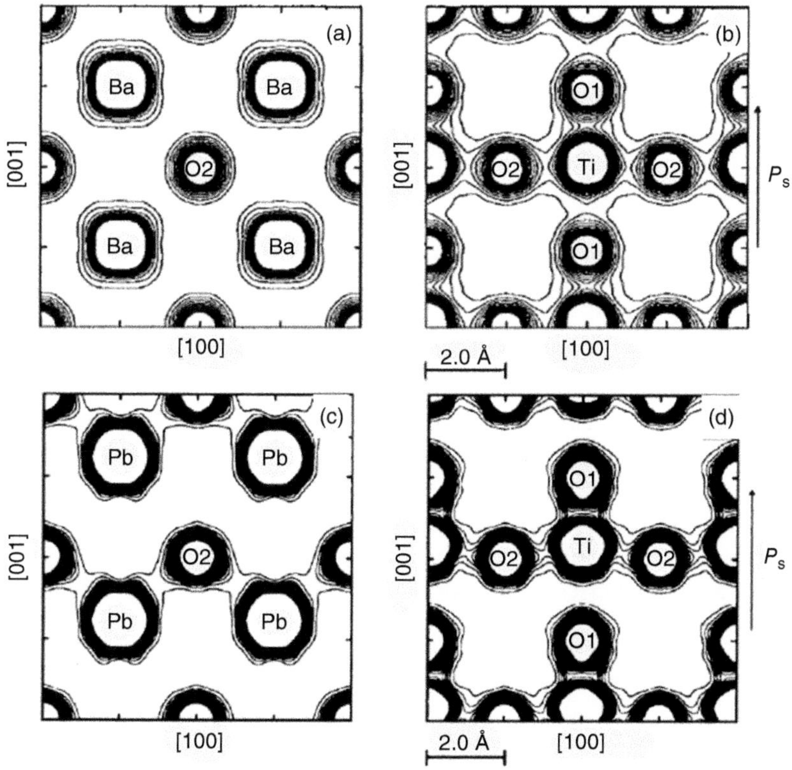

Figure 2.3 Charge-density distributions on (a) the (010) plane (Ba–O plane), (b) the (020) plane (Ti–O plane) of tetragonal $BaTiO_3$ at 300 K, compared to that of tetragonal $PbTiO_3$ at 300 K on (c) the (010) plane (Pb–O plane), and (d) the (020) plane (Ti–O plane). Source: Reprinted with permission from Kuroiwa et al. [20]. Copyright 2001, American Physical Society.

the high performance of Pb-containing piezoceramics. In lead-free counterparts, Saito et al. also found that substituting Nb with Sb, which had higher electronegativity than the former, leading to further piezoelectric property enhancement in $(K,Na)NbO_3$-based ceramics [10]; this is also because of the hybridization of covalency onto ionic bonding. In addition, Bi as a neighbor of Pb has a similar electronic structure with the latter, and can be used a modifier to enhance the contribution of covalent bonding.

2.3 $(K,Na)NbO_3$

$(K,Na)NbO_3$ (shortened as KNN) is a solid solution of $KNbO_3$ and $NaNbO_3$. $KNbO_3$ is a ferroelectric material with $T_c = 435\,°C$, while $NaNbO_3$ is expected to be an antiferroelectric material with $T_c = 640\,°C$ according to the early studies; evidences for the latter's antiferroelectricity are still being investigated. Nevertheless, both compounds can form solid solutions in the full compositional range, analogous to the PZT system that is a solid solution of ferroelectric $PbTiO_3$ and antiferroelectric $PbZrO_3$ [21].

(K,Na)NbO_3 ceramics have been investigated quite early, which was found as a piezoelectric material in 1950s. Jaeger and Egerton [22] fabricated highly dense KNN ceramics using hot pressing and reported that the hot-pressed KNN ceramics possessed high piezoelectric constants and Curie temperature. However, since then there had been only a few reports about piezoelectricity of $NaNbO_3$, $KNbO_3$, or KNN ceramics, probably because of their poor sinterability and high volatility, in addition to their low piezoelectricity compared with the PZT ceramics that have been widely used so far. KNN ceramics have begun to attract special attention again because of a breakthrough made by Saito et al., who obtained high d_{33} (~416 pC/N) in textured KNN-based ceramics [10]. Their work not only revealed the potentials of KNN ceramics as a promising candidate for lead-free piezoelectric ceramics, but also stimulated widespread scientific interest in developing lead-free piezoceramics.

Figure 2.4 shows the phase diagram of the KNN system [23]. It is quite complicated when compared with the PZT system. Similar to $KNbO_3$, the KNN shows the following polymorphisms within a wide composition range: low temperature (less than −10 °C) rhombohedra (R) phase, room-temperature orthorhombic (O) phase,

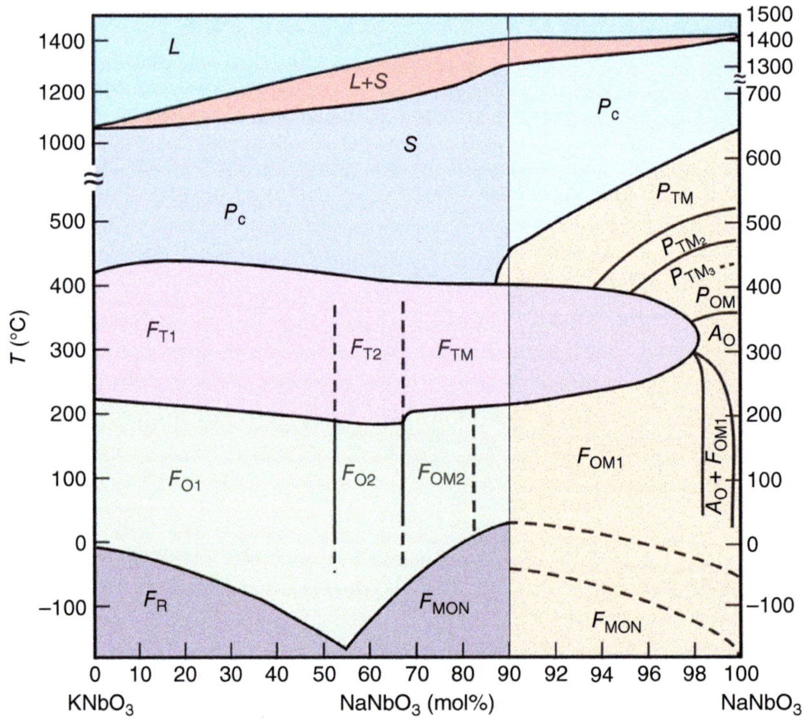

Figure 2.4 Phase diagram of $KNbO_3$–$NaNbO_3$ system. Source: Reprinted with permission from Li et al. [23]. Copyright 2013, John Wiley & Sons, Inc.

high temperature tetragonal (T, 225∼435 °C), and cubic (C, >435 °C) phase. However, the phase transitions of $NaNbO_3$ are even more complicated, exhibiting seven polymorphs across the whole temperature range. The temperature-dependent phase transition of KNN is similar to the case of $BaTiO_3$, but the cubic to non-cubic phase transition temperature in the former, corresponding to the Curie temperature, is much higher than that of $BaTiO_3$, which is a major advantage of KNN. Therefore, KNN-based ceramics are considered as the most promising lead-free piezoceramics with good comprehensive performance (large d_{33} and high T_C).

Significant research progress on KNN-based ceramics has been achieved in the past years [23–26]. The record d_{33} of the textured KNN-based ceramics has been boosted to 704 pC/N in 2018 [27] from the original report of 412 pC/N by Saito et al. in 2004 [10]. Phase boundary engineering is an effective approach to the piezoelectricity enhancement in KNN. Pristine KNN has an orthorhombic phase (O) at room temperature, which transfers to rhombohedral (R) and tetragonal (T) phases at approximately −200 and 200 °C, respectively. Shifting these phase transition temperatures to room temperature by chemical modification has proved to be an effective way to enhance the piezoelectric performance of KNN materials significantly. That is, these two or three phases can be rendered to co-exist near room temperature, thereby leading to enhanced piezoelectricity. Such an enhancement is based on the polymorphic phase transition (PPT) effect, which is different from the classic MPB effect. The high piezoelectricity of KNN-based ceramics deriving from the PPT effect, however, is sensitive to temperature changes [23, 28]; small variations from room temperature can lead to significant drops in the piezoelectricity. Therefore, a simultaneous enhancement of the piezoelectricity and its thermal stability has been one of the focused topics, which is crucial for extending the large-scale applications of KNN-based piezoceramics. An effective strategy to this end is making the phase transition boundary diffused [29–31], which will be described later in the corresponding chapter. In addition to the outstanding overall piezoelectric performance, another advantage of the KNN-based ceramics is their compatibility of co-firing with Ni electrodes that are much cheaper than Ag/Pd electrodes currently used for PZT multilayer piezoelectric ceramics [32].

2.4 $(Bi_{1/2}Na_{1/2})TiO_3$

$(Bi_{1/2}Na_{1/2})TiO_3$ (abbreviated as BNT or NBT) is another important lead-free piezoelectric material with a perovskite structure. Discovered by Smolenskii et al. in 1961 [33], BNT had been considered as a promising lead-free piezoelectric material system [34, 35]. BNT shows strong ferroelectric properties at room temperature, including a large remnant polarization P_r = 38 μC/cm^2 and a coercive field E_c = 73 kV/cm [36]. Due to the poor sintering property of the pure BNT ceramics and the relatively low piezoelectric constant (d_{33} ∼ 100 pC/N), many studies have thus focused on improving the piezoelectric properties through cation substitution at the A and/or B sites of BNT [36–38].

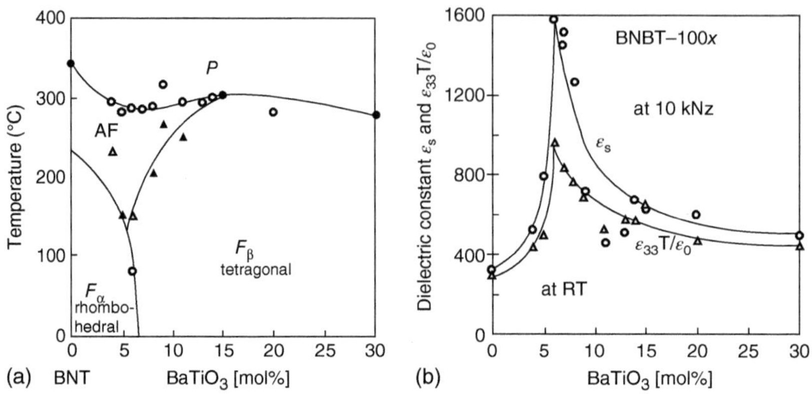

Figure 2.5 (a) Phase diagram of of $(Bi_{1/2}Na_{1/2})TiO_3$–$BaTiO_3$ proposed by Takenaka and (b) compositional dependence of dielectric constant. Source: Reprinted with permission from Takenaka et al. [35]. Copyright 1991, The Physical Society of Japan.

Takenaka and his research group had done extensive investigations about BNT-based lead-free piezoceramics. In 1991, they obtained the phase diagram of the BNT and $BaTiO_3$ solid solution based on the combined measurement results of X-ray diffraction patterns and dielectric and ferroelectric properties [35]. Most importantly, as shown in Figure 2.5, there exists an MPB between BNT and $BaTiO_3$ at the composition close to 6 mol% $BaTiO_3$. Separated by this phase boundary are the rhombohedral and tetragonal phases, similar to that in PZT. Therefore, it is also possible to enhance the piezoelectric properties of BNT-based ceramics by utilizing the MPB effect via compositional modification. However, the BNT–$BaTiO_3$ phase diagram has obtained several newer versions due to the incorporation of other characterization methods [39, 40]. A similar system is the solid solution between $(Bi_{1/2}Na_{1/2})TiO_3$ and $(Bi_{1/2}K_{1/2})TiO_3$ compounds, first reported by Sasaki et al. [41]. The MPB between BNT (rhombohedral) and $(Bi_{1/2}K_{1/2})TiO_3$ (tetragonal) exists in the range of 16–20 mol% $(Bi_{1/2}K_{1/2})TiO_3$ [41].

$(Bi_{1/2}Na_{1/2})TiO_3$ is an interesting system. The structural description of BNT is quite complicated, and some details of the structure are still under debate [36]. Although it is generally agreed that BNT at room temperature is rhombohedral as indicated in the phase diagram, the high-resolution X-ray measurements of single crystals also revealed the hints of a monoclinic phase [42]. Some studies also indicate that BNT has a composite structure at nanoscale, namely, the dispersion of short-range-ordered orthorhombic phase domains within a long-range-ordered rhombohedral or monoclinic phase matrix. The complexity may stem from the fact that there exists two kinds of ions with different valances at the A-site of the perovskite structure. In addition, unmodified NBT and BNT-rich BNT–$BaTiO_3$ ceramics often show relaxor ferroelectric features, although the early phase diagram proposed by Takenaka et al. indicated an antiferroelectric region above a certain temperature as shown in Figure 2.5.

Compared with other lead-free piezoelectric systems, the BNT system exhibits an exceptionally high electric-field-induced strain, which makes it particularly attractive for applications in piezoelectric actuators such as camera focusing, printing nozzle, and impact driving. It has been reported that the poling strain of BNT–$BaTiO_3$ ceramics could reach a high value range of 0.3–0.7% by composition modification [43, 44]. However, one of the drawback of BNT-based ceramics is the existence of a depolarization temperature (approximately 150 °C) below the Curie point, which poses a limit on the application temperature, and its underlying mechanism is still not completely understood.

2.5 $BiFeO_3$

Bismuth ferrite ($BiFeO_3$) possesses a rhombohedrally distorted perovskite structure as shown in Figure 2.6, which can be denoted as a pseudocubic structure with lattice parameter $a = 0.396$ nm and $\alpha = 89.6°$ [45]. $BiFeO_3$ is the only room-temperature multiferroic material with coexisting ferroelectric and magnetic properties in a single compound [46]. Extensive studies have been conducted on this compound in the form of thin films, which are expected to open up new applications [47]. As reported in the epitaxial thin films, $BiFeO_3$ has a high Curie point ($T_c \sim 830$ °C) and large remnant polarization ($P_r \sim 100\,\mu C/cm^2$) at the same time [47]. In addition, the cost of resource is low because of the earth abundance of the constituent elements. Most importantly, $BiFeO_3$ is a lead-free ferroelectric compound where Bi has a similar electronic structure with Pb. Therefore, recently, increasing attention has been paid to the development of $BiFeO_3$-based lead-free piezoelectrics.

However, $BiFeO_3$ and $BiFeO_3$-based ceramics have high leakage current issues, mainly induced by defects and secondary phases, which hinder the poling process required for piezoelectric ceramics [48]. To resolve this problem, extensive investigations have been conducted based on chemical modifications to $BiFeO_3$ [49]. In 2008, Takeuchi and coworkers introduced another rhombohedral–orthorhombic (R–O) MPB that was formed by A-site doping with 14% Sm in $BiFeO_3$ films grown on $SrTiO_3$. The piezoelectric coefficient d_{33} near this MPB could reach 110 pm/V [50]. This report stimulated many studies about Sm-doping on $BiFeO_3$, and the addition of other rare earth elements were also investigated to promote the piezoelectricity and ferroelectricity. For example, a large piezoelectricity ($d_{33} \approx 50$ pC/N) was reported in Sm and La co-doped $BiFeO_3$ ceramics [51]. Higher properties are

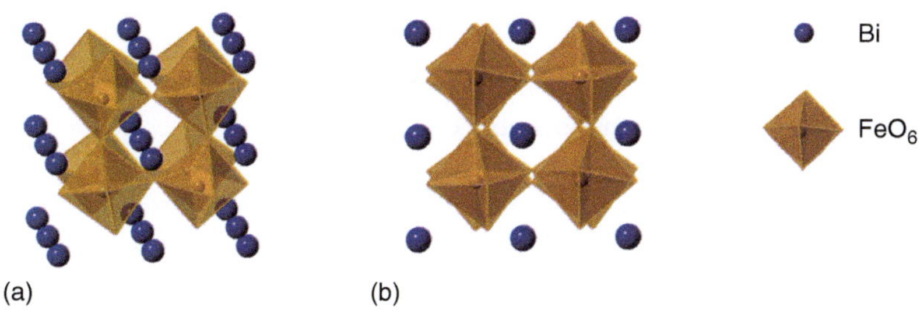

Figure 2.6 Crystalline structure of $BiFeO_3$.

obtained in $BiFeO_3$ ceramics with the addition of ABO_3. In a study by Lee et al., excellent piezoelectric properties with a Curie temperature above 400 °C were reported in $BiFeO_3$–$BaTiO_3$ ceramics with the addition of either $Bi_{1.05}GaO_3$ or $Bi_{1.05}(Zn_{0.5}Ti_{0.5})O_3$ fabricated by a sintering-and-quenching process [52]. Such a special process has proved to effectively suppress the formation of point defects and secondary phase and thereby reduce the high current leakage [53].

The main applications of $BiFeO_3$-based ceramics are considered to be high-temperature piezoelectric sensors, but their lower usage temperatures should be lower as compared with bismuth layer-structured ferroelectric ceramics (BLSFs). Nevertheless, $BiFeO_3$-based ceramics have the potentials to achieve much higher piezoelectric coefficients, so their applications can be extended to broader areas including piezoelectric actuators and transducers at middle and high temperatures.

2.6 Summary

For decades, $Pb(Zr,Ti)O_3$ (PZT)-based ceramics have been market-dominating due to its excellent properties and flexibility in terms of compositional modifications. On the other hand, developing high-performance lead-free piezoelectric ceramics has been one of the most active materials research topics in the last decades. At present, a single replacement for PZT may not be available, but the abovementioned lead-free systems are already applicable for some applications. Among them, KNN appears to be the most promising one for its well-balanced combination of high piezoelectricity and Curie temperature. In addition, as shown in Chapter 3, KNN-based ceramics can be co-fired with base metals like nickel, which is a critical factor for the cost-effective fabrication of multilayered piezoceramic actuators. Certainly, other systems also have certain advantages, such as large piezoelectric strains in BNT-based ceramics and high Curie temperature in $BiFeO_3$-based ceramics. However, all these known lead-free materials are unlikely to be such an "all-round player" that can be modified to exhibit different properties by changing doping elements as in PZT, so there still needs a further long way to develop lead-free piezoceramics that are completely capable of replacing PZT-based ceramics.

References

1 Gagliardi, M. (2019). Lead-free piezoelectric ceramics market projected to grow at much faster pace through 2024. *American Ceramic Society Bulletin* 99: 7.
2 Jaffe, B., Cook, W.R., and Jaffe, H. (1971). *Piezoelectric Ceramics*. Academic Press.
3 Sawaguchi, E. (1953). Ferroelectricity versus antiferroelectricity in the solid solutions of $PbZrO_3$ and $PbTiO_3$. *Journal of the Physical Society of Japan* 8: 615–629.
4 Jaffe, B., Roth, R.S., and Marzullo, S. (1955). Properties of piezoelectric ceramics in the solid-solutions series lead titanate–lead zirconate–lead oxide: tin oxide

and lead titanate–lead hafnate. *Journal of Research of the National Bureau of Standards* 55: 239–254.
5 Noheda, B., Cox, D.E., Shirane, G. et al. (1999). A monoclinic ferroelectric phase in the $Pb(Zr_{1-x}Ti_x)O_3$ solid solution. *Applied Physics Letters* 74: 2059–2061.
6 Noheda, B., Gonzalo, J.A., Cross, L.E. et al. (2000). Tetragonal-to-monoclinic phase transition in a ferroelectric perovskite: the structure of $PbZr_{0.52}Ti_{0.48}O_3$. *Physical Review B* 61: 8687–8695.
7 Wang, H., Zhu, J., Lu, N. et al. (2006). Hierarchical micro-/nanoscale domain structure in M-C phase of $(1-x)Pb(Mg_{1/3}Nb_{2/3})O_{3-x}PbTiO_3$ single crystal. *Applied Physics Letters* 89: 042908.
8 EU-Directive 2002/95/EC (2003). Restriction of the Use of Certain Hazardous Substances in Electrical and Electronic Equipment (RoHS). *Official Journal of the European Union* 46 (L37): 19.
9 Roedel, J. and Li, J.-F. (2018). Lead-free piezoceramics: status and perspectives. *MRS Bulletin* 43: 576–580.
10 Saito, Y., Takao, H., Tani, T. et al. (2004). Lead-free piezoceramics. *Nature* 432: 84–87.
11 Jona, F. and Shirane, G. (1962). *Ferroelectric Crystals*. Oxford, UK: Pergamon Press.
12 Merz, W.J. (1949). The electric and optical behavior of $BaTiO_3$ single-domain crystals. *Physical Review* 76: 1221–1225.
13 Acosta, M., Novak, N., Rojas, V. et al. (2017). $BaTiO_3$-based piezoelectrics: fundamentals, current status, and perspectives. *Applied Physical Review* 4: 041305.
14 Gao, J., Ke, X., Acosta, M. et al. (2018). High piezoelectricity by multiphase coexisting point: barium titanate derivatives. *MRS Bulletin* 43: 595–598.
15 Wada, S., Yako, K., Kakemoto, H. et al. (2005). Enhanced piezoelectric properties of barium titanate single crystals with different engineered-domain sizes. *Journal of Applied Physics* 98: 014109.
16 Ren, X.B. (2004). Large electric-field-induced strain in ferroelectric crystals by point-defect-mediated reversible domain switching. *Nature Materials* 3: 91–94.
17 Karaki, T., Yan, K., Miyamoto, T., and Adachi, M. (2007). Lead-free piezoelectric ceramics with large dielectric and piezoelectric constants manufactured from $BaTiO_3$ nano-powder. *Japanese Journal of Applied Physics Part 2: Letters & Express Letters* 46: L97–L98.
18 Shen, Z.-Y. and Li, J.-F. (2010). Enhancement of piezoelectric constant d_{33} in $BaTiO_3$ ceramics due to nano-domain structure. *Journal of the Ceramic Society of Japan* 118: 940–943.
19 Liu, W.F. and Ren, X.B. (2009). Large piezoelectric effect in Pb-free ceramics. *Physical Review Letters* 103: 257602.
20 Kuroiwa, Y., Aoyagi, S., Sawada, A. et al. (2001). Evidence for Pb–O covalency in tetragonal $PbTiO_3$. *Physical Review Letters* 87: 217601.
21 Shirane, G., Newnham, R., and Pepinsky, R. (1954). Dielectric properties and phase transitions of $NaNbO_3$ and $(Na,K)NbO_3$. *Physical Review* 96: 581–588.
22 Jaeger, R.E. and Egerton, L. (1962). Hot pressing of potassium–sodium niobates. *Journal of the American Ceramic Society* 45: 209–213.

23 Li, J.-F., Wang, K., Zhu, F.-Y. et al. (2013). (K,Na)NbO_3-based lead-free piezoceramics: fundamental aspects, processing technologies and remaining challenges. *Journal of the American Ceramic Society* 96: 3677–3696.

24 Wu, J.G., Xiao, D.Q., and Zhu, J.G. (2015). Potassium–sodium niobate lead-free piezoelectric materials: past, present, and future of phase boundaries. *Chemical Reviews* 115: 2559–2595.

25 Wang, K., Malic, B., and Wu, J.G. (2018). Shifting the phase boundary: potassium sodium niobate derivates. *MRS Bulletin* 43: 607–611.

26 Zhang, Y.C. and Li, J.-F. (2019). Review of chemical modification on potassium sodium niobate lead-free piezoelectrics. *Journal of Materials Chemistry C* 74: 284–4303.

27 Li, P., Zhai, J.W., Shen, B. et al. (2018). Ultrahigh piezoelectric properties in textured (K,Na)NbO_3-based lead-free ceramics. *Advanced Materials* 30: 1705171.

28 Dai, Y.J., Zhang, X.W., and Zhou, G.Y. (2007). Phase transitional behavior in $K_{0.5}Na_{0.5}NbO_3$–$LiTaO_3$ ceramics. *Applied Physics Letters* 90: 262903.

29 Yao, F.-Z., Wang, K., Jo, W. et al. (2016). Diffused phase transition boosts thermal stability of high-performance lead-free piezoelectrics. *Advanced Functional Materials* 26: 1217–1224.

30 Liu, Q., Li, J.-F., Zhao, L. et al. (2018). Niobate-based lead-free piezoceramics: a diffused phase transition boundary leading to temperature-insensitive high piezoelectric voltage coefficients. *Journal of Materials Chemistry C* 6: 1116–1125.

31 Liu, Q., Zhang, Y., Gao, J. et al. (2018). High-performance lead-free piezoelectrics with local structural heterogeneity. *Energy & Environmental Science* 11: 3531–3539.

32 Kawada, S., Kimura, M., Higuchi, Y., and Takagi, H. (2009). (K,Na)NbO_3-based multilayer piezoelectric ceramics with nickel inner electrodes. *Applied Physics Express* 2: 111401.

33 Smolenskii, G., Isupov, V., Agranovskaya, A., and Krainik, N. (1961). New ferroelectrics of complex composition. *Soviet Physics – Solid State* 2: 2651–2654.

34 Roleder, K., Franke, I., Glazer, A.M. et al. (2002). The piezoelectric effect in $Na_{0.5}Bi_{0.5}TiO_3$ ceramics. *Journal of Physics: Condensed Matter* 14: 5399–5406.

35 Takenaka, T., Maruyama, K., and Sakata, K. (1991). $(Bi_{1/2}Na_{1/2})TiO_3$–$BaTiO_3$ system for lead-free piezoelectric ceramics. *Japanese Journal of Applied Physics Part 1* 30: 2236–2239.

36 Paterson, A.R., Nagata, H., Tan, X.L. et al. (2018). Relaxor-ferroelectric transitions: sodium bismuth titanate derivatives. *MRS Bulletin* 43: 600–606.

37 Herabut, A. and Safari, A. (1997). Processing and electromechanical properties of $(Bi_{0.5}Na_{0.5})_{(1-1.5x)}La_xTiO_3$ ceramics. *Journal of the American Ceramic Society* 80: 2954–2958.

38 Takenaka, T. and Nagata, H. (2005). Current status and prospects of lead-free piezoelectric ceramics. *Journal of the European Ceramic Society* 25: 2693–2700.

39 Ma, C., Guo, H.Z., Beckman, S.P., and Tan, X.L. (2012). Creation and destruction of morphotropic phase boundaries through electrical poling: a case study of lead-free $(Bi_{1/2}Na_{1/2})TiO_3$–$BaTiO_3$ piezoelectrics. *Physical Review Letters* 109: 107602.

40 Jo, W., Schaab, S., Sapper, E. et al. (2011). On the phase identity and its thermal evolution of lead free $(Bi_{1/2}Na_{1/2})TiO_3$–6 mol% $BaTiO_3$. *Journal of Applied Physics* 110: 074106.

41 Sasaki, A., Chiba, T., Mamiya, Y., and Otsuki, E. (1999). Dielectric and piezoelectric properties of $(Bi_{0.5}Na_{0.5})TiO_3$–$(Bi_{0.5}K_{0.5})TiO_3$ systems. *Japanese Journal of Applied Physics* 38: 5564.

42 Gorfman, S., Glazer, A.M., Noguchi, Y. et al. (2012). Observation of a low-symmetry phase in $Na_{0.5}Bi_{0.5}TiO_3$ crystals by optical birefringence microscopy. *Journal of Applied Crystallography* 45: 444–452.

43 Zhang, S.-T., Kounga, A.B., Jo, W. et al. (2009). High-strain lead-free antiferroelectric electrostrictors. *Advanced Materials* 21: 4716–4720.

44 Liu, X.M. and Tan, X.L. (2016). Giant strains in non-textured $(Bi_{1/2}Na_{1/2})TiO_3$-based lead-free ceramics. *Advanced Materials* 28: 574–578.

45 Moreau, J.M., Michel, C., Gerson, R., and James, W.J. (1971). Ferroelectric $BiFeO_3$ X-ray and neutron diffraction study. *Journal of Physics and Chemistry of Solids* 32: 1315.

46 Catalan, G. and Scott, J.F. (2009). Physics and applications of bismuth ferrite. *Advanced Materials* 21: 2463.

47 Ramesh, R. and Spaldin, N.A. (2007). Multiferroics: progress and prospects in thin films. *Nature Materials* 6: 21–29.

48 Sun, W., Zhou, Z., Luo, J. et al. (2017). Leakage current characteristics and Sm/Ti doping effect in $BiFeO_3$ thin films on silicon wafers. *Journal of Applied Physics* 121: 064101.

49 Rojac, T., Bencan, A., Malic, B. et al. (2014). $BiFeO_3$ ceramics: processing, electrical, and electromechanical properties. *Journal of the American Ceramic Society* 97 (7): 1993–2011.

50 Fujino, S., Murakami, M., Anbusathaiah, V. et al. (2008). Combinatorial discovery of a lead-free morphotropic phase boundary in a thin-film piezoelectric perovskite. *Applied Physics Letters* 92: 202904.

51 Zheng, T. and Wu, J.G. (2015). Enhanced piezoelectric activity in high-temperature $Bi_{1-x-y}Sm_xLa_yFeO_3$ lead-free ceramics. *Journal of Materials Chemistry* 3: 3684–3693.

52 Lee, M.H., Kim, D.J., Park, J.S. et al. (2015). High-performance lead-free piezoceramics with high curie temperatures. *Advanced Materials* 27: 6976–6982.

53 Chen, J.-Y., Zhang, B.-P., Zhu, L.-F. et al. (2018). Enhanced insulation resistance and electrical properties of $BiFe_{1-x}(Zn_{0.5}Ti_{0.5})xO_3$–$BaTiO_3$ lead-free piezoceramics. *Ceramics International* 44: 8409–8416.

3

(K,Na)NbO_3 System

3.1 Introduction of (K,Na)NbO_3

3.1.1 History of (K,Na)NbO_3

In 1949, some tantalite-based and niobate-based materials were found exhibiting ferroelectricity [1]. In 1954, Shirane et al. [2] found that adding a little amount of ferroelectric $KNbO_3$ into antiferroelectric $NaNbO_3$ could form a new type of ferroelectric material. In 1959, Egerton and Dillon continued to study the properties of $NaNbO_3$–$KNbO_3$ (or (K,Na)NbO_3, hereafter abbreviated as KNN) solid solutions [3]. They found that both piezoelectric and dielectric properties are maximized at the vicinity of $(K_{0.5}Na_{0.5})NbO_3$ [3], which was proposed to be a phase boundary of two different orthorhombic phases [4, 5]. The piezoelectric coefficient d_{33} of $(K_{0.5}Na_{0.5})NbO_3$ was measured to be 80 pC/N and could be promoted to a higher level, i.e. 160 pC/N, in a well-densified ceramic [6–8]. Despite good piezoelectricity and high Curie temperature, KNN did not attract much attention until 2004. Guo et al. [9] firstly discovered that adding Li into the KNN system could enhance d_{33} to 200–235 pC/N along with a high Curie temperature T_C of 452–510 °C, which is almost 100 °C higher than that of the conventional lead zirconate titanate (PZT). In the same year, Saito et al. [10] reported a high piezoelectric coefficient of 300 pC/N in KNN ceramics modified with Li, Ta, and Sb. That value was further enhanced to 416 pC/N by texturing [10]. The study was inspiring to the community, which made scientists believe that KNN could be a promising alternative for replacing the conventional lead-containing PZT [11]. From there on, KNN-based materials attracted much attention [12–15], which can tell from the increasing number of annual publications related to KNN-based materials since 2004. After a decade of research, encouraging performances have been obtained in KNN-based piezoelectric materials, which will be discussed in Section 3.3.

3.1.2 Crystal Structure and Phase Diagram

(K,Na)NbO_3 exhibits an orthorhombic structure (*Amm*2) at room temperature and can transform into tetragonal (*P*4*mm*) and rhombohedral (*R*3*m*) phases at around 200 and −160 °C, respectively. The tetragonal phase transforms into the paraelectric

Lead-Free Piezoelectric Materials, First Edition. Jing-Feng Li.

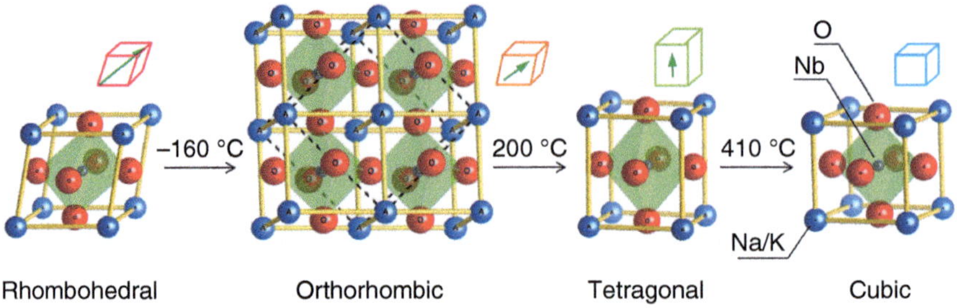

Figure 3.1 The illustration of difference crystal structures of KNN.

cubic phase approximately at 410 °C [2]. Such a series of polymorphic phase transitions (PPTs) inherit from that taking place in pure $KNbO_3$ [16], which is also very similar to that of $BaTiO_3$; however, the phase transition of $NaNbO_3$ is more complicated [17]. Figure 3.1 illustrates the crystal structure of four different phases [14, 18]. Except for the paraelectric cubic phase, all the other ferroelectric phases possess spontaneous polarization of different directions: the tetragonal phase allows dipoles at 6 (100) equivalent lattice directions; the orthorhombic phase allows dipoles at 12 (110) equivalent lattice directions; the rhombohedral phase allows dipoles at 8 (111) equivalent lattice directions.

It is worth noting that the symmetry of the orthorhombic phase needs to be presented with a cell doubling, while the primary unit cell shows a monoclinic symmetry. To understand this better, the orthorhombic structure *Amm*2 of KNN at room temperature is illustrated in Figure 3.2. The primary unit cell, as shown in Figure 3.2a,b, has $c_p = a_p > b_p$ and its interaxial angle β (the angle between b_pc_p and a_pb_p plane) is slightly larger than 90° [18]. Therefore, the primary unit cell is of monoclinic symmetry. In the doubled four-adjacent cell, as shown in Figure 3.2c, the new structure has $a' \neq c' \neq b_p$ and $\beta' = 90°$, which satisfies the orthorhombic structure (*Amm*2 space group). According to the rule of crystallography, the structure should be referred to as a structure of higher symmetry, i.e. the orthorhombic phase. It is worth mentioning that the β' was reported to be also slightly higher than 90°, and there is a marginal difference in length between a_p and c_p [5, 18]. Thus, the phase structure in the doubled four-adjacent cell should also fall into monoclinic symmetry. Nevertheless, such a difference is usually too small to be noticed, and most literature still refers to the room-temperature phase as the orthorhombic phase.

The phase diagram of the KNN system is shown in Figure 2.4. $KNbO_3$ and $NaNbO_3$ can form solid solutions with a perovskite structure regardless of their mixing ratio, but the middle compositions were focused in most previous studies. Certain reports including some early ones considered that there were two different orthorhombic phases (O_1 and O_2) in the vicinity of K/Na = 50/50, indicating there exists a phase boundary similar to the morphotropic phase boundaries (MPBs) observed in the PZT system [8, 19–21]. However, the studies devoted to the investigation of the K/Na ratio dependent phase boundaries are very limited so far. Nevertheless, it is certain that the K/Na ratio can affect the ferroelectric and piezoelectric properties of the pristine KNN and compositionally modified KNN-based ceramics [5, 8, 22, 23],

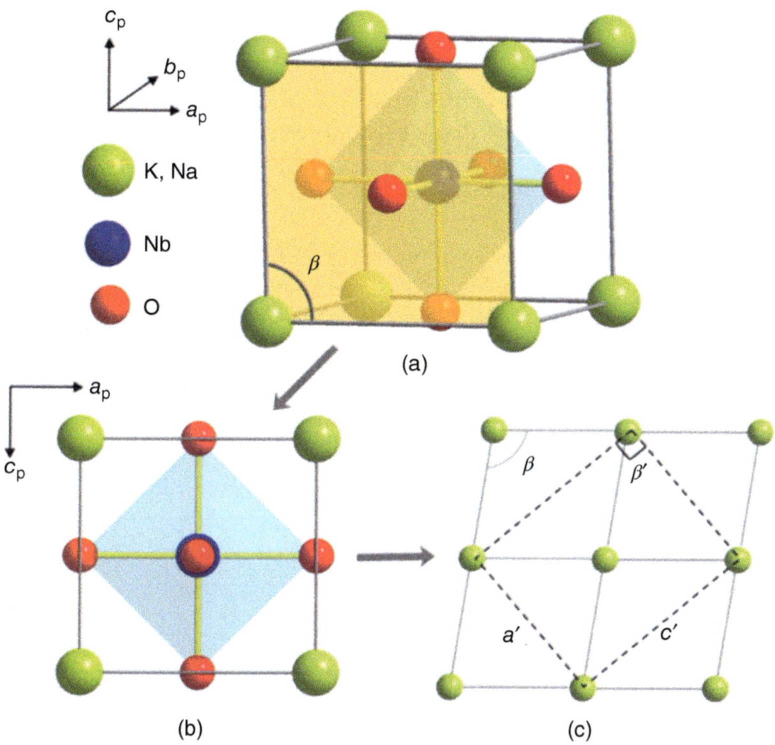

Figure 3.2 The illustration for KNN of the orthorhombic structure at room temperature (a) the perovskite-type ABO_3 primary unit cell. (b) The projection of the primary unit cell along the *b* axis. (c) Four-adjacent unit cell projections in an exaggerated manner to show the geometrical relationship, while omitting Nb and O atoms. Source: Reprinted with permission from Wang and Li [18]. Copyright 2007, AIP Publishing.

whereas the influence of K/Na ratio is not strong as observed in the PZT system that shows clear peaks of electrical properties at the MPB [5, 22–24]. It seems that the optimal K/Na ratio is not exactly 50 : 50 but a little bit Na-rich than this value [8, 20, 23, 24]. It is well-known that the phase transition temperatures in KNN system can be easily shifted by chemical doping, which are important for property enhancement as described later in this chapter. For studying the phase structure, a dielectric permittivity–temperature curve is often used to check such a shifting, in addition to precise analysis of phase structure by X-ray diffraction mostly at room temperature. At the phase transition temperature, the dielectric constant reaches a local maximum, which originates from the increased spontaneous polarization directions, i.e. the summation of 6 (100) and 12 (110) directions. The increased directions will result in a higher susceptibility to the applied electric field and mechanical pressure. Meanwhile, the enhanced dielectric performance has also been explained by the minimization of the energy barrier for polarization rotation. Therefore, the construction of phase coexistence has become a feasible way to improve the performance of piezoelectric materials. It should be noted that KNN shows similar temperature dependence of dielectric permittivity as observed in the $BaTiO_3$ systems (see Figure 2.2) since they possess the same series of temperature-dependent phase transitions.

3.1.3 Current Development of KNN-Based Materials

Extensive studies worldwide have continuously renewed the piezoelectric properties of the KNN-based ceramics. Figure 3.3 shows the top 20% of the highest d_{33} reported annually in KNN-based ceramics [15]. The KNN-based ceramics have exhibited comparable piezoelectric coefficients with commercial PZT-based piezoceramics. The early studies revealed that adding $LiNbO_3$ into KNN was effective for property enhancement, and the first highest d_{33} was 235 pC/N in Li-doped KNN ceramics [9]. Later, much higher d_{33} up to 330 pC/N was achieved also in the Li-doped KNN ceramics when more attention was paid to the fine optimization of sintering temperature and poling processes [25, 26]. To achieve a further higher d_{33}, more and more studies for property enhancement have been conducted on B-site chemical modification [10, 27–31]. Higher d_{33} up to 400 pC/N could be realized by using B-site substitution with Ta and/or Sb [31]. Introducing Sb and (Bi,Na,K)ZrO_3 into KNN can result in an even higher d_{33} of 490–496 pC/N, which is almost comparable with the commercially used soft PZT (PIC 151) [32, 33]. In 2016, Wu and coworkers [34] developed a complex KNN composition with the addition of Sb, $BaZrO_3$, and (Bi,K)HfO_3, achieving an ultrahigh d_{33} of 570 pC/N. In 2018, a similar system, (K,Na)(Nb,Sb)O_3–$CaZrO_3$–(Bi,K)HfO_3, was fabricated with texturing technique, which exhibits a d_{33} of 700 pC/N [35]. This is the highest value achieved in KNN-based ceramics until now. Various other methods have been developed to enhance the piezoelectric properties, such as advanced synthesis process, phase engineering, domain engineering and poling techniques. Despite the focus on the enhancement of d_{33}, it was soon realized that the thermal instability had been one of the major shortages for KNN-based materials compared with the lead-based ones. To have reliable performance, many solutions have been

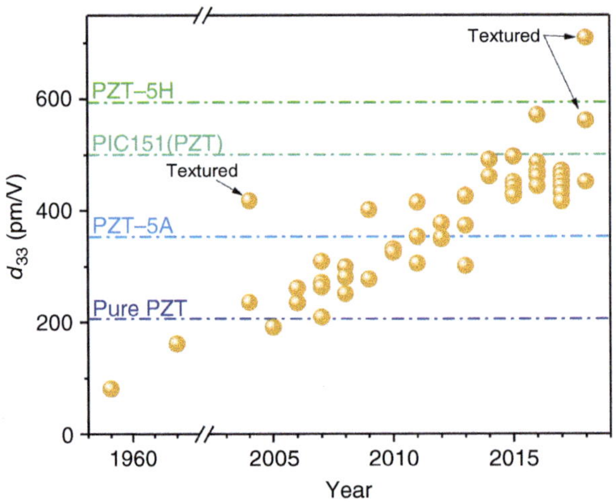

Figure 3.3 The annual top 20% piezoelectric coefficient of KNN-based piezoelectric ceramics reported in the literature. Source: Reprinted with permission from Zhang and Li [15]. Copyright 2019, Royal Society of Chemistry.

proposed to improve the thermal stability, which will be shown later in this chapter. Furthermore, due to the promising performance of KNN bulk ceramics, the studies on the fabrication and characterization of KNN-based thin films also have received increasing attention. At the end of this chapter, the related studies about KNN-based thin films will also be introduced.

3.2 Synthesis

Undoubtedly, a well-designed chemical formula is the key to obtaining high-performance piezoelectric ceramics. However, the processing of ceramics is equally important. Obtaining dense KNN-based ceramics along with good reproducibility is often challenging. The key issues are associated with the alkaline elements in KNN-based materials. The most commonly used alkaline raw materials are carbonates, which are hygroscopic and could easily form hydrates in a moisture atmosphere, causing difficulty for storing and inaccuracy for weighting. Therefore, a drying step at a minimum of 200 °C before weighting is often necessary for guaranteeing the accuracy. It is worth noting that both sodium carbonate and potassium carbonate are soluble in water, which can lead to a discrepancy of compositions. Therefore, organic solvents, e.g. ethanol, are commonly used as the solvent in ball milling. Another problem for alkaline elements is the volatilization during calcination and sintering.

3.2.1 Calcination

Figure 3.4 shows the step-by-step solid-state synthesis of KNN-based powders and bulk materials in a laboratory. Raw materials for making pure KNN include Na_2CO_3, K_2CO_3, and Nb_2O_5. Please note that Na_2CO_3 and K_2CO_3 powders should be kept in an oven to prevent them from moisture. The raw materials are weighted according to chemical formulas and mixed with ethanol during ball milling. Please note that the mixing process needs high-purity ethanol because the carbonate as raw materials can be dissolved in water, which make the final composition (mainly Na and K) deviate from the designed one. For mass-scale production, it may need to develop a process use water as the mixing medium. High-energy planetary ball milling can help achieve similar efficiency (or sometimes even better) within a shorter time (8 hours for this method compared with 24 hours for the traditional ball milling). The ball-milled powder is ground in a mortar after drying in an oven. Finally, the ground powder is calcined in a sealed crucible.

The reaction between alkaline carbonate and niobium oxide would form an intermediate reaction layer of $(K,Na)_2Nb_4O_{11}$ on the surface of Nb_2O_5 particles at the early stage of calcination (approximately 500 °C) [36]. With further increase of temperature, the alkaline species continue to react with $(K,Na)_2Nb_4O_{11}$ layer and Nb_2O_5 core through diffusion, leading to the formation of $(K,Na)NbO_3$ around 600 °C [36]. Since the whole reaction process is dominated by the diffusion of which the diffusion rates are subject to dopants and particle sizes, the optimal calcination

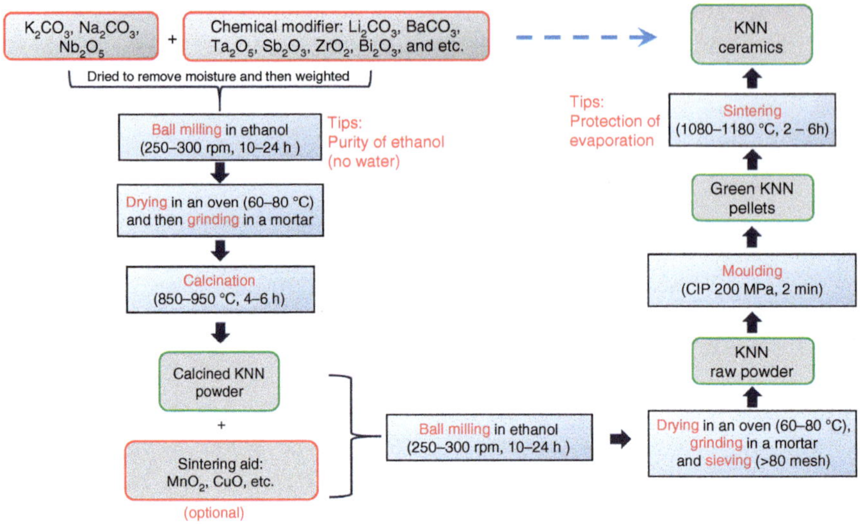

Figure 3.4 Common preparation procedure of KNN-based ceramics on a laboratory scale. Please note that ball milling should use high-purity water and protection of evaporation during sintering should be done.

temperature can vary. Take un-doped KNN, for example, since the diffusion rate of the K^+ is about 1 order of magnitude lower than that of Na^+, the reaction is controlled by the slower species, i.e. K^+. The addition of atomically heavy elements, e.g. Ta and Zr, which have slower diffusion rates, usually requires a higher calcination temperature. Therefore, calcination temperature needs to be optimized for engineering the microstructure of ceramics [37]. Typical calcination temperature for KNN-based ceramics lies between 750 and 950 °C [38]. The two-step calcination, i.e. calcining twice with a grinding step in between, has also been used to synthesize KNN ceramic, which is considered to improve the homogeneity, density, and electric properties [39]. The re-set of CO_2 vapor pressure is likely to contribute to the improvement of repeated calcination [40]. The particle size of the grounded raw powders is important for the calcination: smaller particle size contributes to a larger contact surface area and shorter diffusion paths, which is favorable for calcination.

3.2.2 Sintering

Sintering is an irreversible thermodynamic process in which powder compacts are consolidated into dense polycrystalline ceramics by heating. The reduction of the surface energy of the assembled particles is the main driving force for sintering. Apart from the surface energy, the application of external pressure or chemical reaction can also contribute to the driving force. Sintering process involves densification and grain growth, which determines the microstructures, e.g. density, grain size distribution, and grain morphology. For the details of sintering process and related knowledge, the readers are referred to [41, 42]. It must be additionally concerned that the volatilization of alkali elements is also accompanied during the sintering of KNN ceramics. That is why hot pressing (HP) was first tried to fabricate dense KNN samples at reduced temperatures [6]. Of course, normal sintering or pressureless sintering is most widely used due to its cost efficiency and simplicity, but for the fabrication of highly dense and fine-grained KNN ceramics, spark plasma sintering (SPS), and microwave sintering (MWS) are found more suitable. Sections 3.2.2.1 and 3.2.2.2 summarizes the related studies.

3.2.2.1 Normal Sintering

There is a considerable number of studies devoted to the normal sintering process and related densification mechanism of KNN ceramics [38, 43–50]; see a focused review for a comprehensive understanding [38]. It was found that the densification of KNN-based materials was very sensitive to the fluctuation of sintering temperature [43, 44]. The density of materials increases greatly in a narrow temperature range and starts to decrease when the sintering temperature shifts away from the range. It is worth noting that the optimal piezoelectric properties were not necessarily found at the samples with the highest density if the final composition deviated from the designed one due to the volatilization of alkali elements [43]. The sintering temperature of KNN-based ceramics is relatively high (over 1100 °C) if compared with its melting point (see the phase diagram shown in Chapter 2). Thus, a large degree of loss in alkaline elements and consequent compositional fluctuation is generally unavoidable. From temperature over 1000 °C, serious volatilization can already be detected. The volatilizations of Li, Na, and K are all temperature-dependent but at different rates. In general, the volatilization of Na was found to be more serious than that of K [51, 52], but an opposite result was also reported [40]. Nevertheless, it is likely that the loss of lithium is most serious, particularly at the ~20-μm-thick outmost surface [53], whereas the volatilization of Na and K elements have become less severe compared with the as-sintered surface.

The volatilization problem could seriously deteriorate piezoelectric properties and sometimes even induce secondary phases. Therefore, the sintering parameters are very crucial. It would be ideal if there is an optimal sintering temperature that can densify KNN-based ceramics without volatilization; however, the volatilization starts before the sintering densification temperature [51]. One common strategy to suppress volatilization is to place packing powder (also called sacrificing powder or atmosphere powder), which is the powder of the same composition as the sintered KNN samples, in a sealed crucible to saturate the vapor pressure of alkaline elements [43]. This strategy has been used in PZT to mitigate the volatilization of PbO for a long time. Meanwhile, some researchers added 1–2%-excess alkaline carbonates into the composition to compensate for the potential loss [48].

The sintering temperature of KNN ceramics is close to the solidus line of the liquid and solid area, so a tiny local composition caused by the volatilization and/or temperature change may result in the formation of a liquid phase. This means that liquid phase sintering is available in the KNN ceramics even without intentionally mixing sintering aids [44]. The liquid phase can help promote the densification through particle rearrangement and shortened diffusion paths, as evidenced by a sharp sintering

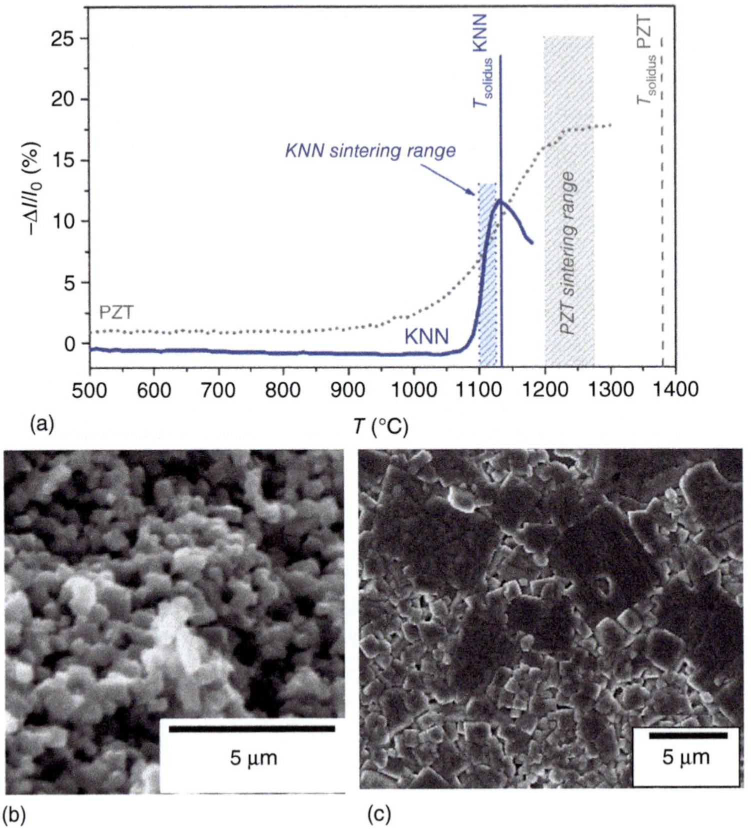

Figure 3.5 (a) Shrinkage vs. temperature curve of KNN and PZT ceramics. (b, c) Scanning electron micrographs of fracture surfaces of Li/Ta-doped KNN ceramics sintered at 1080 and 1100 °C. Note that the densification and grain growth occurred when the sintering temperature was raised by only 20 °C. Source: (a) Malic et al. [38]. (b, c) Li et al. [44].

shrinkage in a narrow temperature and accompanied by significant grain growth, as shown in Figure 3.5 [38, 44].

However, abnormal grain growth (AGG) tends to appear during the sintering, which is associated with the formation of liquid phase [43, 44, 54–57]. The AGG appears as some unexpectedly large grains grow out of a matrix of fine grains, showing a bimodal grain size distribution. As shown in Figure 3.5, only with an increase of 20 °C in sintering temperature did grain coarsening begin within a small part of the grains, which grew up to 5 µm in size. Sometimes, abnormally grown coarse grains with a new core–shell structure could be observed, which affirmed the critical role of liquid phase in facilitating the grain growth. A series of studies indicate that the AGG occurs in the presence of liquid phase due to the anisotropy of grain boundary energy caused by inhomogeneous wetting [55, 56]. In fact, AGG disappears when the sintering temperature is further raised so that more liquid phases are formed [56]. However, the liquid phase also leads to compositional segregation upon cooling according to the phase diagram [50]. In summary, for the densification by normal sintering, liquid phase is indispensable, but the resultant inhomogeneous microstructure is unfavorable in term of property enhancement.

In considering of suppressing volatilization and the formation of inhomogeneous microstructure, low temperature sintering is particularly desired for the KNN ceramics. In this case, liquid phase sintering is intentionally introduced to facilitate the densification by mixing sintering aids that usually have low melting points or produce new compounds with low melting points during sintering. Many studies reported that highly dense KNN ceramics could be fabricated by liquid phase sintering with various sintering aids [58–64]. A good sintering aid must enhance densification and at the same time should not be detrimental to the piezoelectric properties. Excess alkali metal contents (such as Li_2O and Na_2O) have been reported as useful sintering aids, although some of them act as compensation for the volatilization. It should also be noted that controlling the particle size of starting powders, which refers to those obtained after calcination, is quite important. Many milling strategies have been investigated, such as high-energy ball milling, prolonging ball-milling time [65]. Additionally, controlling heating and cooling rate during sintering of KNN-based ceramics requires a special attention to guarantee the final properties, which have been demonstrated by Zhang et al. [47]. A slow cooling rate allows the compositional equilibrium to establish, having K and Na distributed homogeneously to enhance piezoelectric properties [47].

3.2.2.2 Hot Pressing, Spark Plasma Sintering, and Microwave Sintering

Hot pressing (HP) is a pressure-assisted sintering technique that can help achieve high density. In 1962, Jaeger and Egerton reported the fabrication of KNN ceramics by hot pressing [6]. They showed that samples prepared by hot pressing could have a much higher density (above 99% of theoretical density) than the normal-sintered sample. The d_{33} of hot-pressed KNN ceramic was doubled to 160 pC/N [6]. The dielectric constant and mechanical properties were also improved.

SPS is a rapid heating and also pressurized sintering technique, which is increasingly used for the development of new materials [66]. SPS was also applied to the densification of KNN ceramics [7, 67–74], resulting in fine-grained materials with high density. Since SPS usually is conducted using graphite dies in vacuum or inert gas atmosphere, the resultant sample must be annealed in air or oxygen sufficiently to attain the ferroelectric characteristics [66, 69]. For example, pure KNN ceramics have been successfully sintered at a temperature as low as 920 °C, which is about 200 °C lower than that for normal sintering [7]. Up to ~99% of theoretical density was achieved with significantly enhanced d_{33} and ε_r up to 148 pC/N and 559, respectively [7, 67]. Wang et al. compared the microstructure and properties of KNN ceramics prepared by hot pressing and SPS, and found that the SPS-sintered KNN samples show higher room temperature dielectric constant, higher coercive fields, lower remnant polarizations, and lower electromechanical coefficients than that of the hot-pressed counterparts, due to the grain size effects [68]. The usefulness of this technique was further verified by the successful synthesis of KNN ceramics with dopants, such as Li and Ta [69–71]. The SPS-processed ceramics show smaller grain sizes (200–500 nm) as compared with those prepared by normal sintering

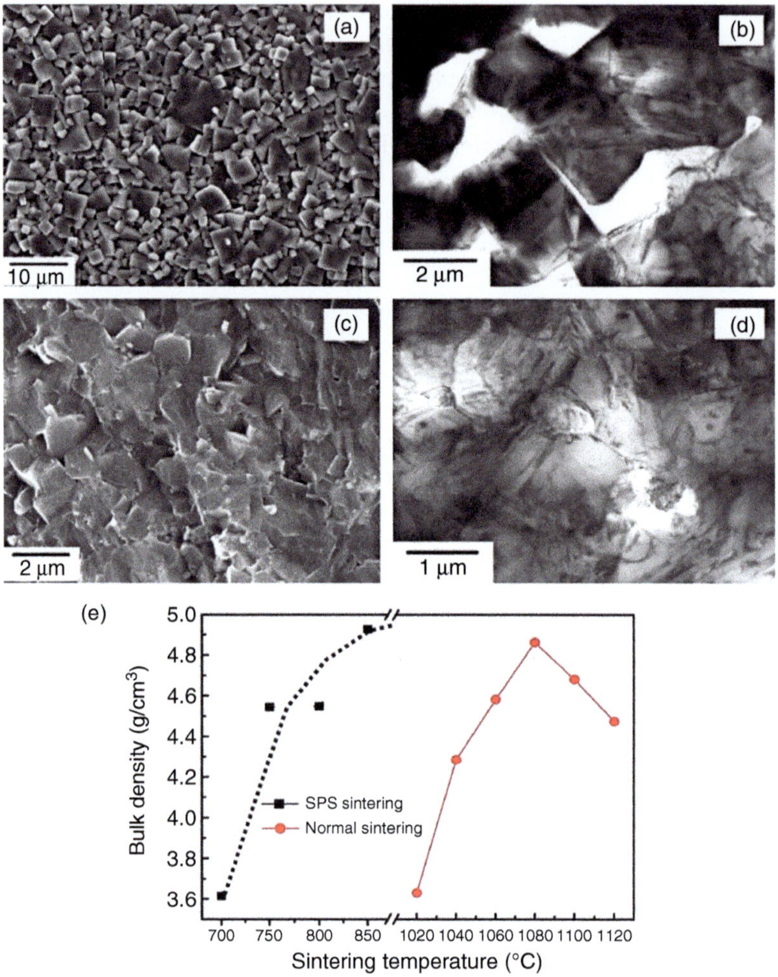

Figure 3.6 The composition of $(Na_{0.535}K_{0.485})_{0.95}Li_{0.05}(Nb_{0.8}Ta_{0.2})O_3$ ceramics: (a) SEM image, (b) TEM image from normally sintered ceramics, (c) SEM image, (d) TEM image of spark plasma sintering (SPS)-sintered ceramics, and (e) bulk density as a function of sintering temperature in comparison with SPS method and conventional solid-state synthesis. Source: Shen et al. [69] and (e) Reprinted with permission from Shen et al. [69]. Copright 2010, John Wiley & Sons, Inc.

at higher temperature, as shown in Figure 3.6. Benefiting from the dense and refined microstructure, the SPS-processed samples have 50% higher mechanical strength than the normally sintered ones [69]; for some applications like actuators, sufficiently high mechanical properties are needed [12].

MWS utilizes electromagnetic energy to heat green bodies. This technology is featured of uniform heating, high heating rate, enhanced diffusion processes, and reduced energy consumption [75]. Despite inconsistent results, this method was reported to have the potential to increase the density, reduce grain size, and lower sintering temperature for KNN-based ceramics [76–82]. Single-phase KNN powders with orthorhombic perovskite structure could be synthesized by microwave-assisted calcination by heating up to 950 °C without soaking time [79]. It was also revealed that microwave-sintered samples exhibited higher piezoelectric constants and improved ferroelectric behavior due to higher density and lower loss of volatile elements as compared with the samples prepared by conventional sintering [80]. Importantly, Lee and coworkers [78] found that MWS could be successfully applied to the co-firing of KNN/AgPd multilayered ceramic actuator with a firing cycle that is 10 times shorter at a lower firing temperature. They also confirmed that microwave-sintered specimens showed electric field-induced strain comparable with those obtained using conventional sintering.

3.2.3 Texturing

Texturing has been proven to be an effective way to obtain high-performance piezoelectric ceramics [83]. A high-performance textured KNN-based ceramic was developed in 2004, having d_{33} of 416 pC/N [10]. Based on high-performance KNN-based compositions, texturing along the (001) directions has been used for obtaining ultrahigh d_{33} [10, 35, 84, 85]. Texturing of ceramics can be realized by hot pressing, reactive templated grain growth (RTGG), templated grain growth (TGG), and multilayer grain growth (MLGG) [83]. For KNN-based ceramics, which are of high-symmetric perovskite structures, the RTGG and TGG techniques are the two promising methods, which are based on tape casting [83]. Both of these two methods are based on tape casting techniques that use the shear force to align the anisotropic templates to realize the texture.

Figure 3.7 shows the preparation of textured KNN by using tape casting [86]. The process begins with the preparation of a slurry, which consists of the ceramic powders, binders, and organic solvents. Then, the slurry will be subjected to ball milling for several hours. After ball milling, crystal templates are now added into the well-mixed slurry. During tape casting, the slurry is slowly cast on a rolling

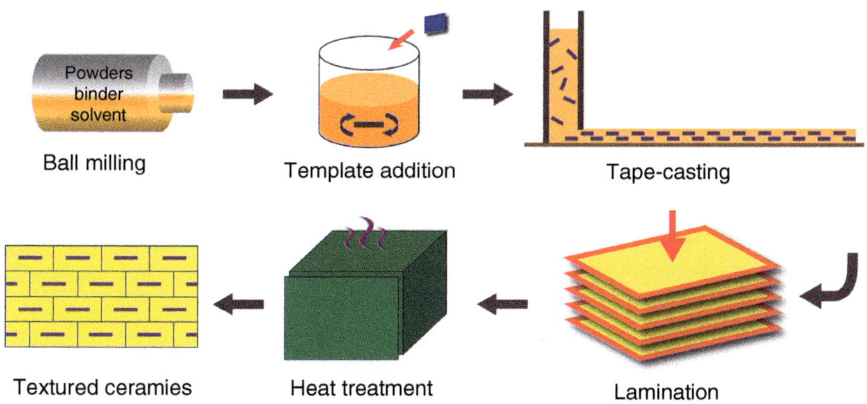

Figure 3.7 Process flow diagram of fabricating textured ceramics by tape casting. Source: Modified from Zhai et al. [86].

supporting film, where the thickness of the casted slurry can be controlled by adjusting the height of the cutting edge. It is worth noting that the templates will align along the rolling direction smoothly, with the help of cutting edge. After drying, the tapes are cut, stacked, and laminated in sequence. The laminated tape is then cut into green bodies before subjected to a series of heat treatments, including de-binding and sintering.

The template plays an important role in the fabrication of textured ceramics. The formation of a textured ceramic via TGG can be mainly divided into three stages, i.e. the nucleation of templates, rapid growth of textured grains, and slow growth of textured grains after impingement [87]. When sufficient thermal activation energy is provided, the heteroepitaxial nucleation of crystallites takes place on the lowest energy surfaces, i.e. $\{001\}_{pc}$ plane, forming into textured grains. Following the nucleation, the textured grains grow rapidly at the expense of small grains. The templates and the matrix should have similar lattice parameters to avoid lattice mismatch. For KNN-based materials, $NaNbO_3$ is the most frequently used template [10, 88]. There are several ways to prepare templates, such as hydrothermal synthesis, molten salt synthesis, and sol–gel synthesis. Saito et al. used two-step molten-salt-synthesized plate-like $NaNbO_3$ templates, which was of 0.5 μm in thickness and 10–15 μm in length and width, to prepare textured KNN-based ceramics [10]. The two-step molten salt synthesis begins with the preparation of a bismuth-layer structure precursor, i.e. $Bi_{2.5}Na_{3.5}Nb_5O_{18}$ (BiNN5), in molten NaCl salt. Then, let BiNN5 react with Na_2CO_3 in the molten NaCl salt, in which Na substitutes into Bi in BiNN5 via topological chemical reactions. After washing away the Bi_2O_3 by-products by using concentrated niobic acid and distilled water, the plate-like $NaNbO_3$ templates can be obtained.

The texturing degree depends on the volume fraction, shape, and distribution of templates. For example, a low volume fraction of templates leads to a poor texturing degree, while excess templates would bring a negative impact to the densification of ceramics and their piezoelectric properties. A good example can be found in the work by Li et al., where the volume of templates in $0.96(K_{0.5}Na_{0.5})(Nb_{0.965}Sb_{0.035})O_3$–$0.01CaZrO_3$–$0.03(Bi_{0.5}K_{0.5})HfO_3$ matrix was optimized at 3 mol% of template addition (d_{33} = 700 pC/N, d_{33}^* = 980 pm/V, and $K_p = 0.76$) [35].

3.3 Approaches to Piezoelectricity Enhancement

The piezoelectric performance of KNN-based materials is determined by many factors, such as grain size, density, composition, phase structure, domain structure, poling condition, and chemical homogeneity. In Section 3.2, we have introduced the synthesis of KNN-based materials. With precise control of processing variables, grain size and density could be easily engineered. In this section, we will introduce the approaches to piezoelectricity enhancement by chemical modification, which is related to the phase structure and domain structure. In terms of the measurement of piezoelectricity, d_{33} is the most frequently used parameters, although d_{13}

and d_{15} could also be useful in certain applications. As shown in Figure 3.3, the d_{33} of KNN-based materials has been greatly improved and become comparable to the PZT-based ones. However, the enhanced d_{33} is usually sensitive to temperature variation, so improvement of the thermal stability is essential and of high priority. The recent advance has revealed a feasible solution that can improve the piezoelectric coefficient and thermal stability simultaneously, i.e. utilizing local heterogeneity. Also, the poling techniques that can further improve the piezoelectric performance will be introduced in this section.

3.3.1 Phase Engineering

3.3.1.1 O–T Phase Boundary

Manipulating phase structure is always one of the effective strategies for piezoelectricity enhancement in ferroelectric materials. As shown in Figure 2.2, the KNN system has a complicated phase diagram, which provides a lot of opportunities to adjust the room temperature phase structure. For pure KNN, the room temperature phase structure is orthorhombic, but it can be changed to tetragonal phase when some substances are incorporated into KNN. That is because the tetragonal–orthorhombic PPT point ($T_{O\text{-}T}$) can be shifted by chemical modification downwards toward room temperature from around 200 °C. Therefore, at room temperature, two phases co-exist in the chemically modified ceramics, leading to enhanced piezoelectricity due to increased polarization directions [9, 18, 89–91]. A lot of chemical elements or compounds have been used to shift the O–T phase transition temperature ($T_{O\text{-}T}$) in the KNN system [12, 13, 15, 90–106], as shown in Figure 3.8 [12, 15]. The evidence of phase coexistence can be manifested by

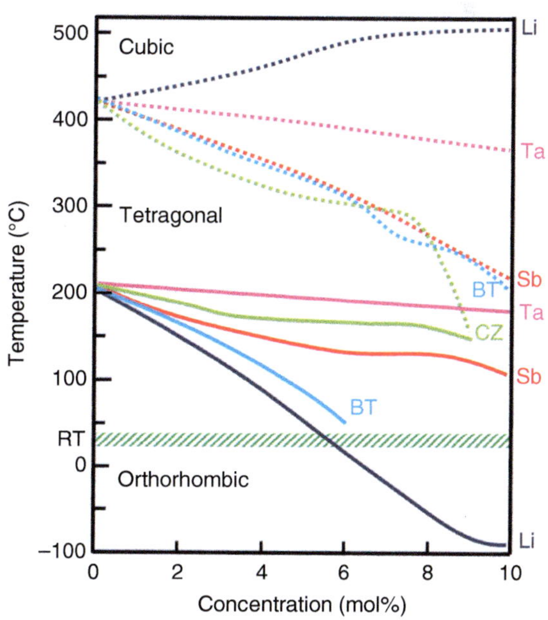

Figure 3.8 Variation of cubic–tetragonal (dots lines) and tetragonal–orthorhombic (solid lines) phase transition temperatures as a function of different doping elements and concentration. Source: Li et al. [12] and Zhang and Li [15].

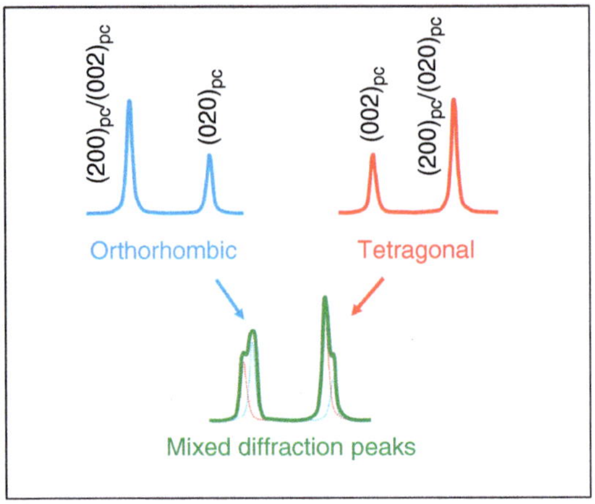

Figure 3.9 Illustration of the 200$_{pc}$ peak splitting in the cases of pure tetragonal phase, pure orthorhombic phase, and orthorhombic–tetragonal phase coexistence. Source: Reprinted with permission from Wang and Li [18]. Copyright 2007, AIP Publishing.

diffraction measurements (e.g. X-ray diffraction or neutron diffraction). Although both *P*4*mm* and *Amm*2 space groups have two peaks at 200$_{pc}$ reflection, the intensity ratio between the two peaks is different, as shown in Figure 3.9 [18]. The *P*4*mm* space group has lattice parameters satisfying $a = b < c$, and the lower angle reflection intensity for (002)$_{pc}$ is half of the higher angle reflection intensity for (200)$_{pc}$ plus (020)$_{pc}$; the *Amm*2 space group has lattice parameters satisfying $a = c > b$, and the lower angle reflection intensity for (200)$_{pc}$ plus (002)$_{pc}$ is double of the higher angle reflection intensity for (020)$_{pc}$. Therefore, when orthorhombic and tetragonal phases coexist, the XRD diffraction pattern would have four peaks at 200$_{pc}$. Sometimes, due to the similarity of lattice parameters between orthorhombic and tetragonal phases as well as the broadening of reflection peaks, the reflections from the two phases may overlap and appear as two peaks. However, phase coexistence can be manifested by judging the ratio of intensities [18]. The state of phase coexistence can be further confirmed by other measurements, such as temperature-dependent or composition-dependent XRD measurement, transmission electron microscopy (TEM) and spectroscopy.

Although all the dopants can shift the T_{O-T} downwards they exhibit distinct efficiencies, as seen in Figure 3.8. The efficiency is mainly associated with the difference in ionic radii, which causes stress [15]. In fact, the phase structure reflects the response of materials to the thermal, chemical, and mechanical pressures. For instance, the Ta^{5+} has a close ionic radius to that of Nb^{5+} and its incorporation does not apparently shift the phase transition temperature while Sb^{5+}, having a much larger radius than Nb^{5+}, exhibits much higher impact. The O–T phase coexistence does not guarantee high d_{33}, e.g. 300 pC/N or even above. Higher d_{33} values >400 pC/N were also reported in Li/Ta/Sb-codoped KNN-based ceramics [107], in which O–T phase coexistence was claimed but the phase structure would be more complicated in term of the compositions. In fact, the most commonly obtained values are among 200–250 pC/N, which could be attributed to several reasons. The first reason is the ratio between the two phases. There is an optimal ratio for the phase fractions to achieve the maximum potential. If the T_{O-T} is

Table 3.1 Effects of dopants on phase transition temperatures. Note: Coordination Number(CN)=12 for A-site, CN=6 for B-site.

Substitutes	Sites	Ionic radii (Å)	T_{R-O}	T_{O-T}	T_C
Li^+	A	1.25*		↓↓↓	↑↑
Ag^+	A	1.54*		↓	↓
Ca^{2+}	A	1.34		↓	↓
Ba^{2+}	A	1.61		↓↓	↓↓
La^{3+}	A	1.36	↑↑	↓↓	↓↓↓
Bi^{3+}	A	1.45*	↑↑↑	↓↓↓	↓↓
Fe^{3+}	B	0.645	↑	↓↓↓	↓↓
Ti^{4+}	B	0.605	↑↑	↓↓	↓↓
Zr^{4+}	B	0.72	↑↑↑	↓↓↓	↓↓
Hf^{4+}	B	0.71	↑↑↑	↓↓↓	↓↓
Sb^{5+}	B	0.60	↑↑	↓↓	↓↓↓
Ta^{5+}	B	0.64	↑	↓	↓

Arrows indicate increase or decrease of phase transition temperatures, and the number of arrows signifies efficiency. The asterisk means that the value is deduced by extrapolation method based on Shannon' s radii.

decreased too fast, the optimal phase constitution is difficult to be controlled at room temperature [105], for example, in the case of Li^+ doping, where the discrepancy in d_{33} among various reports can be as large as around ±150 pC/N. The second one is that the phase structure and properties of KNN-based ceramics are so sensitive to processing conditions [25, 43], such as sintering temperature, that sometimes the reproducibility is not guaranteed. The third factor is associated with a complex stress situation in polycrystalline materials that can influence the phase constitution and piezoelectric properties [106].

The addition of dopants or substances does not only shift T_{O-T} alone but also affect the T_C and the rhombohedral–orthorhombic (R–O) phase transition temperature (T_{R-O}) at the same time, as shown in Table 3.1, which will be described later. It should be pointed out that most dopants would decrease the T_C except the Li^+ and Li^+-containing species, as shown in Figure 3.8. Lithium doping is often indispensable in the KNN-based ceramic to suppress the reduction of T_C, which is usually shifted down with T_{O-T} to create a two-phase coexistence state.

3.3.1.2 R–T Phase Boundary

Another research activity of phase engineering was to shift upwards the R–O phase transition boundary from low temperature (~−160 °C) to room temperature. Wang et al. found that the orthorhombic–rhombohedral (O–R) transition temperature could be moved to room temperature by incorporating 8–15 mol% $BaZrO_3$ into KNN [108]. Interestingly they also found that tetragonal–orthorhombic (T–O) transition temperature can be shifted down if KNN is alloyed with $MTiO_3$ (M = Pb, Ba, Sr, Ca), but incorporating those perovskites with Ti at B-site less affected the O–R phase transition [109]. In addition to Zr, Hf, and Sb, substitutions for Nb were revealed effective to shift upwards the O–R transition temperature [13, 15, 34, 110–115]. Therefore, the phase coexistence of rhombohedral and orthorhombic phases is also possible in KNN-based materials. As both $T_{O\text{-}T}$ and $T_{R\text{-}O}$ continue to be shifted downwards and upwards, respectively, the temperature range for the orthorhombic phase becomes smaller and smaller. Eventually, the orthorhombic phase can no longer be found at room temperature and a new phase transition from rhombohedral–tetragonal (R–T) phase transition occurs [13]. In the XRD pattern, the rhombohedral phase has a single symmetric peak at $\{200\}_{pc}$ while the tetragonal phase has two peaks. The formation of R–T phase coexistence has been confirmed by different characterization techniques, e.g. XRD, convergent beam electron diffraction (CBED), spherical-aberration-corrected TEM, and Raman spectroscopy (Figure 3.10) [34, 114, 115]. However, the complete absence of the orthorhombic phase may not be possible for all compositions and formulations. Therefore, in some literature, the R–O–T phase transition is referred to as an rhombohedral–orthorhombic–tetragonal (R–O–T) phase transition. Constructing the R–T phase transition in the vicinity of room temperature is harder than merely constructing the R–T phase transition. It often requires the corporation of multiple dopants as well as the Na/K ratio and tuning each variable individually.

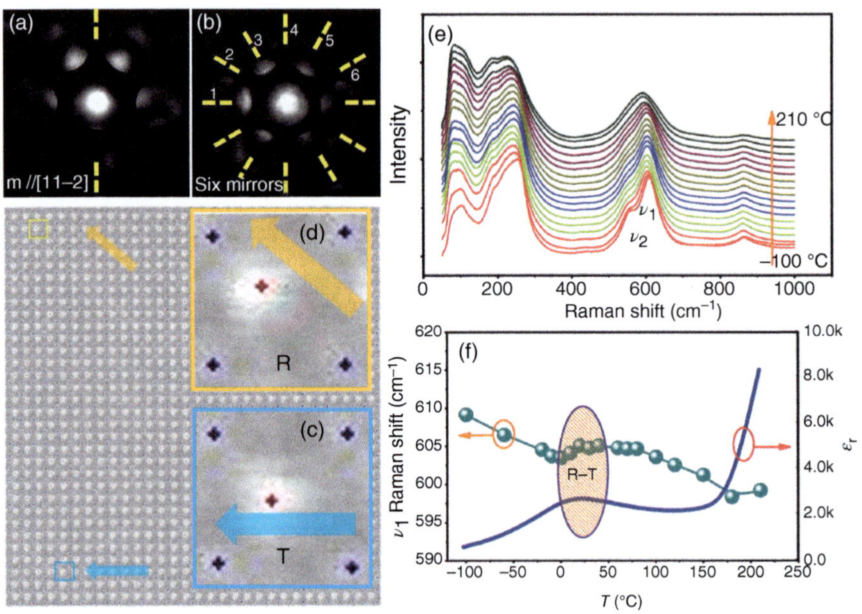

Figure 3.10 (a, b) Convergent beam electron diffraction (CBED) patterns of KNNS–BZ–BKH ceramics from [111] incidence [34]; (c, d) STEM HAADF (high angle annular dark field) lattice images of KNNS–BF–BNZ ceramics [114]; (e, f) temperature dependence of Raman spectra and dielectric constant of KNNS–BNKH ceramics [115]. Source: (a, b) Xu et al. [34] and (e, f) Reprinted with permission from Zheng et al. [115]. Copyright 2017, Royal Society of Chemistry.

Usually, multiple dopants with different effects and efficiencies are used to tune the phase transitions, i.e. R–O and O–T, and merge them precisely around room temperature. Table 3.1 shows some typical substitutions and their effects on each phase transition temperature. It is noted that most of these data are based on temperature-dependent measurements with limited support of diffraction measurements above room temperature. Information on phase transition below room temperature (i.e. R–O) has been rarely reported.

The R–T phase boundary has been proved to greatly increase the piezoelectricity. For example, an ultrahigh d_{33} of 570 ± 10 pC/N was reported in $0.95(K_{0.6}Na_{0.4})(Nb_{0.965}Sb_{0.035})O_3$–$0.02BaZrO_3$–$0.03Bi_{0.5}K_{0.5}HfO_3$ with the R–T phase boundary [34]. Table 3.2 lists the properties of some selected compositions with the R–T or R–O–T phase boundary near or at room temperature [34, 113–120]. In such compositions multiple phases coexist, leading to more significant piezoelectricity enhancement than achieved by manipulating the O–T phase boundary. As shown in Figure 3.11, most compositions with the R–T phase boundary at room temperature can have a d_{33} higher than 400 pC/N [15]. The minimized energy barrier for the polarization rotation between R and T phases has been proposed to be the origin of the enhancement [33, 121–124]. In addition, among three different types of phase boundary, compositions with the R–T phase boundary possesses the lowest coercive field and thus is easy to be poled [125]. It is worth noting that the R–T phase boundary has the lowest T_C. The T_C is influenced by the c/a ratio in tetragonal perovskites. The c/a ratio can be calculated from the splitting of 200_{pc} in the XRD pattern. The larger splitting reflects more distortions apart from the ideal cubic ABO_3 cell. On the other hand, the relationship between c/a ratio and d_{33} is rationalized in that large c/a ratio implies large spontaneous polarization but not necessarily high d_{33}, as smaller c/a ratio favors domain wall motions [12], due to the reduced internal compatibility stress and could contribute to d_{33} as extrinsic contributions. However, it is argued whether the d_{33} should increase monotonously with a reduction of c/a because once spontaneous polarization is too small, it cannot induce an effective piezoelectric response. In some materials like $BiFeO_3$–$PbTiO_3$, there is an optimal value for c/a ratio [126]. Unfortunately, such an optimal value has not been found in KNN-based materials.

3.3.2 Thermal Stability

Although tetragonal and rhombohedral phases can co-exist in KNN ceramics, it is realized by merging the two thermotropic phase boundaries, i.e. O–T and O–T, at high and low temperatures, respectively. Therefore, the R–T phase boundary in KNN is different from the one in PZT, as shown in Figure 3.12. The R–T phase boundary of PZT is sensitive to compositional change and is nearly independent of temperature, termed as MPBs. On the contrary, for KNN-based materials, the phase coexistence is very sensitive to temperature change and is called PPT [12, 127]. At PPT, the phase structure changes from one phase to the other upon an increase in temperature, which means that the high piezoelectric performance is only available at the phase transition point, leading to poor thermal stability. PPT also applies to the O–T and

Table 3.2 Select compositions of R–T phase transition and their properties.

Composition	T_C (°C)	d_{33} (pC/N)	Large signal E_{max} (kV/mm)	Large signal d^*_{33} (pm/V)	Large signal Strain (%)	k_p	ε_r	E_c (kV/cm)	P_r	References
$0.96(K_{0.5}Na_{0.5})_{0.95}Li_{0.05}Nb_{0.93}Sb_{0.07}O_3$–$0.04BaZrO_3$	197	425				0.5	3157	9.9	15.8	[113]
$0.965(K_{0.48}Na_{0.52})(Nb_{0.95}Sb_{0.05})O_3$–$0.035Bi_{0.5}(Na_{0.82}K_{0.18})_{0.5}HfO_3$	220	525	2	465	0.093		2700	5	14.3	[114]
$0.955K_{0.5}Na_{0.5}Nb_{0.96}Sb_{0.04}O_3$–$0.01BaSnO_3$–$0.035Bi_{0.5}Na_{0.5}ZrO_3$	225	480	4	550	0.22	0.5	2600	8.5	17.5	[115]
$0.95K_{0.6}Na_{0.4}Nb_{0.965}Sb_{0.035}O_3$–$0.02BaZrO_3$–$0.03Bi_{0.5}K_{0.5}HfO_3$	190	570				0.49	4200	6	17.4	[34]
$0.964K_{0.6}Na_{0.6}Nb_{0.955}Sb_{0.045}O_3$–$0.006BiFeO_3$–$0.03Bi_{0.5}Na_{0.5}ZrO_3$	237	550				0.53	2641	6	23.2	[116]
$0.93(Na_{0.52}K_{0.48})(Nb_{0.975}Sb_{0.025})O_3$–$0.05BaZrO_3$ –$0.02(Bi_{0.5}Na_{0.5})HfO_3$ + 1 wt% MnO_2	170	370	1	600	0.065	0.52	2700	4.5	13.5	[117]
$0.955K_{0.5}Na_{0.5}NbO_3$–$0.045Bi_{0.5}Na_{0.5}ZrO_3$ + 0.4% MnO	345	300	2.5	430		0.512	1570	12.5	17.3	[118]
$0.96(K_{0.5}Na_{0.5})(Nb_{0.96}Sb_{0.04})O_3$–$0.04Zn_{0.025}(Bi_{0.5}K_{0.5})_{0.975}ZrO_3$	250	442				0.5	2619	8.9	16.9	[119]
$(K_{0.48}Na_{0.52})(Nb_{0.97}Sb_{0.03})O_3$–$0.05(Ca_{0.2}Bi_{0.4}Na_{0.4})ZrO_3$	243	470				0.524	1563	10	16.7	[120]

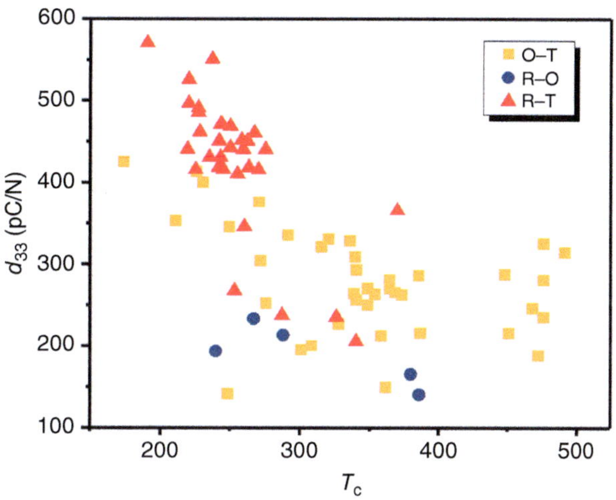

Figure 3.11 The d_{33} and T_C of KNN-based compositions classified according to their phase coexistence types at room temperature. Source: Reprinted with permission from Zhang and Li [15]. Copyright 2019, Royal Society of Chemistry.

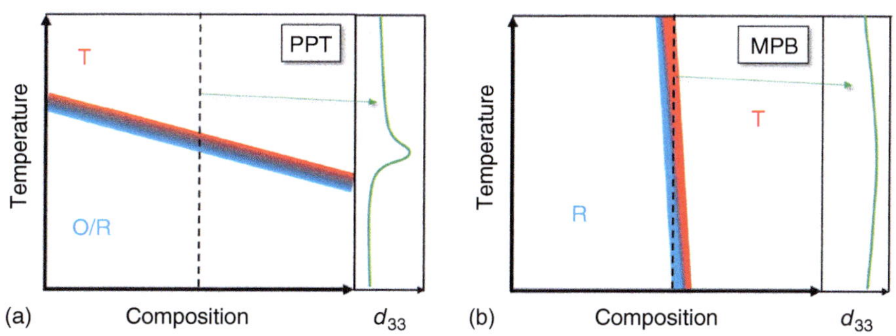

Figure 3.12 Comparison of (a) PPT and (b) MPB including their enhancement on d_{33} for an example composition marked by the dashed line.

R–O phase boundary in the KNN system. The poor thermal stability has also become a significant issue that hinders KNN as well as other lead-free piezoceramics from replacing PZT in large-scale applications [12, 128].

During application, the piezoelectric materials are not only subjected to the external heat from the working environment but also suffer from self-heating during cycling. In fact, the self-heating effect can be significant, e.g. heating up to 80 °C for a bulk material and up to 140 °C for a multilayer actuator [129]. The mechanisms for self-heating include piezoelectric loss, dielectric loss, and mechanical loss. The *in situ* thermal stability is associated with all piezoelectric materials and is critical to many applications, e.g. actuators, which require a stable and accurate strain output. Piezoceramics with PPT would lose the benefit of phase boundary and suffer from a rapid decrease of piezoelectric response if the ambient temperature deviates from the phase transition temperature. Therefore, extra attention should be paid during the design of the phase boundary.

There are several strategies to improve the *in situ* thermal stability of KNN-based materials. The first idea is dealing with the origin of heat. Although the ambient temperature changes are unavoidable, the magnitude of self-heating could be reduced by decreasing the energy loss. The self-heating can be severe when working at high frequency, where the generated heat has insufficient time to be released from the system. Apart from the improvement of the heat-damping system from the engineering point of view, the decrease of dielectric loss would be an alternative option from the chemical modification aspect. Therefore, additives that can increase the mechanical quality factor (Q_m) should be selected for the high-frequency applications [130–132]. The second idea is to shift the phase transition below room temperature and to avoid phase transition. This approach can improve the stability of temperature-dependent piezoelectricity but at the cost of low d_{33} [15, 133–136]. The third idea is to fabricate textured materials [10, 84, 137–139]. This method can significantly improve the thermal stability, though the processing is very complicated. The fourth idea is to broaden the phase transition range of PPT [109, 117, 140]. For example, the phase transition temperature range can be broadened by stacking R–O and O–T phase transition successively to create an R–O–T phase transition, as illustrated in Figure 3.13 [111]. Since the poor thermal stability of KNN ceramics originates from the PPT effect, it would be an ultimate method to resolve it if an MPB like that in PZT can be created by incorporating a new compound into the KNN system. Such an attempt was reported, but the resultant piezoelectricity enhancement was not significant. Of course, excellent thermal stability can be achieved if the phase transients are moved out from the application temperature range [141, 142], but that will sacrifice the two-phase coexisting effect for piezoelectricity enhancement.

It is worth noting that the thermal stability of small-signal d_{33} and large-signal d_{33}^* (normalized piezoelectric strain) may not coincide. Usually, the thermal stability of d_{33}^* is better than that of d_{33}, due to the stimulation of a large electric field [143, 144]. Wang et al. found that the field-induced strain at 4 kV/mm varies less than 10% from room temperature to 175 °C in $CaZrO_3$-modified (Li,K,Na)(Nb,Ta)O_3 (abbreviated KLNNT) ceramics. But, the small-signal d_{33} (piezoelectric coefficient measured by

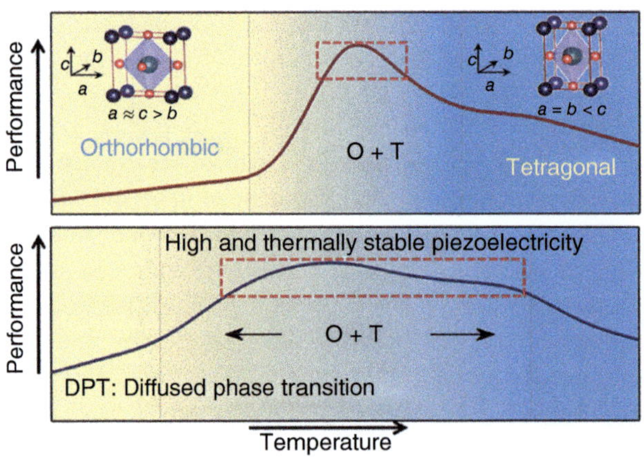

Figure 3.13 Schematic diagram for the enhancement of piezoelectric performance and thermal stability by introducing the diffused phase transition. Source: Reprinted with permission from Liu et al. [111]. Copyright 2018, Royal Society of Chemistry.

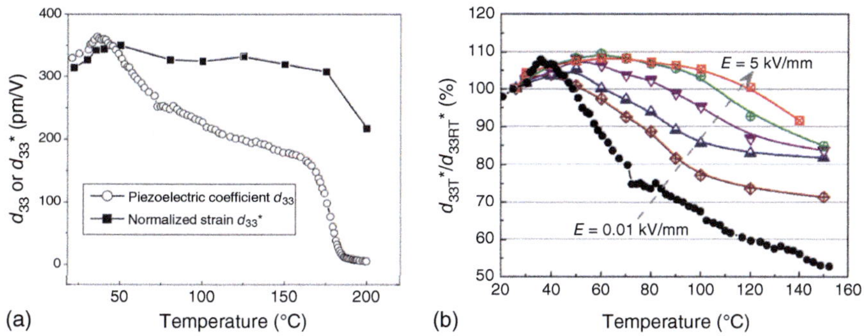

Figure 3.14 (a) Temperature dependence of large signal d^*_{33} measured at 4 kV/mm and small signal d_{33}. (b) d^*_{33} normalized to its room temperature value d_{33RT}* of (K,Li,Na)(Nb,Ta)O_3–$CaZrO_3$ ceramics under electric fields ranging from 0.01 to 5 kV/mm. Source: (a) Reprinted with permission from Wang et al. [143]. Copyright 2013, John Wiley & Sons, Inc. (b) Reprinted with permission from Yao et al. [144]. Copyright 2016, John Wiley & Sons, Inc.

a Berlincourt-type meter) displays a maximum at about 36 °C and then decreases significantly with temperature, as shown in Figure 3.14a [143]. *In situ* synchrotron X-ray diffraction experiments revealed occurrence of the electric-field-induced phase transition from tetragonal to orthorhombic phase, leading to an electrically enhanced diffused polymorphic phase transition (EED-PPT) [144]. As shown in Figure 3.14b, when the applied electric field is lowered gradually, the normalized large-signal d^*_{33} decreases apparently with increasing temperature, and the curve at a low field of 0.01 kV/mm is similar to that of the temperature dependence of small-signal d_{33} shown in Figure 3.14a.

3.3.3 Multiscale Heterogeneity

In reality, chemical heterogeneity can hardly be avoided, and it also plays a critical role in piezoelectric performance. It is worth noting that the heterogeneity should be assessed at different length scales, as shown in Figure 3.15 [15]. A material that is seemingly homogeneous macroscopically can show heterogeneity at the microscopic scale. Also, distorted unit cells may appear to be pseudo-cubic at insufficient resolution, e.g. a relaxor ferroelectric material. The heterogeneity at specific scales can be used to enhance piezoelectric performance, but at other scales may deteriorate the properties.

For KNN-based materials, the un-doped KNN is considered as of the highest homogeneity. The chemical heterogeneity can only be observed at the unit cell scale where the K and Na occupy the A-site alternatively. Despite the different ionic properties of Na and K, the two atoms are elastically coupled, and only a single phase is observed. However, compositional inhomogeneity was even observed in the calcined KNN powders, which was found to be mainly associated with the source of Nb_2O_5 raw materials [145]. Certainly, the addition of dopants will introduce inhomogeneity to unit cells due to different valence states, ionic radii,

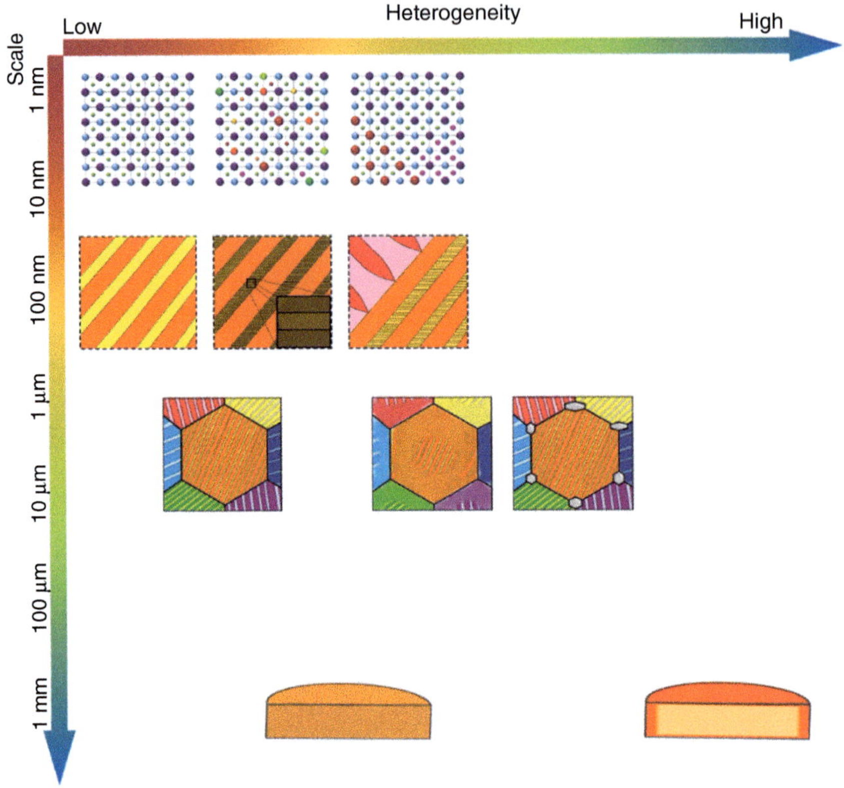

Figure 3.15 Heterogeneity at different length scales. Source: Reprinted with permission from Zhang and Li [15]. Copyright 2019, Royal Society of Chemistry.

and electronegativities [146]. The aliovalent doping introduces vacancies into the matrix; the different ionic radii cause tension and stress; electronegativity of ions varies the bonding fractions and strength (as shown in Table 3.3). Despite elastic coordination, local stress heterogeneities and stress fields can exist and influence the surrounding unit cells. The density and distribution of these stress fields, as well as possible foreign-dopant-induced defects, determine the domain morphology at the next length scale.

In Figure 3.16, most compositions that possess a high d_{33} are doped with a variety of dopants [15]. Particularly, the combination of Bi doping at A-site and Zr and/or Hf doping at B-site was found most effective for the piezoelectricity enhancement [15, 117, 147–150]. However, the increased variety of dopants can significantly reduce the diffusion rates of elements and suppress long-range ordering. Therefore, nano-sized domains are often observed in these materials and appear as a hierarchical domain structure and decreased average domain sizes (Figure 3.17) [151]. The increase of the variety of dopants would increase the density and diversity of stress fields and would be more likely to result in the formation of nano-domains. The decreased domain size leads to an increase in domain wall density, leading to an enhancement of piezoelectricity [117, 151]. Therefore, the design of chemical

Table 3.3 The electronegativity and bonding fraction with oxygen of commonly used elements for piezoelectric ceramics (I = ionic, C = covalent).

Element	Electronegativity	*I*%	*C*%
Li	0.98	77.97	22.03
Na	0.93	79.30	20.70
K	0.82	82.02	17.98
Ca	1	77.43	22.57
Sr	0.95	78.78	21.22
Ba	0.89	80.32	19.68
Pb	1.87	46.00	54.00
Bi	2.02	39.60	60.40
Ti	1.54	59.44	40.56
Zr	1.33	67.14	32.86
Hf	1.3	68.17	31.83
Nb	1.6	57.10	42.90
Ta	1.5	60.97	39.03
Sb	2.05	38.31	61.69

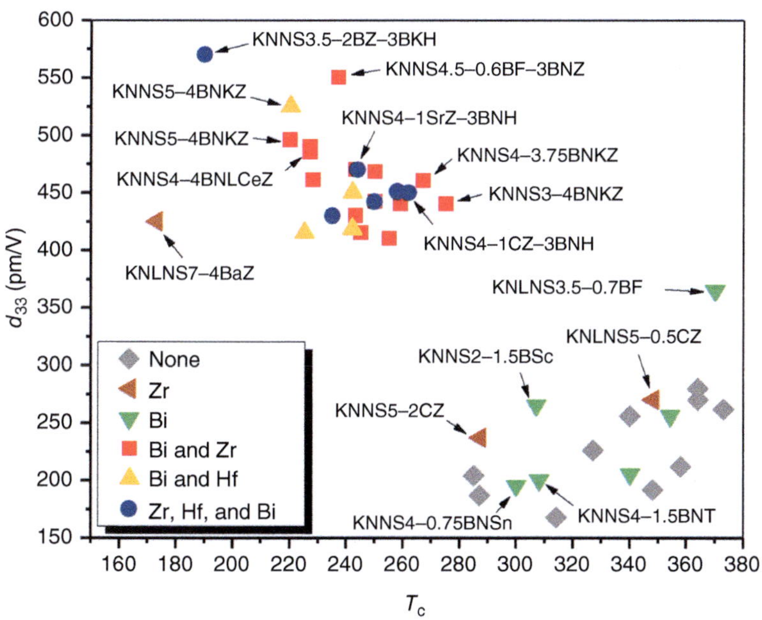

Figure 3.16 d_{33} and Curie temperature of compositions in KNN-system with the constitution of Zr, Bi, and Hf highlighted in color (L = Li, S = Sb, H = Hf, F = Fe, B = Bi, T = Ti, C = Ca, and Z = Zr). Source: Reprinted with permission from Zhang and Li [15]. Copyright 2019, Royal Society of Chemistry.

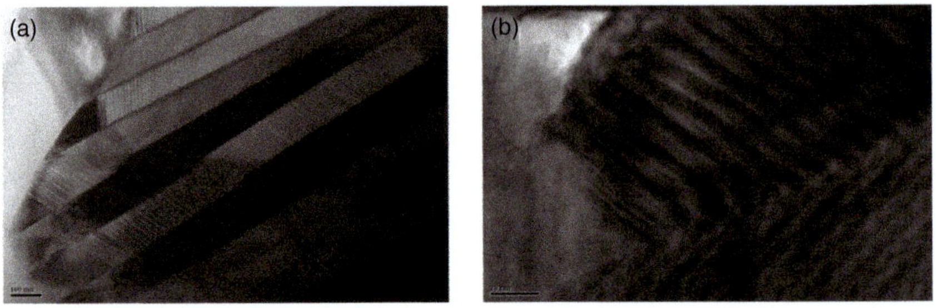

Figure 3.17 Hierarchical domain structure of $BaZrO_3$–(Bi,Na)TiO_3-modified (Li,K,Na)(Nb,Sb)O_3 ceramics: (a) 100 nm domains and (b) 5 nm domains. Source: Liu et al. [151].

compositions plays an important role in achieving high d_{33}. In addition to an appropriate chemical composition, the formation of nano-sized domains is strongly influenced by the raw materials and the fabrication and processing techniques. If the dopants are not evenly distributed in the materials but agglomerate or even segregate, it may result in increased domain size or even phase-structure segregation, i.e. a region in rich of certain phase assemblage [152, 153].

Furthermore, the presence of polar nanoregions (PNRs) also happens on the nanoscale. It is a kind of heterogeneity, but which usually enhances piezoelectricity, especially large-signal d_{33}^*. The PNRs are a region that has different polar orientations from the matrix, usually originating from chemical disorder, random bond/field appearance, and local symmetry lowering. They are often found in relaxor ferroelectric materials and are beneficial to enhance the large-signal d_{33}^* significantly, due to the reduction of remnant strain. The addition of relaxor additives may also introduce the feature of PNRs into KNN-based materials, where the Moiré fringes, an indication for PNRs, has been observed as shown in Figure 3.18 [117]. Compositions with PNRs were found to show enormous large-signal d_{33}^* close to 600 pm/V accompanied by excellent thermal stability [117]. It has been found that

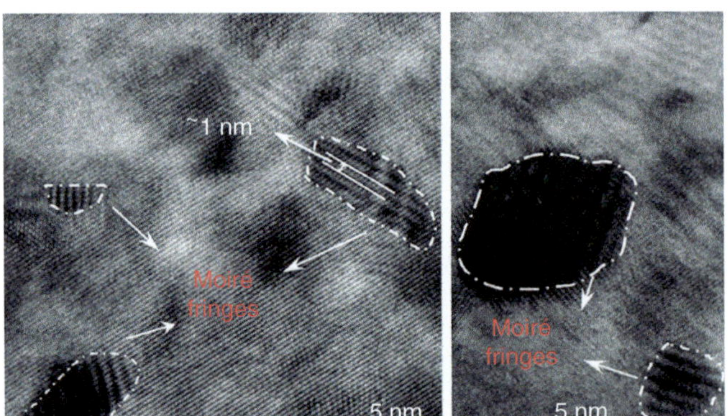

Figure 3.18 Polar nanoregions (PNRs) as Moiré fringes observed in a KNN-based material. Source: Liu et al. [117].

the addition of substances that possess relaxor characteristics [152], e.g. BNT and BNZ [153–157], may also lead to local instability that enhances large-signal d_{33}^*.

3.3.4 Poling Techniques

The as-sintered polycrystalline piezoceramics contain randomly orientated grains. Meanwhile, inside these grains, there are domains with different spontaneous polarization directions, which help retain energetic minimum, and hence no global polarization can be measured. As expected, the as-sintered piezoelectric ceramics do not show piezoelectricity unless they are poled. The purpose of poling is to switch all the randomly oriented domains to one specific direction. The goodness of alignment determines the piezoelectric properties. Therefore, the optimized poling conditions that maximize the alignment of domains are very important to realize the full potential of designed piezoelectric compositions [158]. There are several parameters in the poling process, including poling electric field, duration of poling, and poling temperature [125, 158–160]. Usually, the KNN-based materials are poled with an electric field that is 2–3 times higher than ones' coercive fields mostly at a high temperature (e.g. 120 °C) for 30 minutes or so, followed by field-cooling to room temperature. However, the optimal poling temperature is found to vary depending on the modified composition for phase boundary engineering [161].

One special poling technique is to re-pole samples after aging. The application of this poling technique on an optimized Li-doped KNN composition yields an increase of piezoelectric coefficient d_{33} from 190 to 324 pC/N [26]. It should be noted that this treatment is most effective in the suitable phase structure with orthorhombic–tetragonal phase coexistence, whereas deviations toward either end-members lead to ineffectiveness. The large increment of d_{33} is considered due to the improvement of remnant polarization P_r after the special treatment, which is evidenced by the apparent domain switching deduced from XRD patterns, as shown in Figure 3.19 [26]. This phenomenon is quite unique for KNN that no similar effects have been reported in other systems. One possible mechanism is the change of spontaneous polarization in the domain level, concerning the combined effects of migration of oxygen vacancies and interaction between defect dipoles and spontaneous polarizations inside domains.

3.4 Fatigue and Mechanical Properties

3.4.1 Fatigue

Piezoelectric materials often suffer from the degradation of properties, or even failure, during the mechanical/electrical cycling, so-called fatigue. Fatigue determines the lifetime of materials and thus is another critical factor for commercialization. The degradation of properties could be caused by many reasons, e.g. the damage to the electrode, mechanical cracking or damaged layer, and/or domain pinning by defect dipoles or charge agglomeration. A series of comprehensive investigations

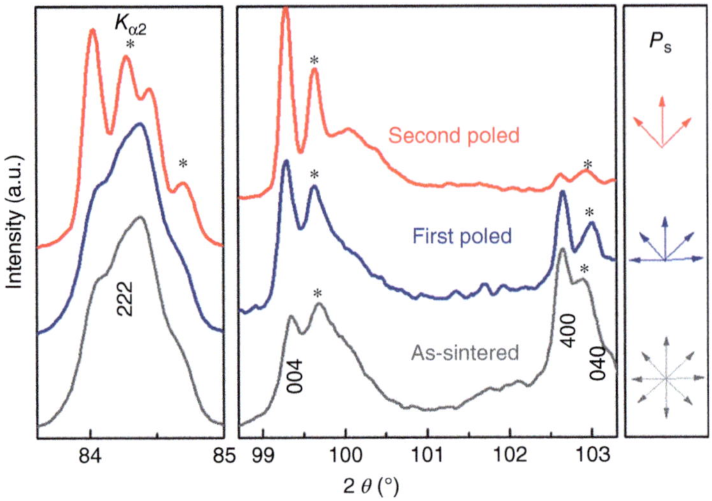

Figure 3.19 X-ray diffraction patterns of (222) and (004) lattice plane series for the sample $0.92(Na_{0.535}K_{0.48})NbO_3-0.08LiNbO_3$, both before and after the first poling, as well as after the second poling. The evolution of P_s inside domains is schematically illustrated in a separate box on the right-hand side, assuming that the electric field applied during poling is parallel to (001) direction. Source: Reprinted with permission from Wang and Li [26]. Copyright 2010, John Wiley & Sons, Inc.

have been conducted about the fatigue behaviors of KNN-based ceramics under different loading conditions [133, 142, 162–169].

The unipolar fatigue behavior of $CaZrO_3$ and MnO_2 co-modified (K,Na,Li)(Nb,Ta)O_3 ceramics, which possess high converse piezoelectricity and enhanced thermal stability, has been investigated systematically [162]. It was found that this ceramic also shows good fatigue resistance by exhibiting only small degradation within 3% at the field amplitude of 2 kV/mm up to 10^7 cycles, being comparable to that of soft commercial PZT, as shown in Figure 3.20a,b [163]. Moreover, the unipolar strain shows a fatigue-free behavior even at elevated temperatures and electric fields. Despite no significant degradation in the strain behavior, the characteristic of unipolar fatigue, namely, the internal bias field, cannot be neglected, particularly at elevated temperatures and electric fields, as shown in Figure 3.20c [165]. The presence of the internal bias field in unipolar fatigued KNN-based piezoceramics is explained by the charge agglomeration near the grain boundary, as illustrated in Figure 3.20d [163]. During electrical cycling, the polarization state in each domain changes continuously. At grain boundaries, the polarization vectors of domains in neighboring grains cannot compensate for each other, due to the crystallographic mismatch. This leads to the occurrence of strong, locally varying depolarization fields (denoted by σ^+ and σ^-). Free charge carriers in the vicinity will be redistributed to compensate for the local depolarization fields. This agglomeration of charges forms a local bias field E_{bias}, which has the same orientation as the cycling field. The superposition of all local bias fields leads to the macroscopically observable shift of the hysteresis loops. It is worth noting that

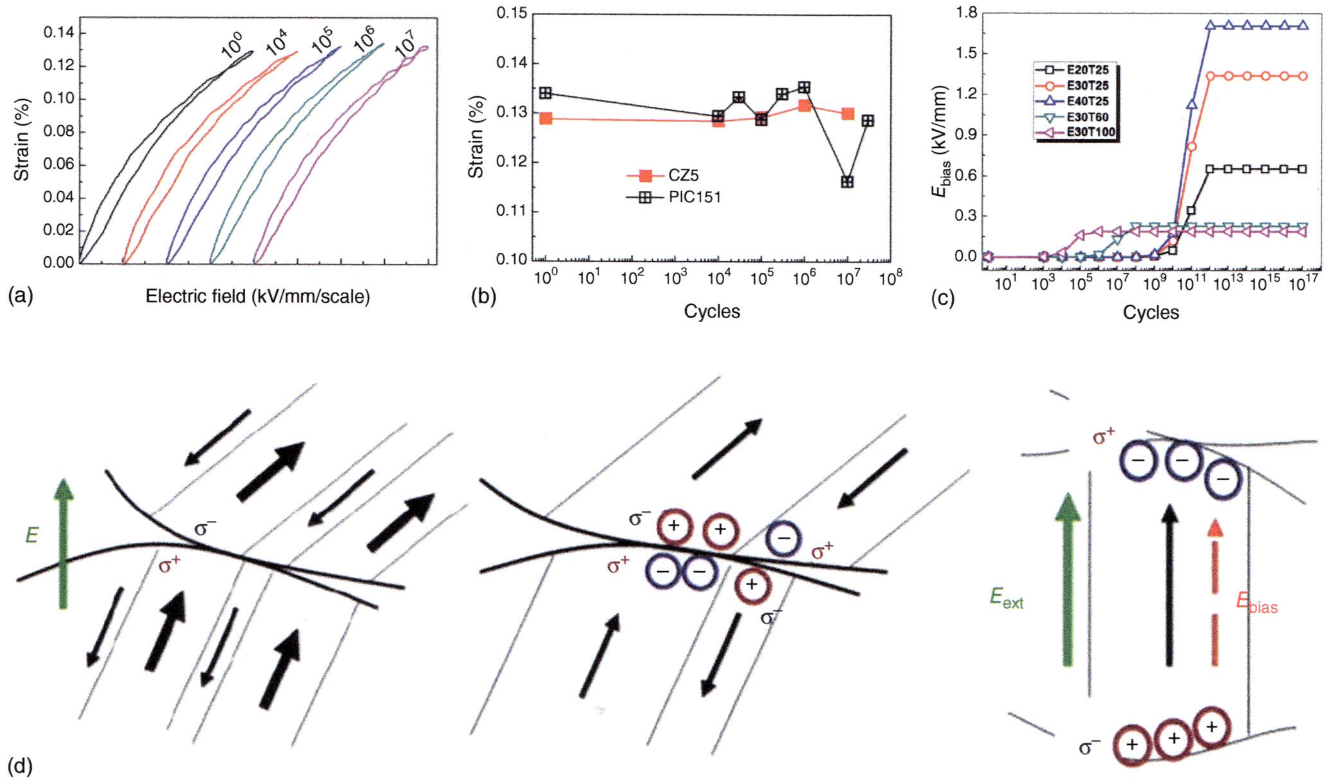

Figure 3.20 (a) Unipolar $S_{33}-E_3$ curves of CZ5 ceramics before and after 10^4, 10^5, 10^6, and 10^7 unipolar fatigue cycles; (b) strain of CZ5 ceramics as a function of cycles with the data of a soft commercial PZT (PIC151) provided for comparison; (c) the calculated internal bias field E_{bias} dependent on different temperatures and cycling fields; and (d) illustration of charge agglomeration near the grain boundary during the fatigue process. Source: Yao et al. [163, 165].

after unipolar cycling, the electrical properties of $CaZrO_3$ and MnO_2 co-modified KNN-based piezoceramics can be fully restored by thermal annealing at the high temperatures (well-above the Curie temperature), because of the absence of dramatic microstructural damage induced by unipolar cycling [163, 165].

In the scenario of bipolar loading, it has been reported that pure KNN could suffer from severe properties degradation after bipolar electric cycling due to the agglomeration of oxygen vacancies and the formation of a damaged layer [157]. However, the bipolar fatigue behavior is sensitive to the types of defects, compositions, and phase structures. Therefore, KNN-based materials with different dopants and fabrication techniques can experience different fatigue behaviors. Since the phase structure can significantly influence bipolar fatigue behavior, various strategies have been adopted to improve the bipolar fatigue resistance of KNN-based systems. The earliest investigation on the fatigue behavior of KNN-based materials studied the bipolar fatigue behavior of $CaTiO_3$ (CT) modified KNLNS. The addition of CT could significantly improve the bipolar fatigue resistance of KNLNS materials, which avoids the PPT while at the cost of piezoelectric performance [133]. The domain wall pinned by defect clusters is significant for KNLNS, due to the O–T phase boundary at room temperature. The 90° domain switching occurs easier than in the materials only with tetragonal phase with CT addition [133]. Similar results were reported in KNLNS doped with Sm_2O_3 or BKT [142, 158]. A study on KNNS–CZ–BNH showed the increasing the T fraction of the R–T phase coexistence could improve the fatigue resistance [169]. However, it is well recognized that the PPT effect is indispensable in enhancing the piezoelectricity of KNN-based ceramics. It seems to be contradictory to simultaneously deliver both promising piezoelectric performance and high bipolar fatigue resistance in the KNN system. A feasible solution to this contradiction has been validated to improve the bipolar fatigue resistance of KNN-based piezoceramics near the PPT by adding $CaZrO_3$, due to the reduced lattice distortion (c/a ratio) and absence of microstructural damage thereof, as shown in Figure 3.21 [164].

In addition to phase engineering, appropriate chemical doping that suppresses the formation of oxygen vacancies is also useful against fatigue [84, 168]. The existence of multiscale nanodomain could also be responsible for high fatigue resistance [166].

3.4.2 Mechanical Properties

The reliability of piezoelectric materials is also associated with their mechanical properties. During electromechanical applications, the piezoelectric materials will experience a large mechanical strain in response of the applied electric field, which can lead to fracture. Therefore, fracture toughness is particularly critical to the lifetime of materials and reliabilities.

An investigation on the fracture toughness of $0.92KNN–0.02Bi_{0.5}Li_{0.5}TiO_3–0.06BaZrO_3$ found that the fracture toughness of an unpoled composition was similar to soft PZT, higher than BCTZ, but lower then BNT–BT [170]. The fracture toughness of the unpoled material was reported to be 0.7–0.95 MPa $\sqrt{m}$, depending on the measurement methods. Meanwhile, a poled $0.92KNN–0.02Bi_{0.5}Li_{0.5}TiO_3–0.06BaZrO_3$

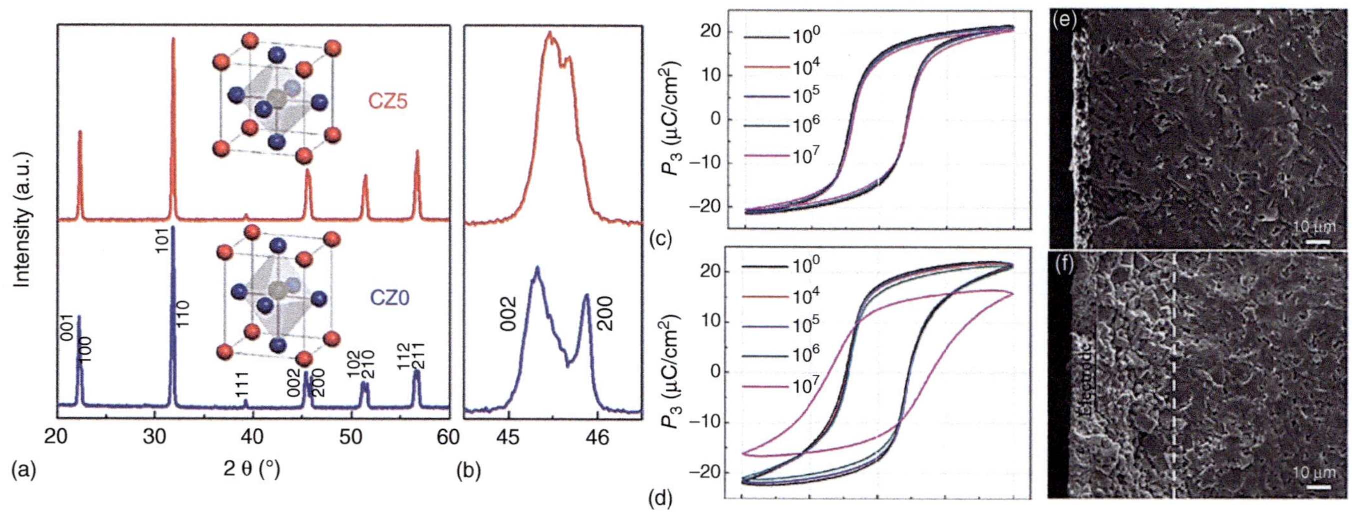

Figure 3.21 (a) XRD patterns of CZ0 and CZ5 ceramics; (b) the enlarged (002) peaks. The inset figures in (a) illustrate the structures of CZ5 and CZ0 ceramics in an exaggerated manner. P_3–E_3 of (c) CZ5 and (d) CZ0 ceramics before and after 10^4, 10^5, 10^6, and 10^7 bipolar fatigue cycles; and SEM images of fracture surfaces of (e) CZ5 and (f) CZ0 ceramics after 10^7 bipolar fatigue cycles. Source: (a–d) Reprinted with permission from Yao et al. [164]. Copyright 2014, AIP Publishing. (e, f) Yao et al. [164].

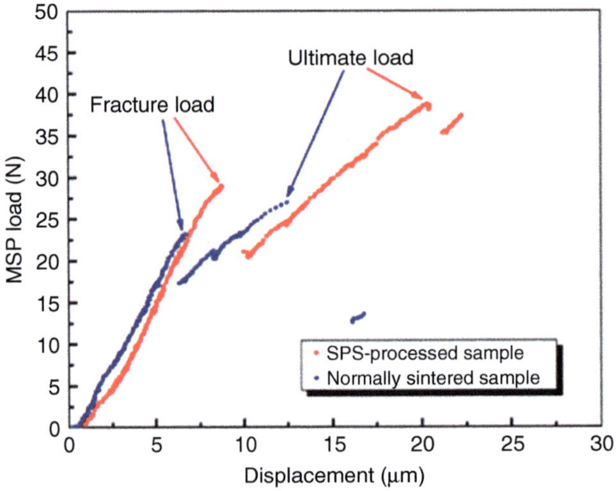

Figure 3.22 Modified small punch (MSP) testing load–displacement curves for SPS-processed and normally sintered $(Na_{0.535}K_{0.485})_{0.95}Li_{0.05}(Nb_{0.8}Ta_{0.2})O_3$ ceramics. Source: Reprinted with permission from Shen et al. [69]. Copyright 2010, John Wiley & Sons, Inc.

was found to have slightly larger fracture toughness than an unpoled one, i.e. 0.88–0.99 MPa $\sqrt{m}$, which is different from the scenario in the PZT. It is also worth noting that the elastic modulus and coercive stress depend on the poling state, crystal structure, and temperatures [171]. Differing from the PZT-based materials, which are nearly independent of the temperature change, the elastic modulus and coercive stress of KNN change with temperature. The fracture strength of KNN-based lead-free piezoceramics is closely related to their microstructure. It is found that dense and fine-grained $(Na_{0.535}K_{0.485})_{0.95}Li_{0.05}(Nb_{0.8}Ta_{0.2})O_3$ ceramics prepared by SPS have considerably higher fracture strength than that fabricated by normal sintering (102.84 vs. 67.9 MPa), as shown in Figure 3.22 [69]. Later, it was demonstrated that chemical modification could also contribute to strength improvement. A tiny amount of $BiFeO_3$ effectively improves the sinterability and densification of (Li,Na,K)(Nb,Ta)O_3 piezoceramics, and hence yielding a significantly enhanced fracture strength of 142 MPa, which is even higher than that of the samples synthesized by SPS as aforementioned [135].

3.5 KNN Thin Films

Driven by the promising properties achieved in the bulk materials, KNN-based films have also drawn increasing attention, because piezoelectric films have been widespread applications for miniaturized devices in microelectromechanical systems (MEMSs) [172–177]. Several methods have been applied to the fabrication of KNN thin films, and main challenges associated with the deposition of high-quality KNN thin films are the composition deviation from stoichiometry and the appearance of the second phase due to the easy volatilization of alkali oxides. Now, the widely used two major methods are the sol–gel process and sputtering, by which high-quality KNN-based films can be fabricated.

3.5.1 Sol–Gel-Processed Films

The sol–gel process has some advantages including simple manufacturing facility and easy control of chemical composition and homogeneity, so it has been well applied to the fabrication of PZT-based films [178, 179]. However, early studies revealed the difficulties of applying this method to the fabrication of KNN films because the source materials containing K and Na are easily hydrolyzed in addition to the easy volatilization during the heat treatment process [180–183]. The volatilization could be compensated by adding excessive raw materials. Meanwhile, Mn doping as well as adding organic agents were treated as an effective way to reduce the leakage current [174, 184–186]. Although the mechanism is still not clear, Mn doping is often applied to the property enhancement in KNN-based bulk materials [187–190]. The preparation of a high-quality and stable sol solution is of primary importance for the sol–gel processing. Special care should be taken to avoid hydrolysis, due to its high alkalinity for the sol–gel processing of KNN films. Figure 3.23 shows the typical processing of the KNN sol solution.

Due to the volatilization of organic compounds, the pyrolysis temperature needs to be optimized to fabricate a sol–gel-processed film of high quality. In addition, various properties depend on the pyrolysis temperature, e.g. crystallinity, surface morphology. It was found that the pyrolysis temperature must be optimized in a range of 400–600 °C [191, 192]. The pyrolysis at lower than 400 °C was found insufficient to complete the decomposition of remnant carbonate phases, which may lead to the formation of pores at the final annealing stage for crystallization. Higher pyrolysis temperatures above 600 °C could cause the start of crystallization, which must happen

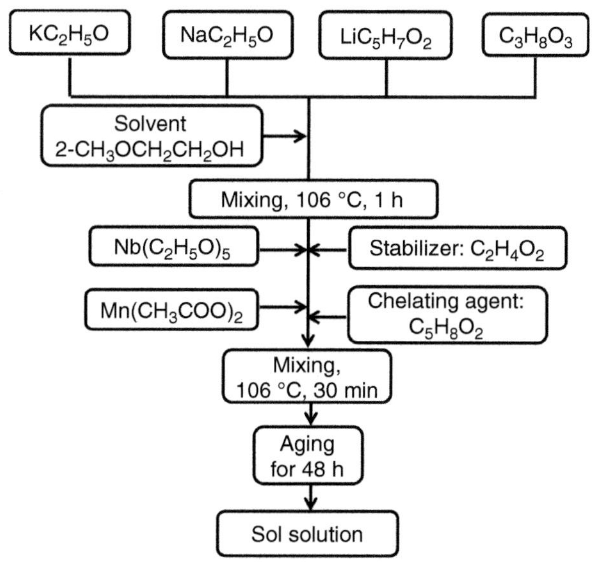

Figure 3.23 The procedure of sol–gel synthesis of the KNN thin film.

at the annealing stage. Zhang et al. found that highly (100)-textured KNN thin films can be fabricated by optimizing the pyrolysis temperature at 550 °C and annealing at 700 °C [192]. For the annealing temperature, a low annealing temperature usually results in cracking and incomplete crystallization with a perovskite-structured phase while a high annealing temperature leads to the volatilization of alkali elements and the formation of secondary phases. The optimal annealing temperature of KNN-based films is in a range of 700–900 °C, depending on compositions [172, 183].

KNN films have inferior piezoelectric and ferroelectric properties compared with bulk ceramics, due to the limited thickness and increased sensitivity to defects. The primary problem for KNN thin films is the high leakage current [193]. Therefore, the reduction of leakage current has been the most critical task for KNN thin films. Similar to bulk ceramics, the addition of an excess amount of K and Na have been found helpful [181], though the magnitude of excess is at an order of 10%, 1 order higher than that of bulk ceramics [194]. The current leakage can also be reduced by introducing other additives, such as Li, Mn, Co, and Ta [172]. For example, the leakage current density of 0.5 mol% Mn-doped KNN film shows the lowest value of $\sim 10^{-7}$ A/cm^2 at 100 V, which is about 4 orders of magnitude lower than that of the undoped KNN thin film [174]. It is evident that the Mn can absorb the free charges left by the volatilization of alkali elements. The Ta doping can inhibit the grain growth resulting in a homogeneous and dense microstructure with little pores [195]. At 150 kV/cm, the leakage is 5 orders lower than the undoped KNN due to the decrease in the number of the pores. The optimization of annealing temperature can also help lower the leakage current [172].

Apart from the quality of the KNN films, the interfaces between the electrodes and KNN films also have a significant impact on the electrical properties, associated with the film orientation, lattice misfit, and residual stress. Conductive perovskite oxide electrodes have lattice constants close to KNN films, which is beneficial for the electrical properties of films. For example, KNN films with $LaNiO_3$ top electrodes are found better ferroelectric and piezoelectric properties compared with that of the films with Pt top electrodes [196]. Another apparent difference between the thin films and bulk ceramics is the use of substrates. For example, KNN films on $SrRuO_3$ buffered $SrTiO_3$ (001) substrate shows improved ferroelectric and current leakage properties and fatigue characteristics [197]. However, the influence of substrate on polycrystalline films is not as significant as the epitaxial films.

The sol–gel process can be applied to fabricate epitaxial KNN films if single crystalline $SrTiO_3$ substrates are used. Yu et al. prepared epitaxial KNN films with three primary orientations of [001], [110], and [111] using Nb-doped $SrTiO_3$ single-crystalline substrates with corresponding cutting directions [198]. As shown in Figure 3.24, a higher remnant polarization value (P_r) of 17.3 μC/cm^2 was obtained along the [110] direction, which is significantly higher than 10.5 μC/cm^2 in [111]-oriented and 10.1 μC/cm^2 in [001]-oriented ones. However, a better piezoelectric response was achieved in the [001]-oriented films with an average local effective piezoelectric coefficient (d_{33}) of 50.5 pm/V, as compared with 45.1 and 39.7 pm/V in [110]- and [111]-oriented films, respectively. This work revealed

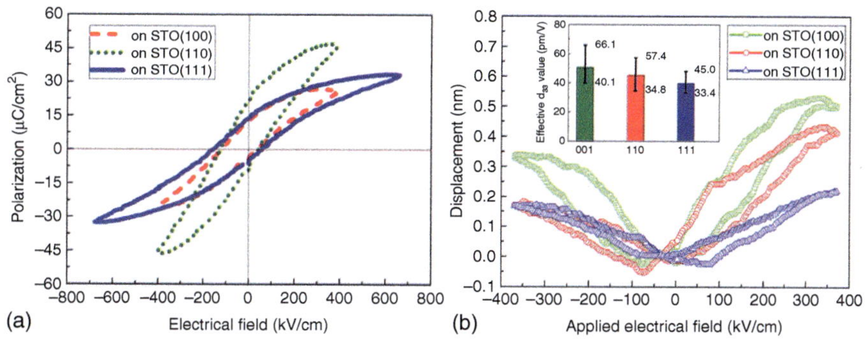

Figure 3.24 (a) Polarization–electric field hysteresis loops and (b) typical displacement–field curves of the KNN films with different orientations. Source: Yu et al. 2013 [198]. Reprinted with permission of AIP Publishing.

the crystallographic orientation dependency of ferroelectric and piezoelectric properties of the KNN system, which is consistent with the spontaneous polarization orientation of orthorhombic phase.

Luo et al. prepared high-quality KNN epitaxial thin films on the (100) Nb:STO substrate and conducted detailed analysis of the phase structure [199, 200]. They found a monoclinic M_C phase at room temperature, which has seldom been experimentally observed in the bulk materials; this is due to strong clamping effect from the substrates in the epitaxial films [201]. The high-quality epitaxial KNN films provided ideal specimens for domain structure analysis by piezoresponse force microscopy (PFM) both at ambient and elevated temperatures. As shown in Figure 3.25, the good epitaxial growth was identified by synchrotron X-ray reciprocal space mapping (RSM) and high-resolution TEM observation of the cross-section. The M_C symmetry has been confirmed by combined analysis of the RSM results and the clear domain structure obtained in the epitaxial film as shown in Figure 3.25. Detailed analysis of the changes in domain structure and local piezoelectricity by PFM during heating and cooling revealed that in the KNN epitaxial films the monoclinic M_C phase transformed into the monoclinic M_A phase around 170 °C, accompanied by the domain structure evolution as shown in Figure 3.25e,f. The local piezoelectricity reaches the highest value at around 170 °C due to the M_C–M_A phase transition [199]. Therefore, engineering the lattice mismatch between the epitaxial film and the substrate can be useful for constructing desirable phase and domain structures.

3.5.2 KNN Films Prepared by Physical Methods

KNN-based films have been prepared by various physical methods, among them pulsed laser deposition (PLD) and radio frequency (RF) magnetron sputtering are the most used ones, which are widely used for the fabrication of a variety of oxide films [202, 203]. It is likely that Grishin and coworkers first demonstrated using PLD to grow a $(K_{0.5}Na_{0.5})NbO_3$ film on Si(111) substrates with a top electrode layer of $Pt_{80}Ir_{20}$ using KrF excimer laser ablation of a stoichiometric ceramic target [204, 205]. The prepared [100]-axis oriented films show a remnant polarization

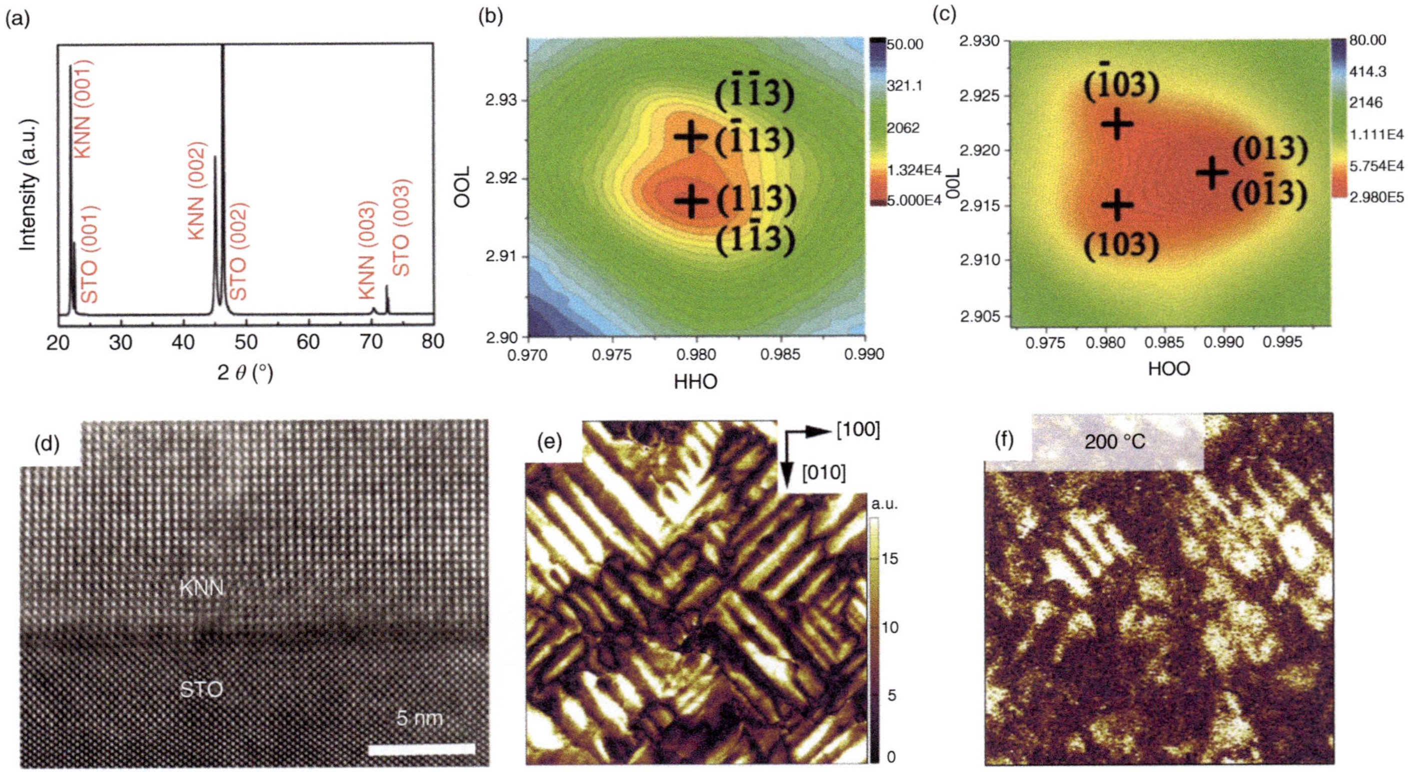

Figure 3.25 (a) XRD pattern of the epitaxial KNN thin film on (001) Nb:SrTO3 substrate. (b, c) Synchrotron X-ray RSMs of $(113)_{pc}$, $(103)_{pc}$ reflections of the film. (d) High-resolution transmission electron microscopy (TEM) image of cross-section. (e, f) Domain at room temperature and 200 °C corresponding to PFM in-plane amplitude. Source: (a) Reprinted with permission from Luo et al. [199]. Copyright 2017, John Wiley & Sons, Inc. (b–f) Luo et al. [199].

P_r of 10 μC/cm^2 and spontaneous polarization P_s of 17.5 μC/cm^2 at 80 kV/cm [205], which are comparable to those of sol–gel processed KNN films [198]. It was found that the deposition temperature and oxygen partial pressure could influence the microstructure and electrical properties of KNN-based thin films. For example, the optimal deposition temperature and oxygen partial pressure for the $(K_{0.5}Na_{0.5})NbO_3$–$LiTaO_3$–$CaZrO_3$ film were reported to be 680 °C and 35 Pa, respectively [206]. Though PLD is a powerful technique, several disadvantages have been noticed, e.g. the formation of interfacial defects during high-energy ion bombardment, the composition deviation from the target, and not suitable for large-scaling production [207].

As compared with PLD, more studies have been conducted on the magnetron sputtering technique for the processing of KNN films [173, 175, 196, 197, 208–210]. As a successful example of R&D of KNN films, a Japanese company realized the mass production of KNN films grown on $Pt/Ti/SiO_2/Si$ substrates using the radio frequency magnetron sputtering technique [173]. Their micrometer-thick KNN films show a high transverse piezoelectric coefficient ($e_{31,f}$) of 14.4 C/m^2, which is comparable with that of commercialized PZT films [173].

3.6 Single Crystals

Despite the limited numbers, the studies about KNN-based single crystals have been conducted. Quite early, KNN single crystals of approximately 1 cm in length with a composition of $0.95(K_{0.5}Na_{0.5})NbO_3$–$0.05LiNbO_3$ were successfully fabricated by Bridgman method by a group in Shanghai Institute of Ceramics, Chinese Academy of Sciences [211]. The (001) plates show good piezoelectric properties with piezoelectric constant d_{33} as high as 405 pC/N, large thickness electromechanical coupling factor $k_t = 61\%$, and low dielectric constant of 185 at room temperature. Later, a large size and high-quality Ta-modified KNN-based single crystal was successfully grown by using the top seeded solution growth (TSSG) technique method with homogeneous composition [212–214]. The crystals show very high electromechanical coupling factors ($k_{33} = 0.827$, $k_t = 0.646$) with low dielectric loss tangent (0.004) (Figure 3.26) [212]. Li-doped KNN single crystals were also fabricated by using a simple low-cost seed-free solid-state crystal growth (SFSSCG) method, which exhibit a higher piezoelectric coefficient d_{33} of 689 pC/N with a high converse piezoelectric coefficient d_{33}^* up to 967 pm/V [215].

At the same time, some studies have been conducted about the synthesis and property characterization of KNN-based nanocrystals, which are expected to be utilized in the fields of nano-electromechanical system (NEMS), energy harvesting devices, photocatalysis, and so on [216]. Single-crystalline KNN nanorods can be synthesized via facile molten-salt reaction [21, 217–219]. Cheng et al. found that it is possible to control the K/Na ratio of KNN nanorods to reach an optimal composition with high piezoelectric output. The hydrothermal method, which is a little bit complicated than the molten-salt reaction method, can be used to fabricate KNN nanocrystals with ordered structures [21]. Xu et al. reported on a hydrothermal process to grow

Figure 3.26 Photograph of the as-grown KNNT single crystal. Source: Zheng et al. [212].

vertically aligned KNN nanorods on $SrTiO_3$ substrate, which had a cross-sectional width about 100 nm and were up to 1–2 μm long along its (100) direction [220]. The vertical arrays of KNN nanorods may find potential applications for energy harvesting nanodevices and nanoscale 1–3 type multifunctional composites.

3.7 Summary

Since 2004 when Saito et al. published their milestone work in Nature, KNN system has been extensively studied, which has made the KNN-based ceramics almost ready for practical applications as lead-free alternatives for PZT-based ceramics. As an apparent and important indicator of the research progress, the piezoelectric charge coefficient (d_{33}) was raised to a high level of 570 pC/N and exceeding 700 pC/N for textured ceramics, which is even higher than some PZT-based ceramics. Despite the narrow sintering window, high-performance KNN-based ceramics can also be produced by the widely used normal sintering process, if special attention is paid to the compositional and microstructural control during the powder calcination and sintering processes. The property enhancement in the KNN system was realized mainly by manipulating the phase boundaries among tetragonal and orthorhombic as well rhombohedral phases. Certainly, the enhanced piezoelectricity lacks thermal stability if compared with the PZT-based ceramics possessing a temperature-insensitive MPB, but several valid approaches have been found to mitigate or solve this problem.

The progress achieved in bulk materials drove the researches of KNN-based thin films, which will find more applications both in the existing and emerging fields. In particular, lead-free piezoelectrics should be more welcome for the new technological applications related to biomedical and mobile electronics. Epitaxial thin films provide good specimens for fundamental studies in addition to specific applications. Of course, more studies about the single crystals are also needed, which should possess higher piezoelectricity than polycrystalline ceramics and also are more suitable for fundamental physical/chemical researches.

References

1 Matthias, B.T. (1949). New ferroelectric crystals. *Physical Review* 75: 1771.

2 Shirane, G., Newnham, R., and Pepinsky, R. (1954). Dielectric properties and phase transitions of $NaNbO_3$ and $(Na,K)NbO_3$. *Physical Review* 96: 581–588.

3 Egerton, L. and Dillon, D.M. (1959). Piezoelectric and dielectric properties of ceramics in the system potassium–sodium niobate. *Journal of the American Ceramic Society* 42: 438–442.

4 Tennery, V.J. and Hang, K.W. (1968). Thermal and X-ray diffraction studies of the $NaNbO_3$–$KNbO_3$ system. *Journal of Applied Physics* 39: 4749–4753.

5 Tellier, J., Malic, B., Dkhil, B. et al. (2009). Crystal structure and phase transitions of sodium potassium niobate perovskites. *Solid State Sciences* 11: 320–324.

6 Jaeger, R.E. and Egerton, L. (1962). Hot pressing of potassium–sodium niobates. *Journal of the American Ceramic Society* 45: 209–213.

7 Li, J.-F., Wang, K., Zhang, B.-P., and Zhang, L.M. (2006). Ferroelectric and piezoelectric properties of fine-grained $Na_{0.5}K_{0.5}NbO_3$ lead-free piezoelectric ceramics prepared by spark plasma sintering. *Journal of the American Ceramic Society* 89: 706–709.

8 Zhang, B.P., Li, J.-F., Wang, K., and Zhang, H.L. (2006). Compositional dependence of piezoelectric properties in $Na_xK_{1-x}NbO_3$ lead-free ceramics prepared by spark plasma sintering. *Journal of the American Ceramic Society* 89: 1605–1609.

9 Guo, Y.P., Kakimoto, K., and Ohsato, H. (2004). Phase transitional behavior and piezoelectric properties of $(Na_{0.5}K_{0.5})NbO_3$–$LiNbO_3$ ceramics. *Applied Physics Letters* 85: 4121–4123.

10 Saito, Y., Takao, H., Tani, T. et al. (2004). Lead-free piezoceramics. *Nature* 432: 84–87.

11 Cross, E. (2004). Materials science – lead-free at last. *Nature* 432: 24–25.

12 Li, J.-F., Wang, K., Zhu, F.-Y. et al. (2013). $(K,Na)NbO_3$-based lead-free piezoceramics: fundamental aspects, processing technologies and remaining challenges. *Journal of the American Ceramic Society* 96: 3677–3696.

13 Wu, J.G., Xiao, D.Q., and Zhu, J.G. (2015). Potassium–sodium niobate lead-free piezoelectric materials: past, present, and future of phase boundaries. *Chemical Reviews* 115: 2559–2595.

14 Roedel, J. and Li, J.-F. (2018). Lead-free piezoceramics: status and perspectives. *MRS Bulletin* 43: 576–580.

15 Zhang, Y.C. and Li, J.-F. (2019). Review of chemical modification on potassium sodium niobate lead-free piezoelectrics. *Journal of Materials Chemistry C* 7: 4284–4303.

16 Shirane, G., Danner, H., Pavlovic, A., and Pepinsky, R. (1954). Phase transitions in ferroelectric $KNbO_3$. *Physical Review* 93: 672–673.

17 Tennery, V.J. (1965). High-temperature phase transitions in $NaNbO_3$. *Journal of the American Ceramic Society* 48: 537–539.

18 Wang, K. and Li, J.-F. (2007). Analysis of crystallographic evolution in (Na,K)NbO$_3$-based lead-free piezoceramics by X-ray diffraction. *Applied Physics Letters* 91: 262902.

19 Wang, K., Zhang, B.-P., Li, J.-F., and Liu, H.-J. (2007). $Na_{1-x}K_xNbO_3$ ($x = 0.2$–0.8) lead-free piezoelectric ceramics prepared by spark plasma sintering. *Key Engineering Materials* 336–338: 224–227.

20 Dai, Y.J., Zhang, X.W., and Chen, K.P. (2009). Morphotropic phase boundary and electrical properties of $K_{1-x}Na_xNbO_3$ lead-free ceramics. *Applied Physics Letters* 94: 042905.

21 Cheng, L.-Q., Wang, K., Li, J.-F. et al. (2014). Piezoelectricity of lead-free (K,Na)NbO$_3$ nanoscale single crystals. *Journal of Materials Chemistry C* 2: 9091–9098.

22 Wu, L., Zhang, J.L., Wang, C.L., and Li, J.C. (2008). Influence of compositional ratio K/Na on physical properties in $(K_xNa_{1-x})NbO_3$ ceramics. *Journal of Applied Physics* 103: 084116.

23 Zhou, J.-J., Li, J.-F., Wang, K., and Zhang, X.-W. (2011). Phase structure and electrical properties of (Li,Ta)-doped (K,Na)NbO$_3$ lead-free piezoceramics in the vicinity of Na/K = 50/50. *Journal of Materials Science* 46: 5111–5116.

24 Kang, Y., Zhao, Y., Huang, R. et al. (2011). Effect of changing Na/K ratio on structure and electrical properties of $(Na_xK_y)(Nb_{0.885}Sb_{0.08})$–$0.035LiTaO_3$ lead-free piezoelectric ceramics. *Journal of the American Ceramic Society* 94: 1683–1686.

25 Zhao, P., Zhang, B.-P., and Li, J.-F. (2007). High piezoelectric d_{33} coefficient in Li-modified lead-free (Na,K)NbO$_3$ ceramics sintered at optimal temperature. *Applied Physics Letters* 90: 242909.

26 Wang, K. and Li, J.-F. (2010). Domain engineering of lead-free Li-modified (K,Na)NbO$_3$ polycrystals with highly enhanced piezoelectricity. *Advanced Functional Materials* 20: 1924–1929.

27 Wang, R.P., Xie, R.J., Hanada, K. et al. (2005). Phase diagram and enhanced piezoelectricity in the strontium titanate doped potassium–sodium niobate solid solution. *Physica Status Solidi A: Applications and Material Science* 202: R57–R59.

28 Zhang, S.J., Xia, R., Shrout, T.R. et al. (2006). Piezoelectric properties in perovskite $0.948(K_{0.5}Na_{0.5})NbO_3$–$0.052LiSbO_3$ lead-free ceramics. *Journal of Applied Physics* 100: 104108.

29 Zhao, P., Zhang, B.-P., and Li, J.-F. (2007). Enhancing piezoelectric d_{33} coefficient in Li/Ta-codoped lead-free (Na,K)NbO$_3$ ceramics by compensating Na and K at a fixed ratio. *Applied Physics Letters* 91: 172901.

30 Wu, J.G., Xiao, D.Q., Wang, Y.Y. et al. (2007). Effects of K/Na ratio on the phase structure and electrical properties of $(K_xNa_{0.96-x}Li_{0.04})(Nb_{0.91}Ta_{0.05}Sb_{0.04})O_3$ lead-free ceramics. *Applied Physics Letters* 91: 252907.

31 Zuo, R., Fu, J., and Lv, D. (2009). Phase transformation and tunable piezoelectric properties of lead-free $(Na_{0.52}K_{0.48-x}Li_x)(Nb_{1-x-y}Sb_yTa_x)O_3$ system. *Journal of the American Ceramic Society* 92: 283–285.

32 Zheng, T., Wu, J.G., Xiao, D.Q., and Zhu, J.G. (2015). Giant d_{33} in nonstoichiometric (K,Na)NbO_3-based lead-free ceramics. *Scripta Materialia* 94: 25–27.

33 Wang, X.P., Wu, J.G., Xiao, D.Q. et al. (2014). Giant piezoelectricity in potassium–sodium niobate lead-free ceramics. *Journal of the American Chemical Society* 136: 2905–2910.

34 Xu, K., Li, J., Lv, X. et al. (2016). Superior piezoelectric properties in potassium–sodium niobate lead-free ceramics. *Advanced Materials* 28: 8519–8523.

35 Li, P., Zhai, J.W., Shen, B. et al. (2018). Ultrahigh piezoelectric properties in textured (K,Na)NbO_3-based lead-free ceramics. *Advanced Materials* 230: 1705171.

36 Feizpour, M., Ebadzadeh, T., and Jenko, D. (2016). Synthesis and characterization of lead-free piezoelectric $(K_{0.50}Na_{0.50})NbO_3$ powder produced at lower calcination temperatures: a comparative study with a calcination temperature of 850 °C. *Journal of the European Ceramic Society* 36: 1595–1603.

37 Wu, W.J., Chen, M., Bo, W. et al. (2017). Modulation of electrical properties of KNNS–BNKZ lead-free ceramics by calcination temperatures. *Journal of Materials Science: Materials in Electronics* 28: 2086–2092.

38 Malic, B., Kuraza, J., Hrescak, J. et al. (2015). Sintering of lead-free piezoelectric sodium potassium niobate ceramics. *Materials* 8: 8117–8146.

39 Wei, N., Wang, J., Li, B. et al. (2015). Improvement of the piezoelectric and ferroelectric properties of $(K_{0.5},Na_{0.5})NbO_3$ ceramics via two-step calcination-milling route. *Ceramics International* 41: 9555–9559.

40 Popovič, A., Bencze, L., Koruza, J., and Malic, B. (2015). Vapour pressure and mixing thermodynamic properties of the $KNbO_3$–$NaNbO_3$ system. *RSC Advances* 5: 76249–76256.

41 German, R.M. (1996). *Sintering Theory and Practice*. New York, NY: Wiley.

42 Kang, S.-J.L. (2005). *Sintering: Densification, Grain Growth, and Microstructure.* Oxford: Elsevier Butterworth-Heinemann.

43 Zhen, Y.H. and Li, J.-F. (2006). Normal sintering of (K,Na)NbO_3-based ceramics: influence of sintering temperature on densification, microstructure, and electrical properties. *Journal of the American Ceramic Society* 89: 3669–3675.

44 Li, J.-F., Zhen, Y.H., Zhang, B.-P. et al. (2008). Normal sintering of (K,Na)NbO_3-based lead-free piezoelectric ceramics. *Ceramics International* 34: 783–786.

45 Shen, Z.-Y., Wang, K., and Li, J.-F. (2009). Combined effects of Li content and sintering temperature on polymorphic phase boundary and electrical properties of Li/Ta Co-doped (Na,K)NbO_3 lead-free piezoceramics. *Applied Physics A* 97: 911–917.

46 Ahn, C.-W., Park, C.-S., Choi, C.-H. et al. (2009). Sintering behavior of lead-free (K,Na)NbO_3-based piezoelectric ceramics. *Journal of the American Ceramic Society* 92: 2033–2038.

47 Zhang, S., Lee, H.J., Ma, C., and Tan, X. (2011). Sintering effect on microstructure and properties of (K,Na)NbO_3 ceramics. *Journal of the American Ceramic Society* 94: 3659–3665.

48 Lee, Y.-H., Cho, J.H., Kim, B.-I., and Choi, D.-K. (2008). Piezoelectric properties and densification based on control of volatile mass of potassium and sodium in $(K_{0.5}Na_{0.5})NbO_3$ ceramics. *Japanese Journal of Applied Physics* 47: 4620–4622.

49 Acker, J., Kungl, H., and Hoffmann, M.J. (2010). Influence of alkaline and niobium excess on sintering and microstructure of sodium–potassium niobate $(K_{0.5}Na_{0.5})NbO_3$. *Journal of the American Ceramic Society* 93: 1270–1281.

50 Fang, J., Wang, X.H., Zuo, R.Z. et al. (2011). Narrow sintering temperature window for (K,Na) NbO_3-based lead-free piezoceramics caused by compositional segregation. *Physica Status Solidi A: Applications and Material Science* 208: 791–794.

51 Wang, K. and Li, J.-F. (2012). (K,Na)NbO_3-based lead-free piezoceramics: phase transition, sintering and property enhancement. *Journal of Advanced Ceramics* 1: 24–37.

52 Song, H.C., Cho, K.H., Park, H.Y. et al. (2007). Microstructure and piezoelectric properties of $(1-x)(Na_{0.5}K_{0.5})NbO_3-xLiNbO_3$ ceramics. *Journal of the American Ceramic Society* 90: 1812–1816.

53 Cheng, L.-Q., Wang, K., Yao, F.-Z. et al. (2013). Composition inhomogeneity due to alkaline volatilization in Li-modified (K,Na)NbO_3 lead-free piezoceramics. *Journal of the American Ceramic Society* 96: 2693–2695.

54 Zhen, Y.H. and Li, J.-F. (2007). Abnormal grain growth and new core–shell structure in (K,Na)NbO_3-based lead-free piezoelectric ceramics. *Journal of the American Ceramic Society* 90: 3496–3502.

55 Jo, W., Kim, D.-Y., and Hwang, N.-M. (2006). Effect of interface structure on the microstructural evolution of ceramics. *Journal of the American Ceramic Society* 89: 2369–2380.

56 Shen, Z.-Y., Zhen, Y.H., Wang, K., and Li, J.-F. (2009). Influence of sintering temperature on grain growth and phase structure of compositionally optimized high-performance Li/Ta-modified (Na,K)NbO_3 ceramics. *Journal of the American Ceramic Society* 92: 1748–1752.

57 Wang, K. and Li, J.-F. (2010). Low-temperature sintering of Li-modified (K,Na)NbO_3 lead-free ceramics: sintering behavior, microstructure and electrical properties. *Journal of the American Ceramic Society* 93: 1101–1107.

58 Matsubara, M., Yamaguchi, T., Sakamoto, W. et al. (2005). Processing and piezoelectric properties of lead-free (K,Na)(Nb,Ta)O_3 ceramics. *Journal of the American Ceramic Society* 2005: 1190–1196.

59 Park, S.H., Ahn, C.W., Nahm, S., and Song, J.S. (2004). Microstructure and piezoelectric properties of ZnO-added $(Na_{0.5}K_{0.5})NbO_3$ ceramics. *Japanese Journal of Applied Physics* 43: L1072–L1074.

60 Park, H.Y., Choi, J.Y., Choi, M.K. et al. (2008). Effect of CuO on the sintering temperature and piezoelectric properties of $(Na_{0.5}K_{0.5})NbO_3$ lead-free piezoelectric ceramics. *Journal of the American Ceramic Society* 91: 2374–2377.

61 Du, H.L., Liu, D.J., Tang, F.S. et al. (2007). Microstructure, piezoelectric, and ferroelectric properties of Bi_2O_3-added $(K_{0.5}Na_{0.5})NbO_3$ lead-free ceramics. *Journal of the American Ceramic Society* 90: 2824–2829.

62 Rubio-Marcos, F., Romero, J.J., Navarro-Rojero, M.G., and Fernandez, J.F. (2009). Effect of ZnO on the structure, microstructure and electrical properties of KNN-modified piezoceramics. *Journal of the European Ceramic Society* 29: 3045–3052.

63 Lin, D.M., Kwok, K.W., and Chan, H.L.W. (2008). Piezoelectric and ferroelectric properties of $K_xNa_{1-x}NbO_3$ lead-free ceramics with MnO_2 and CuO doping. *Journal of Alloys and Compounds* 461: 273–278.

64 Ryu, J., Choi, J.J., Hahn, B.D. et al. (2007). Sintering and piezoelectric properties of KNN ceramics doped with KZT. *IEEE Transactions of Ultrasonics Ferroelectrics and Frequency Control* 54: 2510–2515.

65 Cheng, L.Q., Zhou, J.J., Wang, K. et al. (2012). Influence of ball milling on sintering behavior and electrical properties of $(Li,Na,K)NbO_3$ lead-free piezoceramics. *Journal of Materials Science* 47: 6908–6914.

66 Munir, Z.A., Anselmi-Tamburini, U., and Ohyanagi, M. (2006). The effect of electric field and pressure on the synthesis and consolidation of materials: a review of the spark plasma sintering method. *Journal of Materials Science* 41: 763–777.

67 Zhang, B.-P., Zhang, L.-M., Li, J.-F. et al. (2005). SPS sintering of $NaNbO_3$–$KNbO_3$ piezoelectric ceramics. *Materials Science Forum* 475–479: 1165–1168.

68 Wang, R.P., Xie, R.J., Sekiya, T., and Shimojo, Y. (2004). Fabrication and characterization of potassium–sodium niobate piezoelectric ceramics by spark-plasma-sintering method. *Materials Research Bulletin* 39: 1709–1715.

69 Shen, Z.-Y., Li, J.-F., Wang, K. et al. (2010). Electrical and mechanical properties of fine-grained Li/Ta-modified $(Na,K)NbO_3$-based piezoceramics prepared by spark plasma sintering. *Journal of the American Ceramic Society* 93: 1378–1383.

70 Zhen, Y.H., Li, J.-F., Wang, K. et al. (2011). Spark plasma sintering of Li/Ta-modified $(K,Na)NbO_3$ lead-free piezoelectric ceramics: post-annealing temperature effect on phase structure, electrical properties and grain growth behavior. *Materials Science and Engineering B* 176: 1110–1114.

71 Zhang, H., Wang, X.H., Fang, J. et al. (2013). Piezoelectric properties of Li, Sb, and Ta co-doped $(K,Na)NbO_3$ ceramics with fine grain size sintered by SPS method. *Journal of Electroceramics* 30: 217–220.

72 Bah, M., Giovannelli, F., Schoenstein, F. et al. (2014). High electromechanical performance with spark plasma sintering of undoped $K_{0.5}Na_{0.5}NbO_3$ ceramics. *Ceramics International* 40: 7473–7480.

73 Souza, C.A., Eiras, J.A., and Lente, M.H. (2016). Spark plasma sintering of doped $(K_xNa_{1-x})NbO_3$ piezoceramics. *Ferroelectrics* 499: 47–56.

74 Men, T.L., Yao, F.Z., Zhu, Z.-X. et al. (2016). Piezoelectric properties of $(K_{0.5}Na_{0.5})NbO_3$–$BaTiO_3$ lead-free ceramics prepared by spark plasma sintering. *Journal of Advanced Dielectrics* 6: 1650013.

75 Rybakov, K.I., Olevsky, E.A., and Krikun, E.V. (2013). Microwave sintering: fundamentals and modeling. *Journal of the American Ceramic Society* 96: 1003–1020.

76 Jeong, D.Y., Lee, S.H., Song, H.C. et al. (2011). Hydrothermal synthesis and microwave sintering of $(K_{0.5}Na_{0.5})NbO_3$ ceramics. *Journal of the Korean Physical Society* 58: 663–667.

77 Palei, P., Kumar, P., and Agrawal, D.K. (2012). Structural and electrical properties of microwave processed Ag modified KNN–LS ceramics. *Journal of Microwave Power and Electromagnetic Energy* 46: 76–82.

78 Kang, J.K., Han, H.S., Jeong, S.K. et al. (2013). Microwave and conventional sintering of lead-free $(K,Na)NbO_3$-based piezoelectric ceramic multilayer actuators. *Journal of Ceramic Processing Research* 14: 230–233.

79 Feizpour, M., Bafrooei, H.B., Hayati, R., and Ebadzadwh, T. (2014). Microwave-assisted synthesis and sintering of potassium sodium niobate lead-free piezoelectric ceramics. *Ceramics International* 40: 871–877.

80 Bafandeh, M.R., Gharahkhani, R., Abbasi, M.H. et al. (2014). Improvement of piezoelectric and ferroelectric properties in $(K,Na)NbO_3$-based ceramics via microwave sintering. *Journal of Electroceramics* 33: 128–133.

81 Pang, X.M. (2017). Improved piezoelectricity and luminescence of Er^{3+}/Yb^{3+} co-doped $(K,Na)NbO_3$ ceramics by microwave sintering. *Ferroelectrics* 520: 110–117.

82 Nandinii, R.N., Krishna, M., Suresh, A.V., and Rao, K.N. (2018). Effect of calcination kinetics and microwave sintering parameters on dielectric and pieizo-electric properties of $(Ka_{0.5}Na_{0.5})NbO_3$ ceramics. *Iranian Journal of Materials Science and Engineering* 15: 13–31.

83 Moriana, A.D. and Zhang, S.J. (2018). Lead-free textured piezoceramics using tape casting: a review. *Journal of Materiomics* 4: 277–303.

84 Li, P., Chen, X.Q., Wang, F.F. et al. (2018). Microscopic insight into electric fatigue resistance and thermally stable piezoelectric properties of $(K,Na)NbO_3$-based ceramics. *ACS Applied Materials & Interfaces* 10: 28772–28779.

85 Ge, W.W., Li, J.F., Viehland, D. et al. (2011). Electric-field-dependent phase volume fractions and enhanced piezoelectricity near the polymorphic phase boundary of $(K_{0.5}Na_{0.5})_{1-x}Li_xNbO_3$ textured ceramics. *Physical Review B* 83: 224110.

86 Zhai, J., Liu, B.H., Liu, X. et al. (2017). A review on textured perovskite lead-free piezoelectric ceramics. *Journal of the Chinese Ceramic Society* 45: 1220–1230.

87 Chang, Y., Sun, Y., Wu, J. et al. (2016). Formation mechanism of highly $[001]_c$ textured $Pb(In_{1/2}Nb_{1/2})O_3$–$Pb(Mg_{1/3}Nb_{2/3})O_3$–$PbTiO_3$ relaxor ferroelectric ceramics with giant piezoelectricity. *Journal of the European Ceramic Society* 36: 1973–1981.

88 Chang, Y.F., Poterala, S.F., Yang, Z.P. et al. (2009). <001> textured $(K_{0.5}Na_{0.5})(Nb_{0.97}Sb_{0.03})O_3$ piezoelectric ceramics with high electromechanical coupling over a broad temperature range. *Applied Physics Letters* 95: 232905.

89 Guo, Y.P., Kakimoto, K., and Ohsato, H. (2005). $(Na_{0.5}K_{0.5})NbO_3$–$LiTaO_3$ lead-free piezoelectric ceramics. *Materials Letters* 59: 241–244.

90 Akdogan, E.K., Kerman, K., Abazari, M., and Safari, A. (2008). Origin of high piezoelectric activity in ferroelectric $(K_{0.44}Na_{0.52}Li_{0.04})$–$(Nb_{0.84}Ta_{0.1}Sb_{0.06})O_3$ ceramics. *Applied Physics Letters* 92: 112908.

91 Zhang, J.L., Zong, X.J., Wu, L. et al. (2009). Polymorphic phase transition and excellent piezoelectric performance of $(K_{0.55}Na_{0.45})_{0.965}Li_{0.035}Nb_{0.80}Ta_{0.20}O_3$ lead-free ceramics. *Applied Physics Letters* 95: 022909.

92 Hollenstein, E., Davis, M., Damjanovic, D., and Setter, N. (2005). Piezoelectric properties of Li- and Ta-modified $(K_{0.5}Na_{0.5})NbO_3$ ceramics. *Applied Physics Letters* 87: 182905.

93 Matsubara, M., Yamaguchi, T., Kikuta, K., and Hirano, S. (2005). Effect of Li substitution on the piezoelectric properties of potassium sodium niobate ceramics. *Japanese Journal of Applied Physics* 44: 6136–6142.

94 Wu, J., Wang, Y.Y., Xiao, D.Q. et al. (2007). Piezoelectric properties of $LiSBO_3$-Modified $(K_{0.48}Na_{0.52})NbO_3$ lead-free ceramics. *Japanese Journal of Applied Physics* 46: 7375–7377.

95 Lin, D.M., Kwok, K.W., Lam, K.H., and Chan, H.L.W. (2007). Structure and electrical properties of $K_{0.5}Na_{0.5}NbO_3$–$LiSbO_3$ lead-free piezoelectric ceramics. *Journal of Applied Physics* 101: 074111.

96 Lin, D.M., Kwok, K.W., Tian, H.Y., and Chan, H.L.W. (2007). Phase transitions and electrical properties of $(Na_{1-x}K_x)(Nb_{1-y}Sb_y)O_3$ lead-free piezoelectric ceramics with a MnO_2 sintering aid. *Journal of the American Ceramic Society* 90: 1458–1462.

97 Lin, D.M., Kwok, K.W., and Chan, H.L.W. (2008). Phase transition and electrical properties of $(K_{0.5}Na_{0.5})(Nb_{1-x}Ta_x)O_3$ lead-free piezoelectric ceramics. *Applied Physics A* 91: 167–171.

98 Park, H.Y., Cho, K.H., Paik, D.S. et al. (2007). Microstructure and piezoelectric properties of lead-free $(1-x)(Na_{0.5}K_{0.5})NbO_3$–$xCaTiO_3$ ceramics. *Journal of Applied Physics* 102: 124101.

99 Zhao, P., Zhang, B.-P., and Li, J.-F. (2008). Enhanced dielectric and piezoelectric properties in $LiTaO_3$-doped lead-free (K,Na)NbO_3 ceramics by optimizing sintering temperature. *Scripta Materialia* 58: 429–432.

100 Du, H.L., Zhou, W.C., Luo, F. et al. (2008). Design and electrical properties investigation of $(K_{0.5}Na_{0.5})NbO_3$–$BiMeO_3$ lead-free piezoelectric ceramics. *Journal of Applied Physics* 104: 034104.

101 Liang, W.F., Wu, W.J., Xiao, D.Q., and Zhu, J.G. (2011). Effect of the addition of $CaZrO_3$ and $LiNbO_3$ on the phase transitions and piezoelectric properties of $K_{0.5}Na_{0.5}NbO_3$ lead-free ceramics. *Journal of the American Ceramic Society* 94: 4317–4322.

102 Wu, J.G., Wang, Y.Y., Xiao, D.Q. et al. (2007). Effects of Ag content on the phase structure and piezoelectric properties of $(K_{0.44-x}Na_{0.52}Li_{0.04}Ag_x)(Nb_{0.91}Ta_{0.05}Sb_{0.04})O_3$ lead-free ceramics. *Applied Physics Letters* 91: 132914.

103 Lin, D., Kwok, K.W., and Chan, H.L.W. (2009). Dielectric and piezoelectric properties of $K_{0.5}Na_{0.5}NbO_3$–$AgSbO_3$ lead-free ceramics. *Journal of Applied Physics* 106: 034102.

104 Zhou, J.-J., Wang, K., Li, F. et al. (2013). High and frequency-insensitive converse piezoelectric coefficient obtained in $AgSbO_3$-modified (Li,K,Na)(Nb,Ta)O_3 lead-free piezoceramics. *Journal of the American Ceramic Society* 96: 519–523.

105 Zhang, Q., Zhang, B.-P., Li, H.-T., and Shang, P.-P. (2011). Effects of Sb content on electrical properties of lead-free piezoelectric $[(Na_{0.535}K_{0.480})_{0.942}Li_{0.058}](Nb_{1-x}Sb_x)O_3$ ceramics. *Journal of Alloys and Compounds* 490: 260–263.

106 Zhou, J.-J., Li, J.-F., and Zhang, X.-W. (2012, 2012). Orthorhombic to tetragonal phase transition due to stress release in (Li,Ta)-doped(K,Na)NbO_3 lead-free piezoceramics. *Journal of the European Ceramic Society* 32: 267–270.

107 Gao, Y., Zhang, J.L., Qing, Y.L. et al. (2011). Remarkably strong piezoelectricity of lead-free $(K_{0.45}Na_{0.55})_{0.98}Li_{0.02}(Nb_{0.77}Ta_{0.18}Sb_{0.05})O_3$ ceramic. *Journal of the American Ceramic Society* 94: 2968–2973.

108 Wang, R.P., Bando, H., Katsumata, T. et al. (2009). Tuning the orthorhombic–rhombohedral phase transition temperature in sodium potassium niobate by incorporating barium zirconate. *Physica Status Solidi – Rapid Research Letter* 3: 142–144.

109 Wang, R.P., Bando, H., and Itoh, M. (2009). Universality in phase diagram of (K,Na)NbO_3–$MTiO_3$ solid solutions. *Applied Physics Letters* 95: 092905.

110 Zuo, R.Z. and Fu, J. (2011). Rhombohedral-tetragonal phase coexistence and piezoelectric properties of (NaK)(NbSb)O_3–$LiTaO_3$–$BaZrO_3$ lead-free ceramics. *Journal of the American Ceramic Society* 94: 1467–1470.

111 Liu, Q., Zhang, Y.C., Zhao, L. et al. (2018). Simultaneous enhancement of piezoelectricity and temperature stability in (K,Na)NbO_3-based lead-free piezoceramics by incorporating perovskite zirconates. *Journal of Materials Chemistry C* 6: 10618–10627.

112 Wu, J.G., Tao, H., Yuan, Y. et al. (2015). Role of antimony in the phase structure and electrical properties of potassium–sodium niobate lead-free ceramics. *RSC Advances* 5: 14575–14583.

113 Zhang, B.Y., Wu, J.G., Cheng, X.J. et al. (2013). Lead-free piezoelectrics based on potassium–sodium niobate with giant d_{33}. *ACS Applied Materials & Interfaces* 5: 7718–7725.

114 Wu, B., Wu, H.J., Wu, J.G. et al. (2016). Giant piezoelectricity and high curie temperature in nanostructured alkali niobate lead-free piezoceramics through phase coexistence. *Journal of the American Chemical Society* 138: 15459–15464.

115 Zheng, T., Wu, H.J., Yuan, Y. et al. (2017). The structural origin of enhanced piezoelectric performance and stability in lead free ceramics. *Energy & Environmental Science* 2017: 528–537.

116 Lv, X., Wu, J.G., Yang, S. et al. (2016). Identification of phase boundaries and electrical properties in ternary potassium–sodium niobate-based ceramics. *ACS Applied Materials & Interfaces* 8: 18943–18953.

117 Liu, Q., Zhang, Y.C., Gao, J. et al. (2018). High-performance lead-free piezoelectrics with local structural heterogeneity. *Energy & Environmental Science* 11: 3531–3539.

118 Cen, Z.Y., Wang, X.H., Huan, Y., and Li, L.T. (2018). Temperature stability and electrical properties of MnO-doped KNN-based ceramics sintered in reducing atmosphere. *Journal of the American Ceramic Society* 101: 2391–2407.

119 Wu, W.J., Chen, M., Wu, B. et al. (2017). Lead-free $(K,Na)(Nb,Sb)O_3$–$Zn_{1-x}(Bi_{0.5}K_{0.5})_xZrO_3$ ceramics: phase evolution and electrical properties. *Journal of Alloys and Compounds* 695: 2981–2986.

120 Pan, D., Guo, Y.P., Fu, X.W. et al. (2017). Composition induced rhombohedral–tetragonal phase boundary and high piezoelectric activity in $(K_{0.48},Na_{0.52})(Nb_{1-x}Sb_x)O_3$–$0.05Ca_{0.2}(Bi_{0.5},Na_{0.5})_{0.8}ZrO_3$ lead-free piezoelectric ceramics. *Solid State Communications* 259: 29–33.

121 Zheng, T., Wu, J.G., Xiao, D.Q., and Zhu, J.G. (2018). Recent development in lead-free perovskite piezoelectric bulk materials. *Progress in Materials Science* 98: 552–624.

122 Wang, K., Malič, B., and Wu, J.G. (2018). Shifting the phase boundary: potassium sodium niobate derivates. *MRS Bulletin* 43: 607–611.

123 Rubio-Marcos, F., Lopez-Juarez, R., Rojas-Hernandez, R.E. et al. (2015). Lead-free piezoceramics: revealing the role of the rhombohedral tetragonal phase coexistence in enhancement of the piezoelectric properties. *ACS Applied Materials & Interfaces* 7: 23080–23088.

124 Damjanovic, D. (2009). Comments on origins of enhanced piezoelectric properties in ferroelectrics. *IEEE Transactions on Ultrasonics, Ferroelectrics, and Frequency Control* 56: 1574–1585.

125 Zheng, T. and Wu, J. (2016). Relationship between poling characteristics and phase boundaries of potassium–sodium niobate ceramics. *ACS Applied Materials & Interfaces* 8: 9242–9246.

126 Leist, T., Granzow, T., Jo, W., and Rodel, J. (2010). Effect of tetragonal distortion on ferroelectric domain switching: a case study on La-doped $BiFeO_3$–$PbTiO_3$ ceramics. *Journal of Applied Physics* 108: 014103.

127 Dai, Y.J., Zhang, X.W., and Zhou, G.Y. (2007). Phase transitional behavior in $K_{0.5}Na_{0.5}NbO_3$–$LiTaO_3$ ceramic. *Applied Physics Letters* 26: 262903.

128 Trolier-McKinstry, S., Zhang, S.J., Bell, A.J., and Tan, X.L. (2018). High-performance piezoelectric crystals, ceramics, and films. *Annual Review of Materials Research* 48: 191–217.

129 Zheng, J.H., Takahashi, S., Yoshikawa, S. et al. (1996). Heat generation in multilayer piezoelectric actuators. *Journal of the American Ceramic Society* 79: 3193–3198.

130 Li, E.Z., Kakemoto, H., Hoshina, T., and Tsurumi, T. (2008). A shear-mode ultrasonic motor using potassium sodium niobate-based ceramics with high mechanical quality factor. *Japanese Journal of Applied Physics* 47: 7702–7706.

131 Shen, Z.Y., Xu, Y., and Li, J.F. (2012). Enhancement of Q_m in CuO-doped compositionally optimized Li/Ta-modified $(Na,K)NbO_3$ lead-free piezoceramics. *Ceramics International* 38: S331–S334.

132 Zhou, J.J., Cheng, L.-Q., Wang, K. et al. The phase structure and electric properties of low-temperature sintered $(K,Na)NbO_3$-based piezoceramics modified by CuO. *Ceramics International* 40: 2927–2931.

133 Zhang, S.J., Xia, R., Hao, H. et al. (2008). Mitigation of thermal and fatigue behavior in $K_{0.5}Na_{0.5}NbO_3$-based lead free piezoceramics. *Applied Physics Letters* 92: 152904.

134 Skidmore, T.A., Comyn, T.P., and Milne, S.J. (2009). Temperature stability of ($Na_{0.5}K_{0.5}NbO_3$ (0.93)–$LiTaO_3$ (0.07)) lead-free piezoelectric ceramics. *Applied Physics Letters* 94: 222902.

135 Zhou, J.-J., Li, J.-F., and Zhang, X.W. (2012). $BiFeO_3$-modified (Li,K,Na)(Nb,Ta)O_3 lead-free piezoelectric ceramics with temperature-stable piezoelectric property and enhanced mechanical strength. *Journal of Materials Science* 47: 1767–1773.

136 Zhang, N. and Wu, J.G. (2018). Structure and property of lead-free (K,Na)NbO_3–($Bi_{1/2}Na_{1/2}$)ZrO_3–$CaTiO_3$ piezoelectric ceramics. *Journal of Materials Science: Materials in Electronics* 30: 1663–1669.

137 Liu, B.H., Li, P., Shen, B. et al. (2017). Enhanced piezoelectric properties and temperature-insensitive strain behavior of (001)-textured KNN-based ceramics. *Ceramics International* 43: 8004–8009.

138 Li, P., Liu, B.H., Shen, B. et al. (2018). Mechanism of significantly enhanced piezoelectric performance and stability in textured potassium–sodium niobate piezoelectric ceramics. *Journal of the European Ceramic Society* 38: 75–83.

139 Quan, Y., Ren, W., Niu, G. et al. (2018). Large piezoelectric strain with superior thermal stability and excellent fatigue resistance of lead-free potassium sodium niobate-based grain orientation-controlled ceramics. *ACS Applied Materials & Interfaces* 10: 10220–10226.

140 Liu, Q., Li, J.-F., Zhao, L. et al. (2018). Niobate based lead-free piezoceramics: diffused phase transition boundary leading to temperature-insensitive high piezoelectric voltage coefficient. *Journal of Materials Chemistry C* 6: 1116–1125.

141 Gao, Y., Zhang, J.L., Zong, X.J. et al. (2010). Extremely temperature-stable piezoelectric properties of orthorhombic phase in (K,Na)NbO_3-based ceramics. *Journal of Applied Physics* 107: 074101.

142 Hao, J.G., Xu, Z.J., Chu, R.Q. et al. (2015). Good temperature stability and fatigue-free behavior in Sm_2O_3-modified 0.948($K_{0.5}Na_{0.5}$)NbO_3–0.052$LiSbO_3$ lead-free piezoelectric ceramics. *Materials Research Bulletin* 65: 94–102.

143 Wang, K., Yao, F.Z., Jo, W. et al. (2013). Temperature-insensitive (K,Na)NbO_3-based lead-free piezoactuator ceramics. *Advanced Functional Materials* 23: 4079–4086.

144 Yao, F.Z., Wang, K., Jo, W. et al. (2016). Diffused phase transition boosts thermal stability of high-performance lead-free piezoelectrics. *Advanced Functional Materials* 26: 1217–1224.

145 Thong, H.-C., Xu, Z., Zhao, C. et al. (2019). Abnormal grain growth in (K,Na)NbO_3-based lead-free piezoceramic powders. *Journal of the American Ceramic Society* 102: 836–844.

146 Condurache, O.A., Radan, K., Prah, U. et al. (2019). Heterogeneity challenges in multiple-element-modified lead-free piezoelectric ceramics. *Materials* 12: 4049.

147 Wang, R., Wang, P., Yao, K. et al. (2015). Temperature stability of lead-free niobate piezoceramics with engineered morphotropic phase boundary. *Journal of the American Ceramic Society* 98: 2177–2182.

148 Zhang, M.-H., Wang, K., Zhou, J.-S. et al. (2017). Thermally stable piezoelectric properties of (K,Na)NbO_3-based lead-free perovskite with rhombohedral-tetragonal coexisting phase. *Acta Materialia* 122: 344–351.

149 Zhang, M.-H., Wang, K., Du, Y.-J. et al. (2017). High and temperature-insensitive piezoelectric strain in alkali niobate lead-free perovskite. *Journal of the American Chemical Society* 139: 3889–3895.

150 Qin, Y.L., Zhang, J.L., Yao, W.Z. et al. (2016). Domain configuration and thermal stability of $(K_{0.48}Na_{0.52})(Nb_{0.96}Sb_{0.04})O_3$–$Bi_{0.50}(Na_{0.82}K_{0.18})_{0.50}ZrO_3$ piezoceramics with high d_{33} coefficient. *ACS Applied Materials & Interfaces* 8: 7257–7265.

151 Liu, Q., Zhang, Y.C., Gao, J. et al. (2020). Practical high-performance lead-free piezoelectrics: structural flexibility beyond utilizing multiphase coexistence. *National Science Review* 7: 355–365.

152 Zhu, F.Y., Ward, M.B., Li, J.F., and Milne, S.J. (2015). Core–shell grain structures and dielectric properties of $Na_{0.5}K_{0.5}NbO_3$–$LiTaO_3$–$BiScO_3$ piezoelectric ceramics. *Acta Materialia* 90: 204–212.

153 Zhu, F.Y., Skidmore, T.A., Bell, A.J. et al. (2011). Diffuse dielectric behaviour in $Na_{0.5}K_{0.5}NbO_3$–$LiTaO_3$–$BiScO_3$ lead-free ceramics. *Materials Chemistry and Physics* 129: 411–417.

154 Tao, H., Wu, H.J., Liu, Y. et al. (2019). Ultrahigh performance in lead-free piezoceramics utilizing a relaxor slush polar state with multiphase coexistence. *Journal of the American Chemical Society* 141: 13987–13994.

155 Hao, J.G., Xu, Z.J., Chu, R.Q. et al. (2015). Lead-free electrostrictive $(Bi_{0.5}Na_{0.5})TiO_3$–$(Bi_{0.5}K_{0.5})TiO_3$–$(K_{0.5}Na_{0.5})NbO_3$ ceramics with good thermostability and fatigue-free behavior. *Journal of Materials Science* 50: 5328–5336.

156 Hao, J.G., Xu, Z.J., Chu, R.Q. et al. (2015). Effect of $(Bi_{0.5}K_{0.5})TiO_3$ on the electrical properties, thermal and fatigue behavior of $(K_{0.5}Na_{0.5})NbO_3$-based lead-free piezoelectrics. *Journal of Materials Research* 30: 2018–2029.

157 Lv, X., Wu, J.G., Xiao, D.Q. et al. (2018). Modifying temperature stability of (K,Na)NbO_3 ceramics through phase boundary. *Advanced Electronic Materials* 4: 1800205.

158 Du, H.L., Zhou, W.C., Luo, F. et al. (2007). An approach to further improve piezoelectric properties of $(K_{0.5}Na_{0.5})NbO_3$-based lead-free ceramics. *Applied Physics Letters* 91: 202907.

159 Morozov, M.I., Kungl, H., and Hoffmann, M.J. (2011). Effects of poling over the orthorhombic-tetragonal phase transition temperature in compositionally homogeneous (K,Na)NbO_3-based ceramics. *Applied Physics Letters* 98: 132908.

160 Li, Q., Zhang, M.H., Zhu, Z.X. et al. (2017). Poling engineering of (K,Na)NbO_3-based lead-free piezoceramics with orthorhombic–tetragonal coexisting phases. *Journal of Materials Chemistry C* 5: 549–556.

161 Rubio-Marcos, F., Romero, J.J., Ochoa, D.A. et al. (2010). Effects of poling process on KNN-modified piezoceramic properties. *Journal of the American Ceramic Society* 93: 318–321.

162 Patterson, E.A. and Cann, D.P. (2011). Piezoelectric properties and unipolar fatigue behavior of KNN-based Pb-free piezoceramics. *IEEE Transactions on Ultrasonics Ferroelectrics and Frequency Control* 58: 1835–1841.

163 Yao, F.Z., Glaum, J., Wang, K. et al. (2013). Fatigue-free unipolar strain behavior in $CaZrO_3$ and MnO_2 co-modified (K,Na)NbO_3-based lead-free piezoceramics. *Applied Physics Letters* 103: 192907.

164 Yao, F.Z., Patterson, E.A., Wang, K. et al. (2014). Enhanced bipolar fatigue resistance in $CaZrO_3$-modified (K,Na)NbO_3 lead-free piezoceramics. *Applied Physics Letters* 104: 242912.

165 Yao, F.Z., Wang, K., Shen, Y., and Li, J.-F. (2015). Robust $CaZrO_3$-modified (K,Na)NbO_3-based lead-free piezoceramics: high fatigue resistance insensitive to temperature and electric field. *Journal of Applied Physics* 118: 134102.

166 Yao, F.Z., Wang, K., Cheng, L.-Q. et al. (2015). Nanodomain engineered (K,Na)NbO_3 lead-free piezoceramics: enhanced thermal and cycling reliabilities. *Journal of the American Ceramic Society* 98: 448–454.

167 Namsar, O., Uthaisar, C., and Pojprapai, S. (2018). Ferroelectric fatigue mechanism under bipolar electrical loading in KNN lead free piezoelectric ceramic. *Journal of Materials Science: Materials in Electronics* 29: 7188–7193.

168 Dong, S.S., Du, J., Xu, Z.J. et al. (2016). Enhanced temperature stability and fatigue-resistant behavior in $MgTiO_3$-doped $0.948(K_{0.5}Na_{0.5})NbO_3$–$0.052LiSbO_3$ lead-free ceramics. *Ceramics International* 42: 8051–8057.

169 Lv, X., Wu, J.G., Zhu, J.G. et al. (2018). A new method to improve the electrical properties of KNN-based ceramics: tailoring phase fraction. *Journal of the European Ceramic Society* 38: 85–94.

170 Li, Y.W., Liu, Y.X., Ochsner, P.-E. et al. (2019). Temperature dependent fracture toughness of KNN-based lead-free piezoelectric ceramics. *Acta Materialia* 174: 369–378.

171 Huan, Y., Wang, X.H., Koruza, J. et al. (2016). Inverted electro-mechanical behaviour induced by the irreversible domain configuration transformation in (K,Na)NbO_3-based ceramics. *Scientific Reports* 6: 22053.

172 Zhang, S.-W., Zhou, Z., Luo, J., and Li, J.-F. (2019). Potassium-sodium-niobate-based thin films: lead free for micro-piezoelectrics. *Annalen der Physik* 531: 1800525.

173 Shibata, K., Wang, R., Tou, T., and Koruza, J. (2018). Applications of lead-free piezoelectric materials. *MRS Bulletin* 43: 612–616.

174 Won, S.S., Lee, J., Venugopal, V. et al. (2016). Lead-free Mn-doped $(K_{0.5},Na_{0.5})NbO_3$ piezoelectric thin films for MEMS-based vibrational energy harvester applications. *Applied Physics Letters* 108: 232908.

175 Tsujiura, Y., Suwa, E., Kurokawa, F. et al. (2013). Lead-free piezoelectric MEMS energy harvesters of (K,Na)NbO_3 thin films on stainless steel cantilevers. *Japanese Journal of Applied Physics* 52: UNSP 09KD13.

176 Nguyen, M.D., Dekkers, M., Houwman, E.P. et al. (2016). Lead-free $(K_{0.5}Na_{0.5})NbO_3$ thin films by pulsed laser deposition driving MEMS-based piezoelectric cantilevers. *Materials Letters* 164: 413–416.

177 Wakasa, Y., Kanno, I., Yokokawa, R. et al. (2011). Piezoelectric properties of microfabricated $(K,Na)NbO_3$ thin films. *Sensors and Actuators, A: Physical* 171: 223–227.

178 Schwartz, R.W., Boyle, T.J., Lockwood, S.J. et al. Sol–gel processing of PZT thin-films – a review of the state-of-the-art and process optimization strategies. *Integrated Ferroelectrics* 7: 259–277.

179 Gong, W., Li, J.-F., Chu, X.C. et al. (2004). Preparation and characterization of sol–gel derived (100)-textured $Pb(Zr,Ti)O_3$ thin films: PbO seeding role in the formation of preferential orientation. *Acta Materiallia* 52: 2787–2793.

180 Soderlind, F., Kall, P.O., and Helmersson, U. (2005). Sol–gel synthesis and characterization of $Na_{0.5}K_{0.5}NbO_3$ thin films. *Journal of Crystal Growth* 281: 468–474.

181 Lai, F.P. and Li, J.-F. (2007). Sol–gel processing of lead-free $(Na,K)NbO_3$ ferroelectric films. *Journal of Sol-Gel Science and Technology* 42: 287–292.

182 Lai, F.P. and Li, J.-F. (2007). Sol–gel processing and characterization of $(Na,K)NbO_3$ lead-free ferroelectric films. *Ferroelectrics* 358: 1063–1069.

183 Seog, H.J., Ullah, A., Ahn, C.W. et al. (2018). Recent progress in potassium sodium niobate lead-free thin films. *Journal of the Korean Physical Society* 72: 1467–1483.

184 Abazari, M. and Safari, A. (2010). Leakage current behavior in lead-free ferroelectric $(K,Na)NbO_3$–$LiTaO_3$–$LiSbO_3$ thin films. *Applied Physics Letters* 97: 262902.

185 Lee, S.Y., Ahn, C.W., Ullah, A. et al. (2011). Effect of Mn substitution on ferroelectric and leakage current characteristics of lead-free $(K_{0.5}Na_{0.5})(Mn_xNb_{1-x})O_3$ thin films. *Current Applied Physics* 11: S266–S269.

186 Wang, L.Y., Ren, W., Shi, P., and Wu, X.Q. (2014). Structures, electrical properties, and leakage current behaviors of un-doped and Mn-doped lead-free ferroelectric $K_{0.5}Na_{0.5}NbO_3$ films. *Journal of Applied Physics* 115: 034103.

187 Rubio-Marcos, F., Marchet, P., Vendrell, X. et al. (2011). Effect of MnO doping on the structure, microstructure and electrical properties of the $(K,Na,Li)(Nb,Ta,Sb)O_3$ lead-free piezoceramics. *Journal of Alloys and Compounds* 509: 8804–8811.

188 Liu, Q., Zhu, F.Y., Zhao, L. et al. (2016). Further enhancing piezoelectric properties by adding MnO_2 in $AgSbO_3$-modified $(Li,K,Na)(Nb,Ta)O_3$ lead-free piezoceramics. *Journal of the American Ceramic Society* 99: 3670–3676.

189 Yao, F.Z., Zhang, M.H., Wang, K. et al. (2018). Refreshing piezoelectrics: distinctive role of manganese in lead free perovskites. *ACS Applied Materials & Interfaces* 10: 37298–37306.

190 Wang, L.Y., Ren, W., Ma, W.H. et al. (2015). Improved electrical properties for Mn-doped lead-free piezoelectric potassium sodium niobate ceramics. *AIP Advances* 5: 097120.

191 Yu, Q., Li, J.-F., Chen, Y. et al. (2014). Effect of pyrolysis temperature on sol–gel synthesis of lead-free piezoelectric (K,Na)NbO$_3$ films on Nb:SrTiO$_3$ substrates. *Journal of the American Ceramic Society* 97: 107–113.

192 Zhang, S.-W., Luo, J., Zhou, Z., and Li, J.-F. (2019). Sol–gel processed highly (100)-textured (K,Na)NbO$_3$-based lead-free thin films: effect of pyrolysis temperature. *Journal of the American Ceramic Society* 102: 2696–2705.

193 Lai, F.P., Li, J.-F., Zhu, Z.-X., and Xu, Y. (2009). Influence of Li content on electrical properties of highly piezoelectric (Li,K,Na)NbO$_3$ thin films prepared by sol–gel processing. *Journal of Applied Physics* 106: 064101.

194 Nakashima, Y., Sakamoto, W., Maiwa, H. et al. (2007). Lead-free piezoelectric (K,Na)NbO$_3$ thin films derived from metal alkoxide precursors. *Japanese Journal of Applied Physics Part 2: Letters & Express Letters* 46: L311–L313.

195 Ahn, C.W., Seog, H.J., Ullah, A. et al. (2012). Effect of Ta content on the phase transition and piezoelectric properties of lead-free $(K_{0.48}Na_{0.48}Li_{0.04})(Nb_{0.995-x}Mn_{0.005}Ta_x)O_3$ thin film. *Journal of Applied Physics* 111: 024110.

196 Li, T., Wang, G.S., Li, K. et al. (2013). Influence of LNO top electrodes on electrical properties of KNN/LNO thin films prepared by RF magnetron sputtering. *Journal of the American Ceramic Society* 96: 787–790.

197 Li, T., Wang, G.S., Li, K. et al. (2014). Electrical properties of lead-free KNN films on SRO/STO by RF magnetron sputtering. *Ceramics International* 40: 1195–1198.

198 Yu, Q., Li, J.-F., Sun, W. et al. (2013). Electrical properties of $K_{0.5}Na_{0.5}NbO_3$ thin films grown on Nb:SrTiO$_3$ single-crystalline substrates with different crystallographic orientations. *Journal of Applied Physics* 113: 024101.

199 Luo, J., Sun, W., Zhou, Z. et al. (2017). Monoclinic (K,Na)NbO$_3$ ferroelectric phase in epitaxial films. *Advanced Electronic Materials* 3: 1700226.

200 Luo, J., Sun, W., Zhou, Z. et al. (2017). Domain evolution and piezoelectric response across thermotropic phase boundary in (K,Na)NbO$_3$-based epitaxial thin films. *ACS Applied Materials & Interfaces* 9: 13315–13322.

201 Zhou, M.-J., Wang, J.-J., Chen, L.-Q., and Nan, C.-W. (2018). Strain, temperature, and electric-field effects on the phase transition and piezoelectric responses of $K_{0.5}Na_{0.5}NbO_3$ thin films. *Journal of Applied Physics* 123: 154106.

202 Dinescu, M. (2010). *PLD of Piezoelectric and Ferroelectric Materials*, Springer Series in Materials Science. Springer.

203 Kelly, P.J. and Arnell, R.D. (2000). Magnetron sputtering: a review of recent developments and applications. *Vacuum* 56: 159–172.

204 Cho, C.-R. and Grishin, A. (1999). Self-assembling ferroelectric $Na_{0.5}K_{0.5}NbO_3$ thin films by pulsed-laser deposition. *Applied Physics Letters* 75: 268–270.

205 Cho, C.R., Katardjiev, I., Grishin, M., and Grishin, A. (2002). $Na_{0.5}K_{0.5}NbO_3$ thin films for voltage controlled acoustoelectric device applications. *Applied Physics Letters* 80: 3171–3173.

206 Abazari, M. and Safari, A. (2009). Effects of doping on ferroelectric properties and leakage current behavior of KNN–LT–LS thin films on SrTiO$_3$ substrate. *Journal of Applied Physics* 105: 094101.

207 Zhu, M.M., Shang, X.Z., Chang, G. et al. (2013). Pulsed laser deposition of single-phase lead-free NKLNST thin films with K- and Na-excess targets. *Journal of Alloys and Compounds* 567: 97–101.

208 Margolin, A.M., Surovyak, Z.S., Zakharchenko, I.N. et al. (1988). (K,Na)NbO_3 ferroelectric-films formed under HF cathode sputtering. *Zhurnal Tekhnicheskoi Fiziki* 58: 2355–2359.

209 Blomqvist, M., Koh, J.H., Khartsev, S. et al. (2002). High-performance epitaxial $Na_{0.5}K_{0.5}NbO_3$ thin films by magnetron sputtering. *Applied Physics Letters* 81: 337–339.

210 Lee, H.J., Kim, I.W., Kim, J.S. et al. (2009). Ferroelectric and piezoelectric properties of $Na_{0.52}K_{0.48}NbO_3$ thin films prepared by radio frequency magnetron sputtering. *Applied Physics Letters* 94: 092902.

211 Chen, K., Xu, G.S., Yang, D.F., and Wang, X.F. (2007). Dielectric and piezoelectric properties of lead-free $0.95(K_{0.5}Na_{0.5})NbO_3$–$0.05LiNbO_3$ crystals grown by the Bridgman method. *Journal of Applied Physics* 101: 044103.

212 Zheng, L.M., Huo, X.Q., Wang, R. et al. (2013). Large size lead-free (Na,K)(Nb,Ta)O_3 piezoelectric single crystal: growth and full tensor properties. *CrystEngComm* 15: 7718–7772.

213 Tian, H., Meng, X., Hu, C. et al. (2016). Origin of giant piezoelectric effect in lead-free $K_{1-x}Na_xTa_{1-y}Nb_yO_3$ single crystals. *Scientific Reports* 6: 25637.

214 Huo, X.Q., Zhang, R., Zheng, L.M. et al. (2015). (K,Na,Li)(Nb,Ta)O_3:Mn lead-free single crystal with high piezoelectric properties. *Journal of the American Ceramic Society* 98: 1829–1835.

215 Yang, J., Zhang, F.Q., Yang, Q.B. et al. (2016). Large piezoelectric properties in KNN-based lead-free single crystals grown by a seed-free solid-state crystal growth method. *Applied Physics Letters* 108: 182904.

216 Cheng, L.Q. and Li, J.-F. (2016). A review on one dimensional perovskite nanocrystals for piezoelectric applications. *Journal of Materiomics* 2: 25–36.

217 Xu, S.Y. and Li, J.-F. (2011). Synthesis and piezoelectricity of single-crystalline (K,Na)NbO_3 nanobars. *Journal of the American Ceramic Society* 94: 3812–3818.

218 Cheng, L.-Q., Wang, K., and Li, J.-F. (2013). Synthesis of highly piezoelectric lead-free (K,Na)NbO_3 one-dimensional perovskite nanostructures. *Chemical Communications* 49: 4003–4005.

219 Cheng, L.-Q., Wang, K., Yu, Q., and Li, J.-F. (2014). Structure and composition characterization of lead-free (K,Na)NbO_3 piezoelectric nanorods synthesized by molten-salt reaction. *Journal of Materials Chemistry C* 2: 1519–1524.

220 Xu, Y., Yu, Q., and Li, J.-F. (2012). A facile method to fabricate vertically aligned (K,Na)NbO_3 lead-free piezoelectric nanorods. *Journal of Materials Chemistry* 22: 23221–23226.

4

$(Bi_{1/2}Na_{1/2})TiO_3$ System

4.1 Introduction of BNT System

$Na_{1/2}Bi_{1/2}TiO_3$ (often abbreviated as BNT or NBT), a perovskite piezoelectric material with rhombohedral (R) symmetry at room temperature, was firstly discovered by Smolenskii et al. in 1961 [1]. BNT has a high Curie temperature $T_C = 320\,°C$ with an additional phase transition at around 200 °C. A relatively large remnant polarization $P_r = 38\,\mu C/cm^2$ and a high Curie temperature make the BNT ceramic a promising lead-free substitution [2, 3]. In fact, it is well known that KNN-based materials have been a hot research topic since 2004 [4]. BNT had been extensively studied even earlier, especially led by Takenaka's group [5, 6].

Early studies found that the pure BNT ceramic has a large leakage current and a high coercive field, which makes it difficult to be polarized and exhibit desirable performances [5, 7]. In addition, the BNT ceramics need to be sintered at rather high above 1200 °C, which further enhances the volatilization of inevitably A-site elements [8]. Nevertheless, the d_{33} of pristine BNT ceramic is also relatively low, generally less than 100 pC/N. To overcome these issues and enhance the piezoelectricity, another perovskite is introduced to construct a BNT-based solid solution with morphotropic phase boundary (MPB). The representative one is $Na_{1/2}Bi_{1/2}TiO_3$–$BaTiO_3$ (BNT–BT) solid solution system, which is the major topic of this chapter. Other well-studied solid solution systems with MPB are $(Bi_{1/2}Na_{1/2})TiO_3$–$(Bi_{1/2}K_{1/2})TiO_3$ (BNT–BKT) [7, 9–11] and $Na_{1/2}Bi_{1/2}TiO_3$–$(K,Na)NbO_3$ (BNT–KNN) [12].

If compared with KNN-based ceramics, the piezoelectric charge coefficient d_{33} reported in the BNT-based ceramics is not high. However, this system is characteristic of high electric-field-induced strain, which makes themselves potential to be applied in piezoelectric actuators such as the camera focusing system, printing nozzle, and impact driving. It has been reported that the electric-field-induced strain of BNT–BT ceramics could reach 0.5% by specific compositional modification [13–15]. In addition to the large electric-field-induced strain, many unusual behaviors have been found in BNT-based ceramics, such as frequency dispersion observed in the temperature-dependent dielectric spectrum, temperature-dependent strain loops, and double-loop *P*–*E* hysteresis. After decades of research, these abnormal behaviors and large electric-field-induced strain are believed to be related to composition-, temperature-, and electric field-dependent phase transition [16–22].

Lead-Free Piezoelectric Materials, First Edition. Jing-Feng Li.

Many recent investigations found that the phase structures of BNT-based materials at room temperature could be more complicated than what we knew [23]. It has been suggested that the composition at MPB could be a relaxor ferroelectric or an antiferroelectric with polar nanoregions (PNRs), instead of a traditional ferroelectric like lead zirconate titanate (PZT) [22, 24, 25]. Furthermore, some works reported that the space group of $(Bi_{1/2}Na_{1/2})TiO_3$ is *Cc*, rather than the pre-supposed *R3c* [26–30]. There are still many controversies and unclear structure details in BNT-based ceramics.

BNT-based ceramics are featured with a high mechanical quality factor (Q_m), originating from its high coercive electric field (E_c). Since its stability in the high-power working state is much superior to hard PZT, it has become the most promising candidate for high-power piezoelectric applications. Many works have been focusing on the optimization of BNT-based ceramics for applications.

In this chapter, we will make a review and discussions on some issues that have been highly concerned in the BNT-based piezoelectric system, including the composition-/temperature-dependent phase structure, the origin of the large electric-field-induced strain and high-power application of BNT-based piezoelectrics.

4.2 Extensive Research on Phase Diagram of $(Bi_{1/2}Na_{1/2})TiO_3$–$BaTiO_3$ System

4.2.1 Relaxor or Antiferroelectric?

As mentioned earlier, BNT–BT is a representative BNT-based system with high piezoelectric performance. When it was first fabricated by Takenaka et al. in 1991, the BNT–BT system was considered to be a solid solution with *R*–*T* MPB at the composition around 0.94BNT–0.06BT [5]. It was also found that the BNT–BT ceramics manifest double *P*–*E* hysteresis at high temperature (Figure 4.1) and, therefore, the high-temperature phase is suggested to be an antiferroelectric phase.

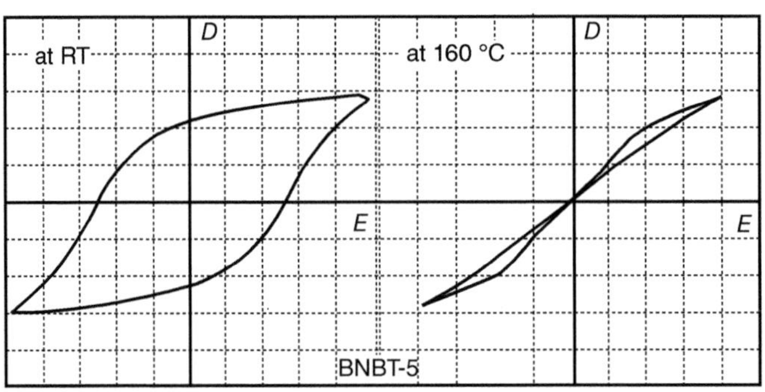

Figure 4.1 The *P*–*E* loops of BNT–5BT at room temperature and 160 °C. Source: Reprinted with permission from Takenaka et al. [5]. Copyright 1991, The Physical Society of Japan.

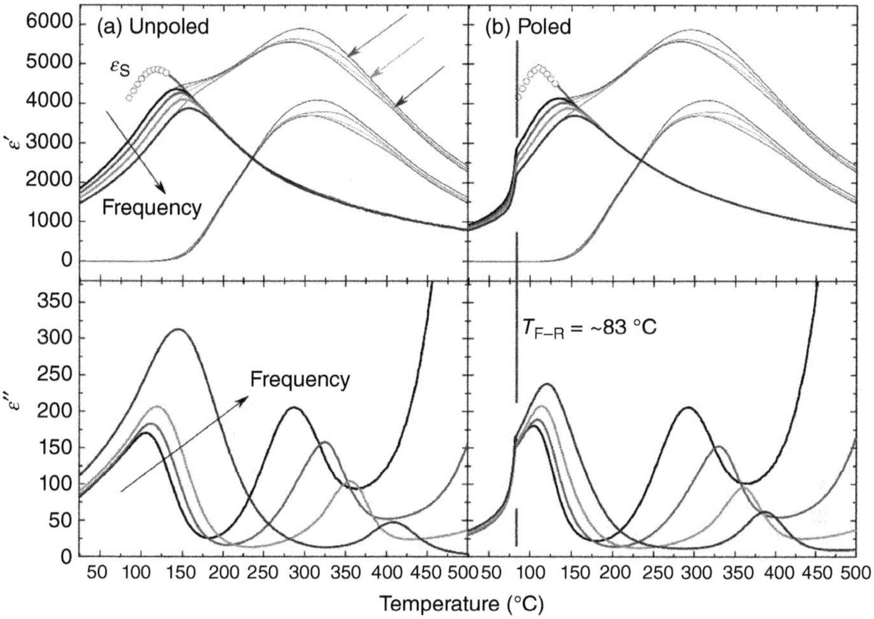

Figure 4.2 The temperature-dependent dielectric permittivity curves of (a) unpoled and (b) poled BNT–BT ceramics. Source: Jo et al. 2011 [20]. Reprinted with permission of AIP Publishing.

Following this work, many researches have been focusing on characterizing the exact MPB composition and to verify the presence of the antiferroelectric phase structure at high temperature. From the temperature-dependent dielectric (ε–T) spectra measured at different frequencies, however, a dielectric spectrum that is different from the traditional ferroelectrics was observed, as shown in Figure 4.2 [20].

The ε–T spectra of a traditional ferroelectric show a sharp peak phase transition temperature without frequency dispersion. Whereas, the ε–T spectra of the BNT–BT ceramics exhibit not only diffused phase transition but also significant frequency dispersion. In addition, two dielectric anomalies can be observed in poled BNT–BT ceramics, which occur at depolarization temperature (T_d) and the temperature where ε_r reaches a maximum value (T_m). Such ε–T spectra have been observed in the relaxor ferroelectrics, such as lead magnesium niobate $(1-x)Pb(Mg_{1/3}Nb_{2/3})O_3$–$xPbTiO_3$ (PMN–PT) and lanthanum-doped PZT $(Pb_{1-x}La_x)(Zr_{0.65}Ti_{0.35})O_3$ (PLZT) [31, 32].

The dielectric relaxation behavior observed in relaxors has been attributed to the presence of PNRs [33, 34]. The PNRs are considered as individual thermally activated dipoles giving rise to the orientational local spontaneous polarization [25]. As a result, the measured relaxor peak should be related to the thermally activated reorientation of dipole moments of PNRs. The appearance of PNRs is believed to originate from the chemical disorder. e.g. the disorder arrangement of Mg^{2+} and Nb^{5+} cations

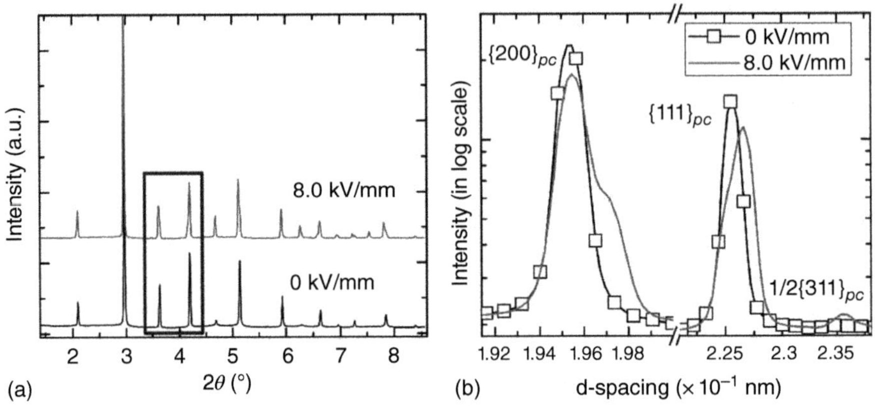

Figure 4.3 (a) In situ X-ray diffraction pattern of BNT-6BT at its virgin statein comparison with that at 8 kV/ mm. (b) An expanded view of $\{200\}_{pc}$ and $\{111\}_{pc}$ reflections. To pronounce the $1/2\{311\}_{pc}$ superlattice reflection, the intensity of which is considerably weak compared to that of Bragg reflections, the intensity of reflections are plotted on a logarithmic scale. Source: Modified from Jo et al. [20].

in PMN–PT ferroelectrics [35, 36]. In BNT-based materials, the coexistence of two A-site ions of Bi^{3+} and Na^{+} is believed to provide the basis for the chemical disorder.

Over the past decade, more and more evidence has been shown to support the presence of relaxor behavior in BNT–BT [16, 20, 37–39]. One of the most important pieces of evidence is that BNT–6BT shows a pseudocubic perovskite structure [16, 20]. All the X-ray diffraction (XRD) peaks show the symmetric single peak shape, which was not consistent with XRD results of the antiferroelectric structure. Significant rhombohedral and tetragonal distortions are visible with the application of the electric field over a threshold value. The XRD result of BNT–6BT under an applied electric field is shown in Figure 4.3 [40].

It is worth noting that even after the electric field is unloaded, the rhombohedral and tetragonal distortions still remain, which indicates an irreversible electric-field-induced phase transition from a macro-non-polarized phase to a ferroelectric phase has occurred. Jo et al. suggested that the pre-assumed "pseudocubic" phase or rhombohedral phase could be a slightly distorted "non-polar" phase decorated with two types of PNRs, having rhombohedral *R3c* and tetragonal *P4bm* symmetries, embedded in a cubic matrix [20]. This hypothesis was examined by the selected area electron diffraction (SAED) taken along $[130]_{pc}$ zone axis. The result shows that each grain has a superlattice pattern of 1/2{*ooo*} and 1/2{*ooe*}, which correspond to rhombohedral *R3c* and tetragonal *P4bm* symmetries, respectively. Meanwhile, the confirmed domain structure of BNT–6BT is different from the traditional ferroelectrics, where no ferroelectric domain can be observed by bright-field transmission electron microscope (TEM).

Ma et al. further investigated the domain structure and crystal symmetry of the BNT–*x*BT system by using TEM [24, 25]. In the bright-field TEM images two different kinds of nanodomains were observed in BNT–BT, one of which is the "complex domain" with domain walls aligned along {100} planes but no sharp contrast. And

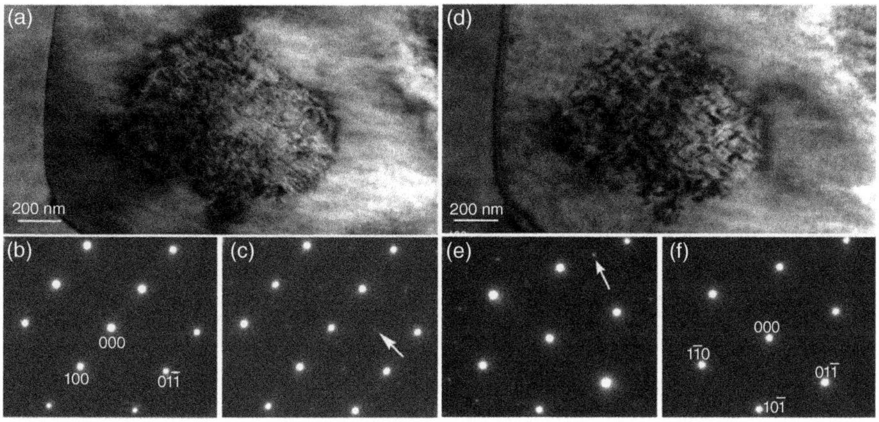

Figure 4.4 (a) Transmission electron microscope bright-field image of a typical grain with core–shell structure for the BNT–6BT along its [011] zone axis; (b) diffraction pattern of shell along [011] zone axis; (c) diffraction pattern of the core along [011] zone axis; (d) TEM bright-field image of the same grain tilted to its [111] zone axis; (e) diffraction pattern of shell along [111] zone axis; and (f) diffraction pattern of the core along [111] zone axis. Source: Ma et al. [25].

the other kind is the nanodomains with faint contrast. In BNT–6BT, a core–shell domain structure was found. As shown in Figure 4.4, about 40% of the grains display a core–shell structure. The core exhibits a "complex domain" structure consisting of differently oriented nano-domain in size around 100 nm, while the shell consists of nanodomains with faint contrast.

The rest of the grains consist of nanodomains only. A $R3c$ symmetry with the $a^-a^-a^-$ oxygen octahedron tilt is evidenced by the SAED patterns of the "core", with strong $1/2(ooo)$ superlattice diffraction spots in SAED patterns of the [011] zone axis and the absence of superlattice spots in those of the [111] or [−111] zone axes. The SAED patterns of the "shell" exhibited no superlattice spots along the [011] zone axis, and the $1/2(ooe)$ spots were observed along the [111] zone axis, which characterizes the antiferroelectric $P4bm$ symmetry with $a^0a^0c^+$ oxygen octahedron tilt. Other TEM works also revealed similar results that the presence of $P4bm$ nanometer-scale platelets embedded in the $R3c$ matrix in pure BNT [41]. On the other hand, it is worth noting that the diffraction signal of $1/2(ooe)$ superlattice spots of BNT is really weak compared with the traditional antiferroelectrics. Therefore, it is believed that the antiferroelectric nanodomains in BNT–xBT are embedded in a matrix of undistorted cubic lattice, as illustrated in Figure 4.5 [25].

The dynamic fluctuation of the cation displacements along (001) direction causes the diffusion and frequency dispersion in the ε–T spectra, similar to the case in traditional relaxors. In consequence, a new name "relaxor antiferroelectric" has been proposed to describe the BNT–BT, to differentiate it from traditional relaxors and antiferroelectrics [25]. It is interesting to be mentioned that the concept "antiferroelectric relaxor" or "antiferroelectric PNRs" was first prompted in the classic relaxor $Pb(Mg_{1/3}Nb_{2/3})O_3$ (PMN), to explain the contradiction between pseudocubic crystal structure observed in XRD and superlattice spots observed in SAED [25].

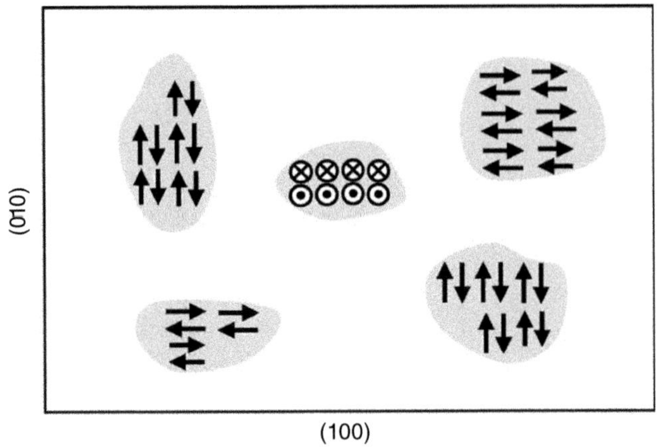

Figure 4.5 Schematic diagram of the nanostructure of BNT–BT. The AFE nanodomains (the gray islands in the diagram) with the *P4bm* symmetry are embedded in the undistorted cubic matrix (the white background in the diagram). Source: Ma et al. 2010 [25]. Reprinted with permission of AIP Publishing.

The coexistence of the *R3c* and *P4ma* PNRs in BNT–BT materials, including ceramics and single crystals, has been verified by many groups [20, 25, 42]. Neutron scattering and high-energy XRD are two efficient techniques that have been utilized to investigate the coupling of two different oxygen tilt symmetries, *R3c* and *P4ma*. Delaire and coworkers used these methods to characterize the BNT–BT single crystals at MPB composition [43]. They found that the *R3c* and *P4ma* PNRs have a similar temperature-independent average size but the stacking sequence is not regular. Luo and coworkers conducted neutron scattering on a single-crystal BNT–BT, and their results show that the diffuse streaks in BNT originate from the PNRs induced by Bi^{3+} displacement along $(001)_{pc}$, which is a mechanism different from the lead-based B-site relaxor ferroelectrics [44]. Addition of BT makes the diffuse streaks rotate from $(001)_{pc}$ to $(011)_{pc}$, with Bi^{3+} displacement rotating to $(111)_{pc}$. A large correlation length was observed in the MPB composition, which was consistent with the aforementioned PNRs stacking model.

With the advancement of characterization techniques, BNT–BT at MPB has been confirmed to be a relaxor ferroelectric that consists of *R3c* and *P4ma* PNRs. *P4ma* is an antiphase tetragonal symmetry and can remain its original structure with increasing temperature, which is the main reason for the antiferroelectric properties.

4.2.2 MPB and Complex Phase Structure

Although BNT–6BT is assumed to be the MPB composition, the actual MPB region in the BNT–BT system remains vague. Until now, numerous studies have been conducted to explore its nature and impact on the electrical properties [5, 20, 24, 25, 30, 45–48], and at least six different phase diagrams of BNT–BT have already been proposed. Due to the coexistence of PNRs and ferroelectric

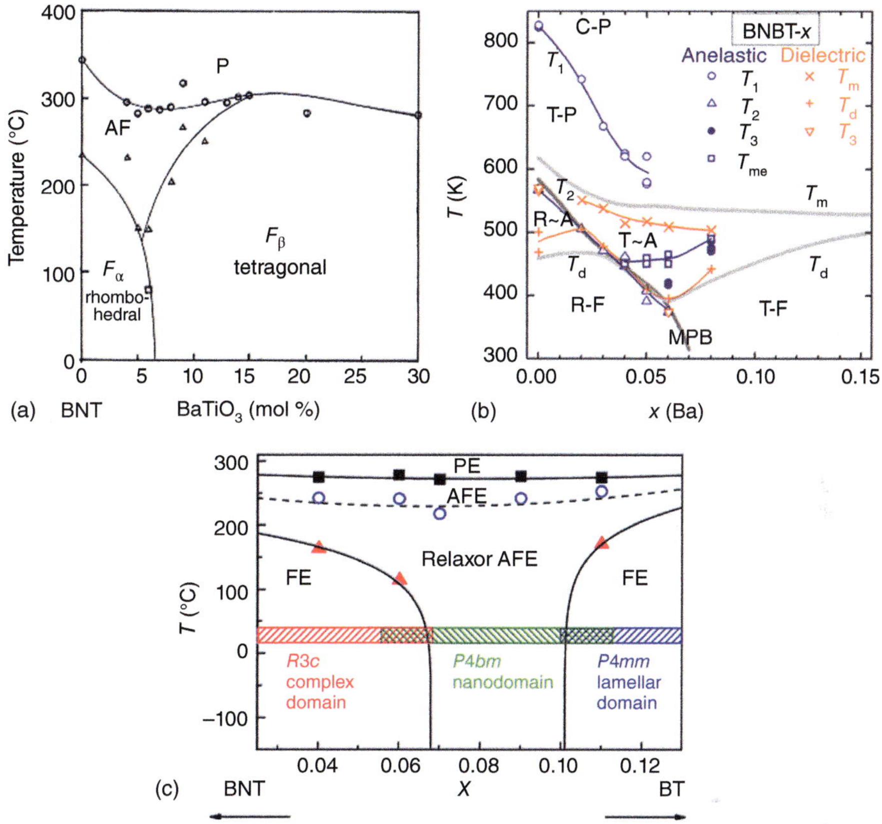

Figure 4.6 (a) The first diagram of BNT–BT, (b) the diagram of BNT–BT based on the results of the anelastic and dielectric measurements, and (c) the phase diagram of BNT–BT based on the TEM measurements. Source: (a) Reprinted with permission from Takenaka et al. [5]. Copyright 1991, The Physical Society of Japan; (b) Reprinted with permission from Cordero et al. [45]. Copyright 2010, American Physical Society; (c) Ma et al. 2010 [25]. Reprinted with permission of AIP Publishing.

domains, the phase structures of the BNT–BT solid solutions are much more complicated than the other traditional ferroelectrics.

One of the earliest and the most recognizable diagrams of BNT–BT is shown in Figure 4.6a showing that the MPB is at around BNT–6BT composition, which separates the rhombohedral phase and tetragonal phase for BNT–xBT ($x < 6$) and BNT–xBT ($x > 6$), respectively. Later, a more accurate phase diagram was proposed according to some new scientific results obtained by anelastic, dielectric, and other electric measurements, as shown in Figure 4.6b [45]. However, little inaccuracies have been revealed recently in this phase diagram, to which many modifications have been added. There are two major parts of modification. The first modification is based on the idea that MPB is a composition region rather than a specific composition point. The second one is that the pure BNT was found to be a

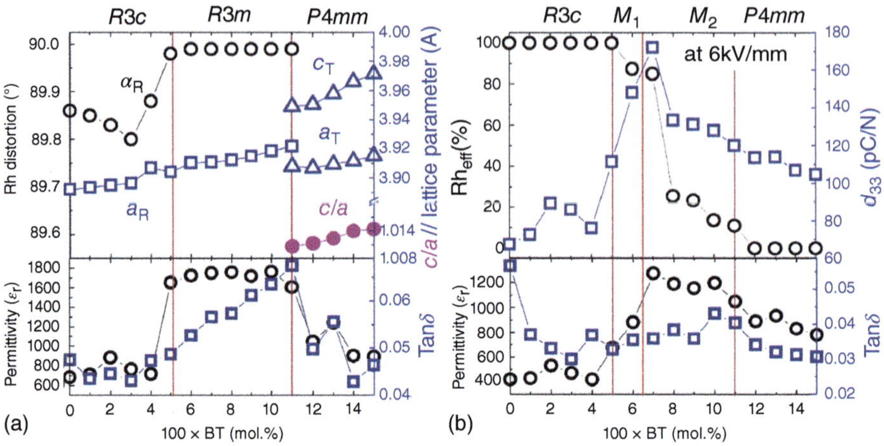

Figure 4.7 Phase diagram and dielectric properties of BNT–BT ceramic (a) before and (b) after poling. Source: Jo et al. 2011 [46]. Reprinted with permission of AIP Publishing.

monoclinic (M) phase rather than an R phase and that a new M–R phase boundary was suggested to be at the composition around BNT–3BT.

Ma et al. used TEM to investigate the phase structures of unpoled BNT–xBT ceramics [24, 25]. They proposed a new phase diagram of unpoled BNT–xBT as shown in Figure 4.6c, which is different from the original phase diagram of poled samples as shown in Figure 4.6a,b. In this diagram, a new phase with *P4bm* symmetry was added at the BNT–xBT compositions for $x = 0.07–0.09$.

Jo et al. implemented *in situ* synchrotron XRD and dielectric measurement to investigate the phase structures of poled and unpoled BNT–xBT ceramics and subsequently obtained a phase diagram as shown in Figure 4.7 [46].

Two-phase boundaries can be significantly observed at the compositions around BNT–5BT and BNT–11BT, as indicated by the abrupt changes of rhombohedral distortion, lattice parameters, and permittivity. The positions of these phase boundaries were almost identical to the ones proposed by Ma and Tan [24]. The space groups determined by the Rietveld refinements are also similar to those derived from the TEM results, except for the compositional region between BNT–BT5 and BNT–BT11. In these works, the space group was thought to be of *R3m* symmetry rather than *P4bm* symmetry. The differences could be rationalized by the different testing scales between XRD and SAED (i.e. XRD gives the macroscopic information of the average structure, while the SAED reveals the lattice structure in a nano-sized region). These findings support that the relaxor phase is composed of *P4bm* PNRs embedded in a pseudocubic matrix.

Although pure BNT has long been considered as a rhombohedral *R3c* phase, recently many groups found that it should be a monoclinic (*M*) *Cc* phase [26, 29, 30, 49, 50]. Gorfman and Thomas applied the high-resolution X-ray to get the reciprocal-space map (RSM) for characterizing the phase structure of a BNT single crystal. RSM is an effective method to identify the monoclinic phase in the perovskite ferroelectrics [49]. The *M* and *R* phases have different splitting in the

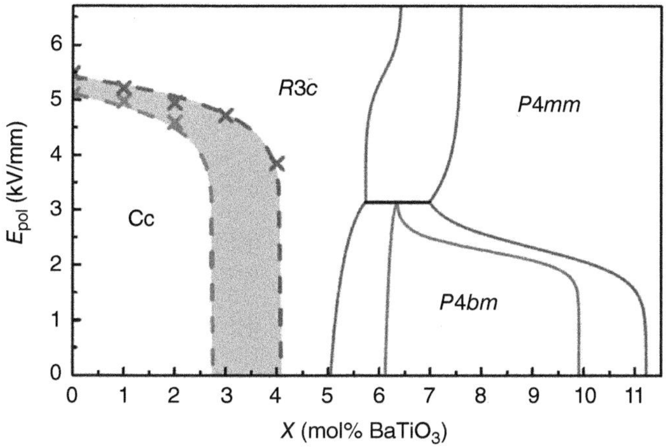

Figure 4.8 The phase diagram for polycrystalline $(1-x)(Bi_{1/2}Na_{1/2})TiO_3-xBaTiO_3$ ceramics under electric poling. Source: Reprinted with permission from Ma et al. [30]. Copyright 2013, John Wiley & Sons, Inc.

peaks with some special indices, such as (114) and (222). The Rietveld refinement results of the high-resolution X-ray pattern suggested that the *Cc* symmetry fits better than the *R*3*c* symmetry [26]. Similar results were obtained in the study conducted on sintered BNT ceramics by Levin and Reaney [29]. Following these results, electron diffuse scattering together with high-resolution TEM was utilized to characterize the octahedral tiling and cation displacements of BNT ceramics [29]. BNT was found to consist of nanoscale twin domains that are featured with $a^-a^-c^+$ tilting, exhibiting an average monoclinic distortion $a^-a^-c^-$ tilting system and supporting that the *Cc* symmetry is an averaged "best fit" cell. Recently, a new method was proposed to distinguish the *Cc* and *R*3*c*, which was based on SAED of two adjacent (101) axes within the domain wall plane [30]. It has been found that the *Cc* phase dominates in BNT–*x*BT ($x \leq 3$), while the *R*3*c* dominates in BNT–*x*BT ($x \geq 4$).

Further investigations have indicated that the reported global *Cc* distortion was not a thermodynamic equilibrium state but might originate from the presence of local structural and strain heterogeneities. Poling could remove the structural inhomogeneities and irreversibly induce a stable *R*3*c* phase [27, 28].

Combining all the structure analysis earlier, the most accurate phase diagram of the BNT–BT solid solution, considering both the *P*4*bm* relaxor region and the *Cc* monoclinic phase, is shown in Figure 4.8. However, this diagram is not the endpoint but a new stepping-stone for further research on BNT–BT systems.

4.3 High Converse Piezoelectricity

Among the lead-free piezoelectrics, BNT-based system is the only system that has large electric-field-induced strain (S_E) that is comparable with lead-based

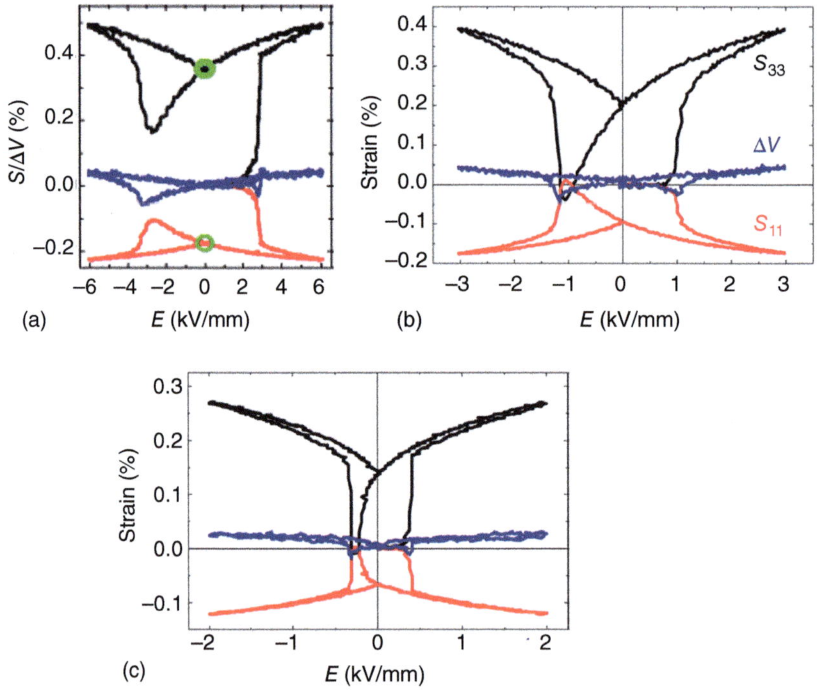

Figure 4.9 Axial and radial strains measured during the bipolar poling cycle on (a) BNT−6BT [16], (b) soft PZT, and (c) relaxor PLZT 8/65/35 [51]. Source: Reprinted with permission from Ma et al. [30]. Copyright 2013, John Wiley & Sons, Inc.

piezoelectrics. In this part, the high converse piezoelectricity of BNT-based system and the origin of their outstanding stain-related performances will be discussed in detail. In 2009, Rödel and coworkers systematically studied this material system [16, 18, 19], which indicated that BNT–BT ceramics have a large strain during the first poling cycle, while the strain at the second poling cycle is reduced, as shown in Figure 4.9a.

The result indicates that the large strain in the first cycle is irreversible. When the temperature rises to a threshold value, the strain curve gradually becomes reversible, with a maximum electric-field-induced strain. Thus, at this temperature, the intrinsic large S_E is completely released. This temperature is almost the same as the depolarization temperature T_d. It was further found that this transition temperature can be adjusted by doping KNN into the BNT–BT binary system, achieving high reversible strain at room temperature for application. Zhang et al. obtained a large S_E of 0.45% in the 0.92BNT–0.06BT–0.02KNN ceramic, and a large-signal inverse piezoelectric constant d_{33}^* ~560 pm/V [13]. The high performance of the BNT-based system under a large electric field soon became a hot research topic. Many interesting properties including electric-field-induced phase transition and temperature/doping dependent nonergodic (NR) to ergodic relaxor (ER) phase transition were discovered, revealing a more detailed depiction of the structure of BNT-based piezoelectrics.

4.3.1 Electric-Field-Induced Phase Transition

As aforementioned, the irreversible electric-field-induced strain of BNT–6BT shown in Figure 4.9a is an abnormal phenomenon that is different from the first *S–E* loop of the traditional ferroelectrics, such as soft PZT and relaxor PLZT, as shown in Figure 4.9b,c. The irreversible electric-field-induced strain was found in not only the MPB composition, but also many other compositions. Figure 4.10 shows the first loop of BNT–*x*BT ($x = 1, 2, 3, \ldots, 15$) ceramics.

The irreversible strain can be found in the compositions from pure BNT up to BNT–12%BT–12BT. In these *S–E* loops, two features are indicated. One is that the presence of remnant volume (V_{rem}) that decreases monotonically with increasing BT content, and the other is the non-zero strain level measured on the negative side of the bipolar strain curve, which will be referred to as the "unrecoverable" strain (S_{irr}). These loops reveal a phase transition during the first cycle of an applied electric field, accompanied by a change in the molar volume of the phase. Since the *P–E* loop of the poled BNT–BT behaves like the traditional ferroelectric hysteresis, the phase after the first cycle of an applied electric field is considered to be a ferroelectric phase.

The *in situ* XRD was used to characterize the structural evolution of the as-sintered BNT–BT ceramics upon an applied electric field [16, 18]. Figure 4.11 shows the XRD

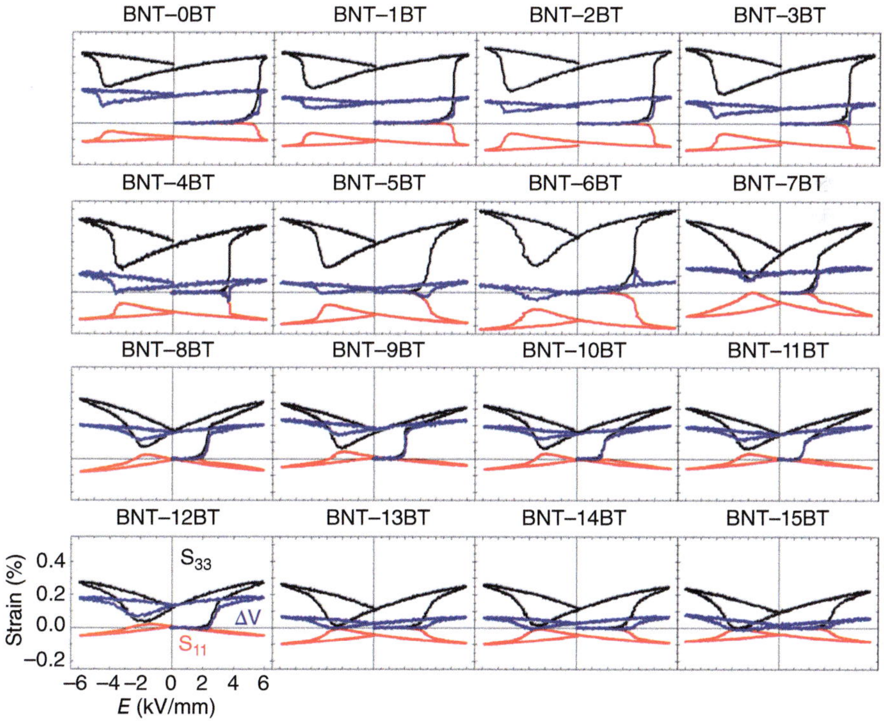

Figure 4.10 Electric-field-induced stain and volume changes of BNT−100*x*BT ($0 \leq x \leq 0.15$) during the bipolar poling cycle. Source: Jo and Rödel 2011 [51]. Reprinted with permission of AIP Publishing.

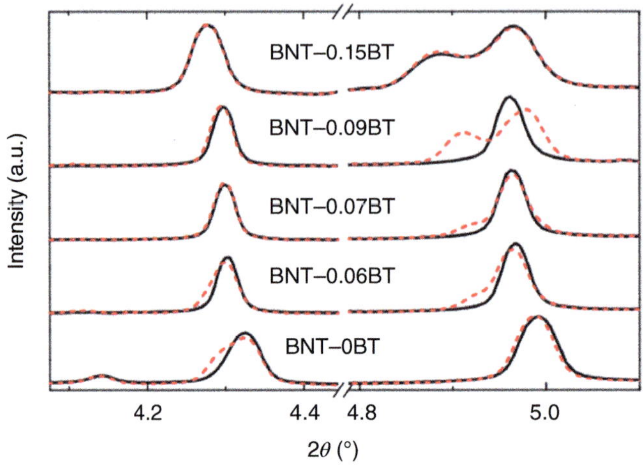

Figure 4.11 X-ray diffraction patterns of BNT–BT showing $(111)_{pc}$ and $(200)_{pc}$. The solid black lines and broken red lines refer to the diffraction patterns collected at 0 and 6 kV/mm, respectively. Source: Reprinted with permission from Jo et al. [16]. Copyright 2012, Springer Nature.

patterns of $(111)_{pc}$ and $(200)_{pc}$ peaks of selected compositions in the BNT–BT system at 0 kV/mm and at 6 kV/mm.

A series of compositions from BNT–5BT to BNT–10BT displayed cubic symmetry within the resolution limit of the XRD. All phase structures evolved into non-cubic symmetries under a sufficiently high electric field, representing a relaxor with PNRs embedded in cubic matrix to ferroelectric phase transition.

Even the aforementioned *R*3*c* phase and *Cc* phase compositions from pure BNT to BNT–4BT were found to undergo a phase transition with the change in the lattice parameter and the lattice distortion degree before and after electrical poling [30]. These results are consistent with the irreversible *S*–*E* loops in the compositions from BNT–0BT to the BNT–12BT. As a contrast, the BNT–15BT, which is a *T* phase ferroelectric composition, shows no change in the XRD pattern before and after poling. Similar phase transition evidence is also found by using an area detector to measure the diffraction patterns with the scattering beam aligned in different directions. The setup is shown in Figure 4.12A.

Both the synchrotron XRD and the neutron diffraction results show similar results [16, 40]. The diffraction pattern at 90° shows no difference with the origin diffraction pattern with a pseudocubic structure. The diffraction pattern in other directions shows significant peak splitting in (002) diffraction peak and other peak transformation from a pseudocubic to a T structure arising from an electric-field-induced phase transition in BNT–7BT as shown in Figure 4.12B and from a pseudocubic to a *T*/*R* mixing structure in BNT–6BT as shown in Figure 4.12C, with a strong domain texture aligning preferentially parallel to the electric field.

The *in situ* TEM was performed on the BNT–6BT sample to investigate the domain and phase structure on the application of electric field [52]. The bright-field TEM

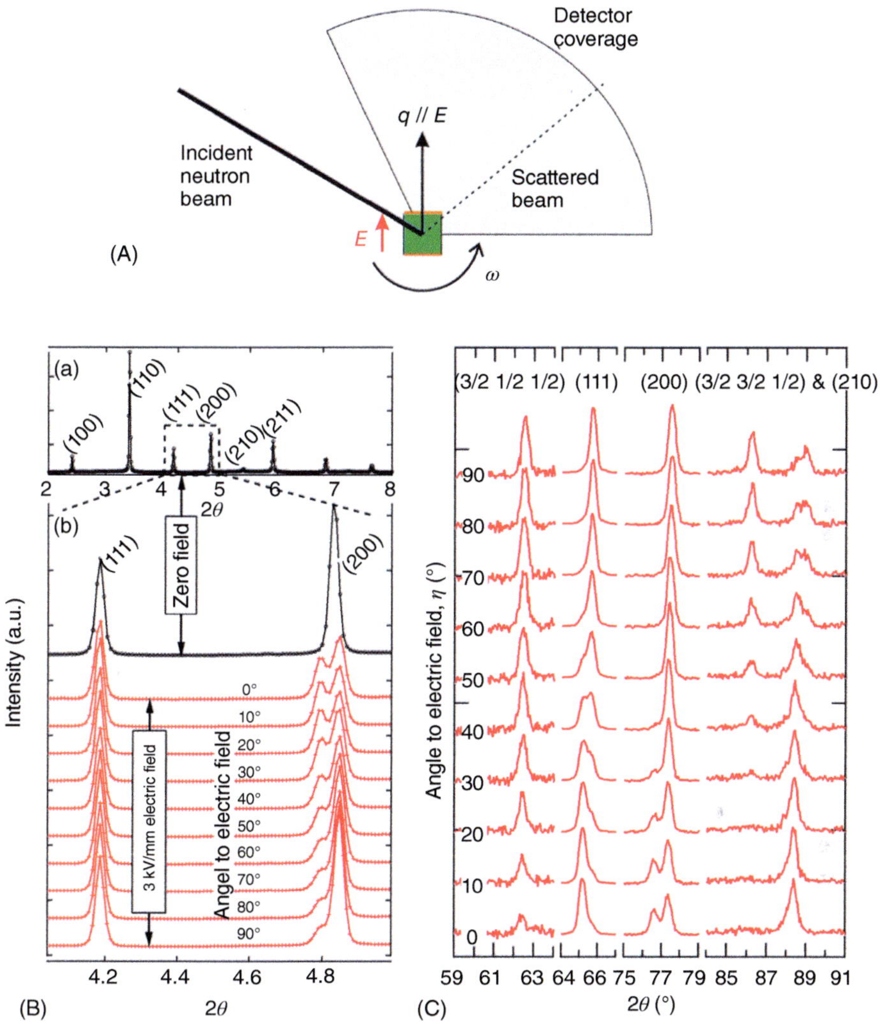

Figure 4.12 (A) The measurement setup; (B) the XRD pattern of the BNT−7BT; and (C) neutron diffraction result of the BNT−6BT. Source: (a, b) Daniels et al. 2009 [18]. Reprinted with permission of AIP Publishing; (a, c) Simons et al. 2011 [40]. Reprinted with permission of AIP Publishing.

image and corresponding SAED patterns of a specific grain under different electric fields are shown in Figure 4.13.

At zero field, about 60% of the grains exhibit only *P4bm* nanodomains with faint contrast while the rest exhibit mixed *P4bm* and *R3c* nanodomains. The *P4bm* nanodomain is characterized by the presence of 1/2(*ooe*) and the absence of 1/2(*ooo*) superlattice diffraction spots (Figure 4.13e and, on the contrary, the *R3c* nanodomain is distinct by the presence of 1/2(*ooo*) and the absence of 1/2(*ooe*) spots (Figure 4.13f). The selected grains are the ones with only *P4bm* nanodomains originally (Figure 4.13a). As the applied electric fields increase, these nanodomains

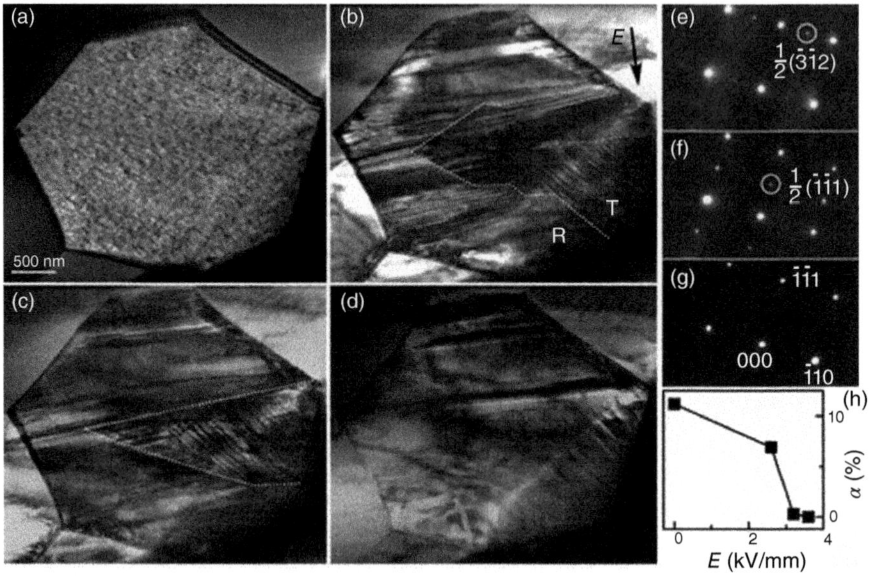

Figure 4.13 Transmission electron microscope observations along the [53] zone axis of the selected grain of BNT−66BT at (a) 0 kV/mm, (b) 3.2 kV/mm, (c) 3.6 kV/mm, and (d) 4.0kV/mm. Representative selected area electron diffraction patterns are displayed in (e) from nanodomains at 0 kV/mm, (f) from region R at 3.2 kV/mm, and (g) from region T at 3.6 kV/mm and (h) the evolution of the intensity ratio with increasing applied fields. Source: (a–g) Ma et al. [52]; (h) Reprinted with permission from Ma et al. [52]. Copyright 2012, American Physical Society.

turn into lamellar domains. At 3.2 kV/mm, a large part of the grain is taken by the wedge-shaped domains that have *R*3*c* symmetry, while the remaining lamellar domains go through a *P*4*bm*-to-*P*4*mm* transition illustrated by the sudden decrease in the 1/2(*ooe*) intensity toward zero as shown in Figure 4.13h. With the further increasing electric field, the region consisting of *R*3*c* wedge-shaped domains expands and finally occupies the entire grain at 4 kV/mm (Figure 4.13c,d). Thus, the phase transition of BNT–6BT under an applied electric field can be described as the following process: relaxor of *R*3*c*/*P*4*bm* PNRs → ferroelectric with *R*3*c*/*P*4*mm* macrodomains → ferroelectric with *R*3*c* macrodomains. It is believed that the irreversible transition from PNRs to ferroelectric macrodomains accounts for the S_{irr} in the composition between BNT–0BT and BNT–5BT, and the irreversible transition from *P*4*bm* PNRs to ferroelectric *P*4*mm* macrodomains accounts for the S_{irr} in the composition between BNT–7BT and BNT–11BT. For BNT–6BT, the MPB composition, both kinds of domain transitions, contribute to the S_{irr}.

4.3.2 Ergodic and Nonergodic Relaxor

According to the classical theory, a relaxor transforms from high-temperature non-polar paraelectric (P_E) phase into the ER state with PNRs in randomly distributed directions of dipole moments at Burns temperature (T_B). Under an applied electric field, an *ER* will be reversibly poled into a ferroelectric phase that will return to an ER when the electric field is removed. Therefore, the *P–E* loop manifests a double ferroelectric hysteresis loop with almost zero P_r and an *S–E* loop with zero negative strain (S_{neg}). At the temperature close to T_B, the PNRs are randomly fluctuating and their behavior is ergodic. Upon cooling, their dynamics slow down and at a low enough temperature, freezing temperature (T_f), the PNRs will be frozen into a NR state, while remaining an average cubic symmetry. Different from the ER, the NR can be irreversibly transformed into an FE state by an applied electric field. Thus, a ferroelectric type hysteresis and an *S–E* loop with S_{irr} are the characteristics of the NR. Upon heating the FE phase transforms to the ER one at the temperature T_C, which is very close to T_f. However, in many relaxors, the spontaneous transition from the ER to FE instead of an NR phase occurs at T_C and thus the NR state does not exist. Cryogenic investigations have already shown the presence of an NR state in the PMN [32, 54].

The concept of NR and ER has been extensively used in the BNT system to explain the irreversible electric-field-induced phase transition and the temperature-dependent phase transition [20, 37, 55]. The BNT–6BT that suffers an irreversible relaxor-ferroelectric phase transition under sufficient large electric field is considered to be an NR. When a suitable amount (2–3%) of KNN is doped into the BNT–6BT, the materials manifest electric properties of an ER, a pinched double-loops ferroelectric hysteresis, and a sprout-shaped *S–E* loop without S_{neg} [16]. Figure 4.14 compares the *P–E* loops and *S–E* loops together with the volume changes of (a) BNT–6BT, (b) BNT–6BT–1KNN, and (c) BNT–6BT–3KNN.

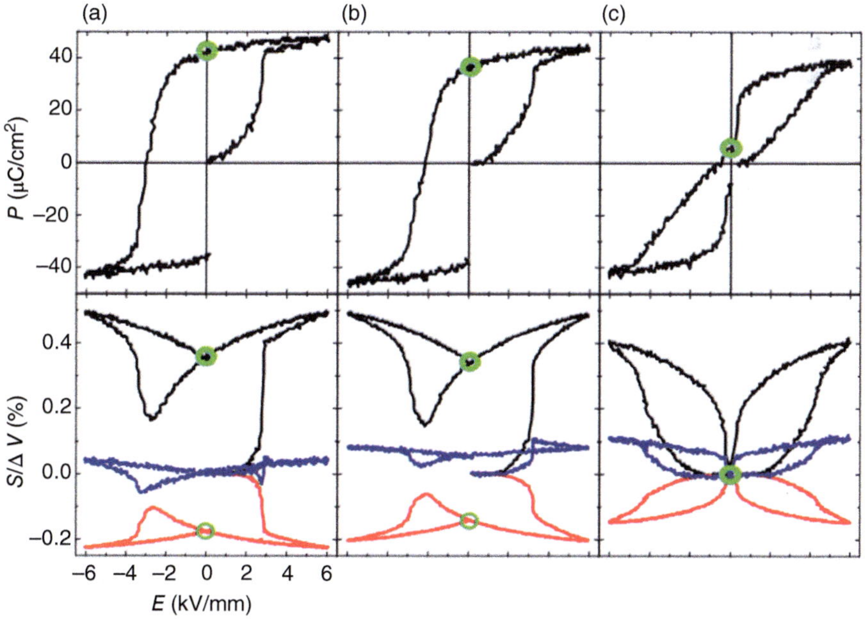

Figure 4.14 *P–E* and *S–E* loops or volume changes in (a) BNT–6BT, (b) BNT–6BT–1KNN, and (c) BNT–6BT–3KNN. Source: Reprinted with permission from Jo et al. [16]. Copyright 2012, Springer Nature.

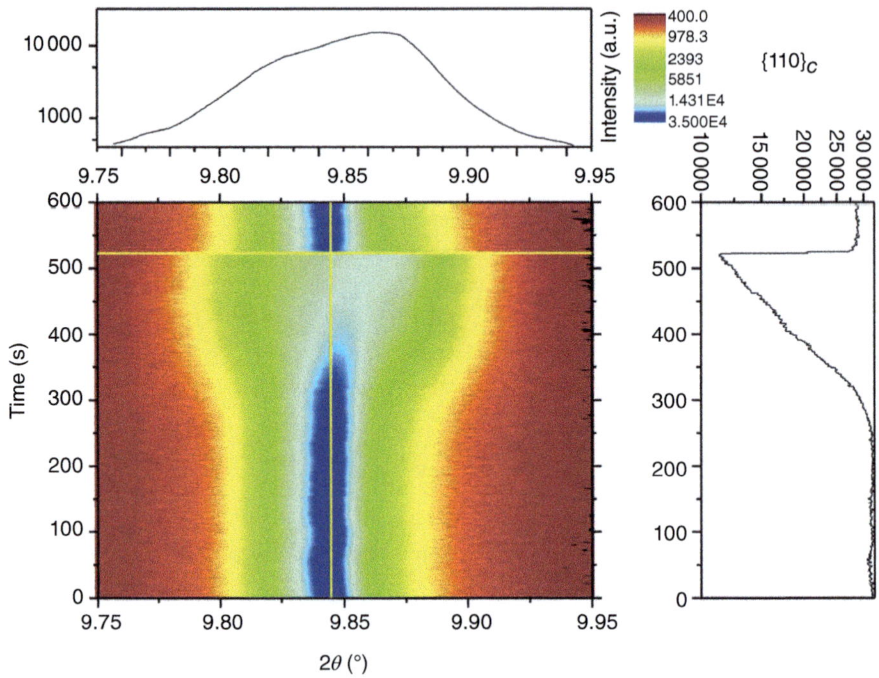

Figure 4.15 Contour and perspective image of electric-field-dependent XRD on BNT–6BT–2KNN. Source: Reprinted with permission from Hinterstein et al. [56]. Copyright 2010, International Union of Crystallography.

However, to fully confirm the ER phase structure of the BNT–BT–KNN, two key points should be strictly verified: one is the reversible phase transition under an electric field, and another is the essence of the relaxor rather than antiferroelectric. The electric field-induced reversible phase transition was confirmed by the *in situ* XRD/neutron scattering techniques and *in situ* TEM [42, 56, 57]. In the *in situ* XRD experiment as shown in Figure 4.15, the initially highly symmetric $(110)_{pc}$ reflection begins to split at ~2.2 kV/mm and disappears upon the removal of the applied electric field. The pattern returns back to its initial state, confirming the reversible phase transition.

Figure 4.16 shows the *in situ* TEM result. Neither ferroelectric nor antiferroelectric domain was observed in the bright-field TEM image, as shown in Figure 4.16a. At a nominal field of 25 kV/cm, the small domains appear at the bottom right side in Figure 4.16b with a bending contrast and a lamellar domain configuration is clearly visible in the $[110]_c$ zone axis as shown in Figure 4.16c. After the removal of the applied electric field, the contrast of domains disappeared, with only strain bending contrast and dislocations can still be observed, as shown in Figure 4.16d. Piezoresponse force microscopy (PFM) was also applied to characterize the domain evolution of BNT–BT–KNN ceramics in nanoscale under an applied electric field. The litho-PFM, which can pole the selected area on the sample by using

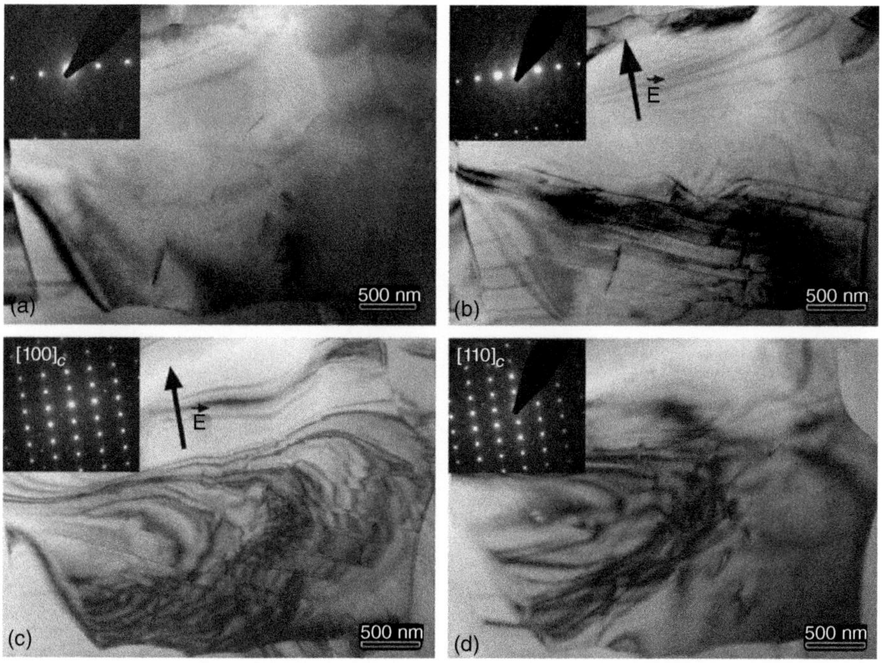

Figure 4.16 The *in situ* TEM results: (a) the grain before applying electric field and (b) at nominal 25 kV/cm (actual field about 50 kV/cm). (c) The same grain oriented in $[110]_c$ direction at nominal 25 kV/cm (actual field about 50 kV/cm) and (d) at 0 kV/cm (after the removal of electric field). Source: Kling et al. [57].

a conducting probe, is implemented to compare the different domain switching behaviors in BNT–BT ceramics. Figure 4.17 provides the out-of-plane (OP) PFM result of selected areas directly after poling and after 66 minutes (4000 s) for (b) BNT–6BT, (c) BNT–6BT–1KNN, and (d) BNT–6BT–3KNN.

The relaxation function of the average PFM signal in the switched area is also depicted in Figure 4.17a. The piezoresponse of the switched domains in BNT–6BT shows no decay, whereas the one in BNT–6BT–1KNN decreases a little and asymptotically saturated to a certain value. The domains switched in BNT–6BT–3KNN relaxed rapidly to zero and totally disappear after 30 minutes, demonstrating reversible electric-field-induced ferroelectric phase transition [22]. Similar to the BNT–BT ceramics, the frequency dispersion of the ε–T spectra in the BNT–BT–KNN system is the most decisive feature in the relaxor ferroelectric. All the measurements, including electrical measurement, *in situ* XRD, neutron diffraction, TEM, and PFM mapping, support that the BNT–BT–KNN is an ER. The different electric-field-induced phase transition mechanisms observed in BNT–6BT and BNT–6BT–2KNN are illustrated in Figure 4.18 demonstrating how ER and NR evolve under an applied electric field.

It was recently proposed that the non-ergodicity in BNT-based materials may be related to the Bi–O hybridization that provided a long-range non-cubicity [58].

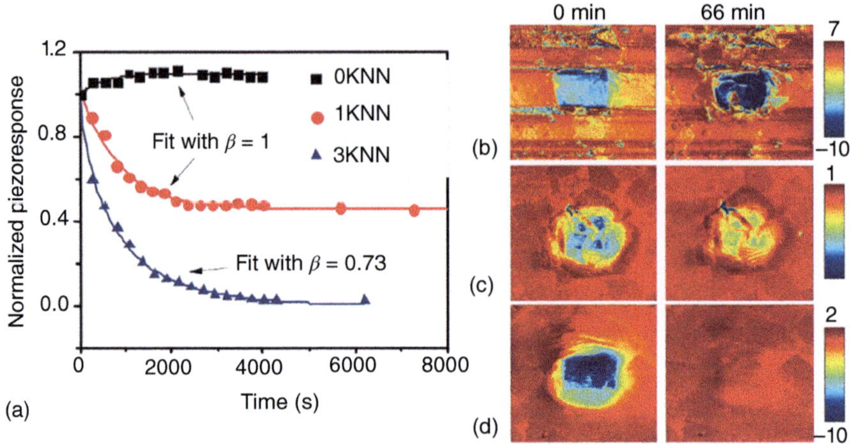

Figure 4.17 (a) Relaxation of the average PFM signal in the switched area and OP PFM images after poling and after 4000 s for (b) 0KNN, (c) 1KNN, and (d) 3KNN. Source: (a) Reprinted with permission from Dittmer et al. [22]. Copyright 2012, John Wiley & Sons, Inc.; (b–d) Dittmer et al. [22].

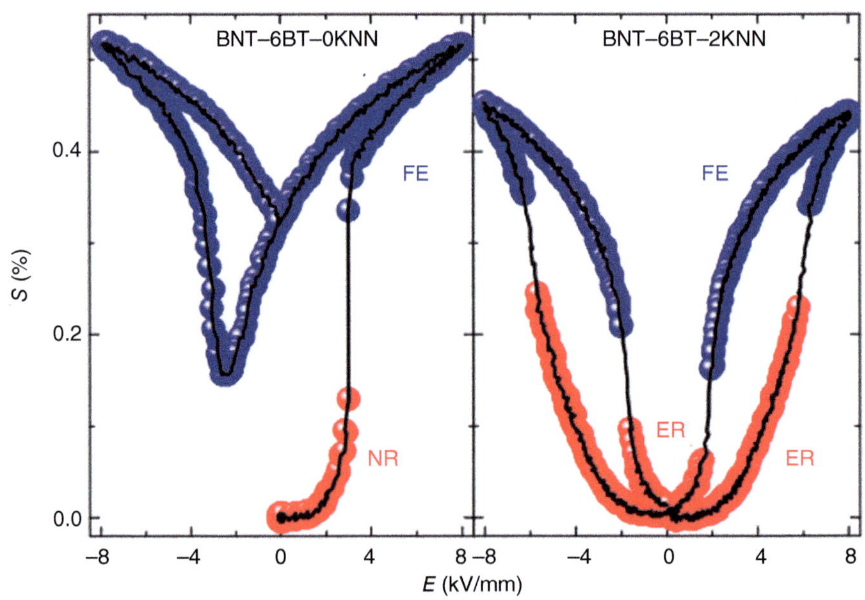

Figure 4.18 The identity of the electric-field-induced phase transition observed in BNT–6BT and BNT–BT–2KNN. Source: Reprinted with permission from Jo et al. [16]. Copyright 2012, Springer Nature.

4.3.3 Modulation of Depolarization Temperature

The temperature-dependent piezoelectric performance of BNT-based materials is closely related to their temperature-dependent evolution of the phase structure. The depolarization temperature (T_d) is defined as the temperature where the inflection point of the associated decaying polarization curve occurs [59]. In the BNT system, a ferroelectric-to-relaxor phase transition happens in the poled sample as the temperature increases, and this process is accompanied by a shrink of classic *P–E* loop to a pinched *P–E* loop that leads to a rapid decrease in P_r. However, the transition does not happen at a specific temperature as in classic ferroelectrics, but in a temperature range. Thus, T_d is always at a temperature close to the ferroelectric-to-relaxor phase transition temperature ($T_{F\text{-}R}$). On one hand, the depolarization leads to an abrupt decrease in d_{33}, which also means the T_d should be as high as possible. On the other hand, it is found that the S_E reaches its maximum around T_d and it becomes reversible at T_d. Therefore, for large-signal applications, the T_d should be modulated to room temperature. In BNT–6BT, the T_d is at around 100 °C, which is neither high enough for the temperature stability for d_{33} application nor low enough for the room-temperature large-signal S_E application. Thus, many works have been conducted to modulate the T_d of BNT-based ceramics to a different temperature region, depending on the application purpose. There are mainly three different approaches for T_d adjustment: (i) the compositional modification approach achieved by doping, (ii) the composite approach achieved by adding other BNT-based materials, and (iii) the internal stress approach.

4.3.3.1 Compositional Modification Approach

By doping an appropriate amount of KNN into BNT–6BT, a pinched *P–E* loop can be achieved at room temperature. In other words, the T_d of the BNT–6BT can be shifted to room temperature by the doping of KNN. The addition of KNN can disrupt the long-range ferroelectric phase. Many other dopants were also found to be effective to shift the T_d of BNT-based materials to room temperature and, consequently, achieve a large reversible strain at room temperature. The effective dopants include the B-site donor ions Nb^{5+}, Sb^{5+}, and Ta^{5+}, which tend to destabilize ferroelectric order [60–62], and compounds such as $KNbO_3$, $SrTiO_3$, $BaZrO_3$, and $BiMeO_3$ (Me are metals such as Sc, Mn, and Al) [17, 63–66]. The room-temperature S_E and large-signal d_{33}^* of some representative compositions in BNT-based ceramics are summarized in Table 4.1.

Unlike the B-site donor doping that could decrease the T_d, the B-site acceptor dopants like Mn^{2+}, Fe^{3+}, and Zn^{2+} can stabilize the long-range ferroelectric order and thus increases the T_d [76]. The B-site acceptor dopants can cause the formation of defect complexes, such as $(Fe'_{Ti}–V_O^{\cdot\cdot})^{\cdot}$ and $Mn''_{Ti}–V_O^{\cdot\cdot}$, that will pin the movement of the domain walls and stabilize the long-range ferroelectric order [77, 78]. It was also reported that oxygen vacancy may exist in the quenched samples and consequently pins the domain walls, resulting in significant enhancement of depolarization temperature [79].

Binary $Bi_{1/2}Na_{1/2}TiO_3$–$Bi_{1/2}K_{1/2}TiO_3$ (BNT–BKT) solid solutions with the *R–T* MPB at around BNT–2BKT have a relatively high T_d = 150–190 °C and d_{33} around

Table 4.1 Room-temperature S_E and large-signal d_{33}^* of some representative compositions in BNT-based ceramics.

Composition	S_E (%)	d_{33}^* (pm/V)	References
$(1-x)$BNT–xKNbO$_3$, $x = 0.08$	0.40	498	[67]
$(1-x)$BNT–xBaZrO$_3$, $x = 0.055$	0.40	500	[68]
$(1-x)$BNT–xBa(Al$_{0.5}$Sb$_{0.5}$)O$_3$, $x = 0.045$	0.48	532	[69]
0.975(0.94BNT–0.06BT)–0.025LiNbO$_3$	0.6	857	[70]
0.98(0.94BNT–0.06BT)–0.02BaSnO$_3$	0.4	669	[71]
$Bi_{0.5}(Na_{0.82}K_{0.18})_{0.5}(Ti_{0.97}Nb_{0.03})O_3$	0.47	641	[61]
$(Bi_{0.5}Na_{0.385}K_{0.09}Li_{0.025})(Ti_{0.975}Ta_{0.025})O_3$	0.436	727	[72]
0.98(0.935BNT–0.065BT)–0.02SrZrO$_3$	0.39	722	[73]
$(Bi_{0.5}Na_{0.5})_{0.943}(Bi_{0.2}Sr_{0.7}\square_{0.1})_{0.02}Ba_{0.055}TiO_3$	0.428	713	[74]
$(Bi_{0.5}Na_{0.42}K_{0.08})_{0.96}Sr_{0.04}Ti_{0.975}Nb_{0.025}O_3$	0.70	1400	[75]

Table 4.2 Representative BNT-based solid solutions with high T_d and large d_{33}.

Composition	T_d(°C)	d_{33} (pC/N)	References
0.79BNT–0.14BKT–0.07BT	162	170	[80]
0.88BNT–0.08BKT–0.04BT	224	135	[80]
0.77BNT–0.23BKT–excessive Bi	330	207	[11]
0.92BNT–0.08Ba(Hf$_{0.05}$Ti$_{0.95}$)O$_3$	180	136	[81]
0.91BNT–0.09B(Ti$_{0.942}$Zr$_{0.058}$)O$_3$	244	147	[82]
0.83BNT–0.17BT–0.2 wt% CuO	218	148	[83]
0.9BNT–0.1BT–1 at %MnO$_2$, single crystal	150	287	[84]
0.925BNT–0.075BT, (001)-textured	140	202	[85]

190 pC/N, which is preferred for the small-signal application. Some BNT-based compositions with high T_d and large d_{33} are summarized in Table 4.2.

4.3.3.2 Composite Approach

In recent years, the composite approach has been an important approach for T_d modification. Mainly two different kinds of materials were dispersed into the BNT matrix to construct the BNT-based composites.

The first kind of means is introducing simple binary oxides, such as ZnO. Zhang et al. constructed a 0–3 ceramic composite by introducing semiconductor ZnO particles into BNT–BT ceramics [86]. It is found that the introduction of ZnO can increase T_d. When the addition amount of ZnO reaches 30 mol%, the depolarization phenomenon completely disappears. At the same time, the ZnO addition improves the electrical properties of BNT–BT. It was proposed that ZnO can capture a certain amount of charges and thus compensates for the depolarization after the electric field is removed. While recent research proposed that the thermal residual stresses caused by the introduction of the ZnO second phase can also cause the shifting of T_d [70]. This research opens up a new idea for the improvement of temperature stability of BNT-based piezoelectric ceramics, e.g. other oxides, such as Al_2O_3, was also found to be able to delay the depolarization process. Additionally, by using the composite approach, the T_d can also be raised to achieve high small-signal d_{33} and good temperature stability, simultaneously.

The second kind of means is constructing relaxor–ferroelectric composites, consisting of a ferroelectric or NE phase (seed), embedded in a non-polar or ER phase (matrix). This composite can effectively overcome some shortcomings in the BNT-based materials, such as the relatively high operational electric field, large hysteresis, and poor temperature stability. Many investigations have been conducted on the BNT–BT–KNN/BNT–BT composite, where BNT–BT and BNT–BT–KNN act as the seed and the matrix, respectively [87, 88]. It was found that the composite that contains 20% seed shows the highest strain and the largest d^*_{33}, while the best temperature stability is achieved when the seed ratio is 40%. The result indicates that the composite approach could meet the requirements for different applications by adjusting the seeding content to an appropriate value. Recently, Zhang et al. added 30 vol% of BNT–7BT into $0.91Bi_{1/2}Na_{1/2}TiO_3$–$0.06BaTiO_3$–$0.03AgNbO_3$ to construct a 0–3 composite [89]. An extremely large strain, $d^*_{33} = 824$ pm/V at room temperature, is achieved, accompanied by a relatively low operation electric field of 4 kV/mm. It was found that both the polarization coupling and the strain coupling play critical roles in the reduction of the RE–FE phase transition field.

4.3.3.3 Stress Approach

The stress approach is a widely accepted method for adjusting the phase transition temperature in the traditional ferroelectrics. The mechanism of stress approach is based on the Ginzburg–Landau theory, in which the free energy of a ferroelectric material can be written as a Taylor expansion in terms of the order parameter, P, while the external electric field and stress count for the other two components of the free energy. Thus, the stress plays an important role in the phase transition, and, as a result, can be manipulated for modulating the phase transition temperature. It has been reported that the isostatic stress often leads to the formation of a higher symmetric phase while uniaxial stress could lead to the formation of a lower symmetric phase [90, 91].

However, limited works are focusing on the effect of stress on the phase transition temperature of BNT-based materials. Webber and coworkers have done a systematic work about the effect of stress on BNT-based ceramics [91–94]. It was found that the applied uniaxial compressive stress can shift the relaxor-ferroelectric phase transition to high temperature, as shown in Figure 4.19a, which indicates the stabilization of the long-range ferroelectric order and the enhancement of the T_d.

Based on the results, a phase diagram was depicted, as shown in Figure 4.19b. The stress–temperature phase diagram is similar to the reported electric

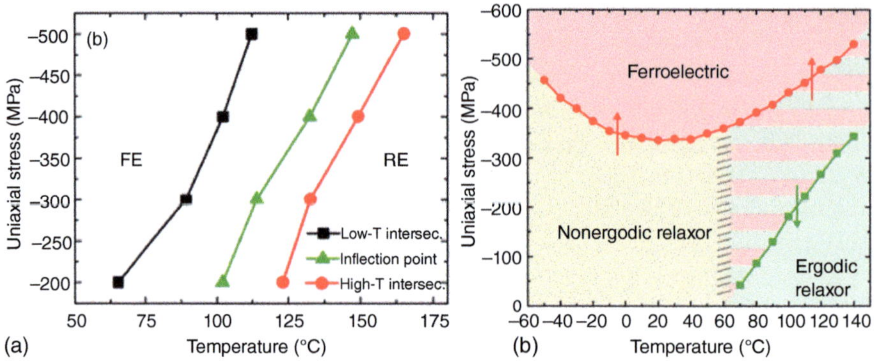

Figure 4.19 (a) Relaxor-ferroelectric phase transition induced by the uniaxial compress stress derived from the $\varepsilon-T$ spectra; (b) a stress and temperature phase diagram. Source: (a) Modified from Schader et al. [93]; (b) Reprinted with permission from Schader et al. [93]. Copyright 2016, American Physical Society.

field-temperature phase diagram [46], exhibiting the parallel influence of the electrical field and mechanical field to the long-range ferroelectric order in BNT-based materials, as predicted in Ginzburg–Landau theory. However, high uniaxial stress also suppresses the d_{33} of the BNT-based ceramics and decreases the frequency dispersion in the $\varepsilon-T$ and $d_{33}-T$ spectra simultaneously. These effects could be explained by the domain wall and PNR clamping caused by applied stress.

Note that it is not easy to apply stress on the ceramics during application. While for film materials, stress can be easily controlled by selecting different substrates because of the lattice mismatch. An enhancement in both T_d and $d_{33,eff}$ has been observed in (001)-textured BNT–6BT thin film on platinized Si substrate [95]. It is suggested that the in-plane tensile stress perpendicular to the texture orientation, which is caused by a larger expansion coefficient of BNT–6BT compared with that of Si, can stabilize the long-range ferroelectric order and the (001)-textured structure prompted the piezoelectric performance. For comparison, the T_d of randomly oriented BNT–6BT thin film grown on Si substrate was about 100 °C, which is the same as the BNT–6BT ceramics.

4.4 Thin Films

There has been increasing studies on the lead-free piezoelectric thin film over the last decade, following the extensive research on BNT–BT bulk materials. Research on the films has been driven by the widespread applications of the miniaturized devices in microelectromechanical systems (MEMSs). Besides, thin films are also an idea sample for fundamental studies, particularly for the investigation on phase structure, domain structure, and their related physical mechanisms. Different from bulk ceramics, one can easily introduce mechanical stress onto thin films by manipulating the difference among samples and substrates or buffer layers, such as thermal expansion coefficients and lattice parameters. The mechanical stress was found to be able to modify the phase structure and depolarization temperature (T_d) of ferroelectrics.

The BNT-based thin films can be fabricated via three major approaches, including magnetron sputtering, sol–gel processing, and pulsed laser deposition (PLD). A research group that belongs to a Japan company has been extensively working on the development of high-performance BNT–BT thin films prepared by magnetron sputtering [96–101], although the earliest work was done by another group from Singapore [102]. Epitaxial BNT–BT thin films could be grown on a single-crystalline MgO substrate with Pt surface electrodes by radio frequency (RF) magnetron sputtering. It was found that the BNT–BT thin films grown on (110) substrates show high piezoelectricity, whose transverse piezoelectric coefficient d_{31}^* and piezoelectric stress coefficient e_{31}^* reached −221 pC/N and −20.6 C/m^2, respectively [96, 97]. It should be noted that the phase structures of the epitaxial BNT–BT thin films are different from their bulk counterparts, due to the compressive thermal strain caused by the difference in thermal expansion among the BNT–BT samples and MgO substrates [98]. Highly *c*-axis-oriented polycrystalline BNT–BT thin film was prepared on a $LaNiO_3$-buffered Si substrate by RF magnetron sputtering. It is important that the temperatures at maximum permittivity (T_m) of the BNT–BT films were significantly enhanced by 250 °C than that of the BNT–BT bulks, accompanied by the disappearance of T_d. The resulting enhancement of temperature stability in BNT–BT films is beneficial for applications [99].

Pulsed laser deposition is a handy processing technique for thin film, especially in research laboratories. Several groups have used PLD to fabricate BNT–BT films [103–108]. For example, Cheng et al. fabricated a series of BNT–BT films with a wide range of compositions from 0% to 44% of BT on the (001)-$LaAlO_3$ substrates by using PLD in conjunction with a combinatorial methodology [103]. An obvious MPB was found at about 9% of BT, which deviates from what has been reported in bulk materials, probably due to the substrate clamping effect. Safari and coworkers prepared BNT–BT thin films on a $SrRuO_3/SrTiO_3$ substrate and investigated their dielectric and piezoelectric properties. However, it seems that the obtained properties are inferior to those reported in the magnetron sputtered BNT–BT thin films [104, 105]. Nevertheless, a giant electric energy density up to 154 J/cm^3 was realized in pulsed laser deposited La/Zr-codoped BNT–0.0582BT epitaxial thin films [108]. This value is roughly three times the highest value of the lead-based systems and five times the value of the best dielectric/ferroelectric polymer, which is attributed to the high epitaxial quality, great relaxor dispersion, and the coexistence of the ferroelectric/antiferroelectric phases [108].

The sol–gel process has been increasingly used for the fabrication of thin films because of its simplicity and controllability of chemical compositions. Recently, there were quite a few studies about the sol–gel-processed BNT-based thin films [53, 95, 109–115]. For example, Guo et al. prepared highly (100)-oriented BNT–0.06BT thin films on $LaNiO_3/\gamma\text{–}Al_2O_3/Si$ substrates [109]. They found that ethanolamine is a very effective coordinating ligand of Ba^{2+}, resulting in a transparent and stable precursor chemical solution, which is critically important for the sol–gel processing. Zhou et al. prepared BNT–0.06BT films by using sodium acetate, bismuth acetate, and barium acetate as starting precursors, which dissolved in 2-methoxyethanol and acetic acid [53]. They found that the phase structure and electrical properties of the resultant films could be greatly affected by the pyrolysis temperature. High-quality thin films could be obtained only with good control over the temperature-dependent crystallization behavior. The best ferroelectric and piezoelectric properties were achieved when the film was subjected to the pyrolysis at around 400 °C. They also fabricated highly (001)-textured BNT–0.06BT thin films on platinized Si wafers with a $LaNiO_3$ buffer layer [95]. The effective piezoelectric coefficient $d_{33,\text{eff}}$ is increased by 65% to 43 pm/V in the textured films from 26 pm/V in the non-textured counterparts. Figure 4.20 shows the surface and cross-sectional morphologies of the samples, together with the piezoresponse curves [95]. In addition, the texturing also raises the T_d to approximately 140 °C, which is about 40 °C higher than that in the non-textured films. The enhancement is attributed to the in-plane tensile thermal stress and the (001)-textured structure, which is again favorable for applications.

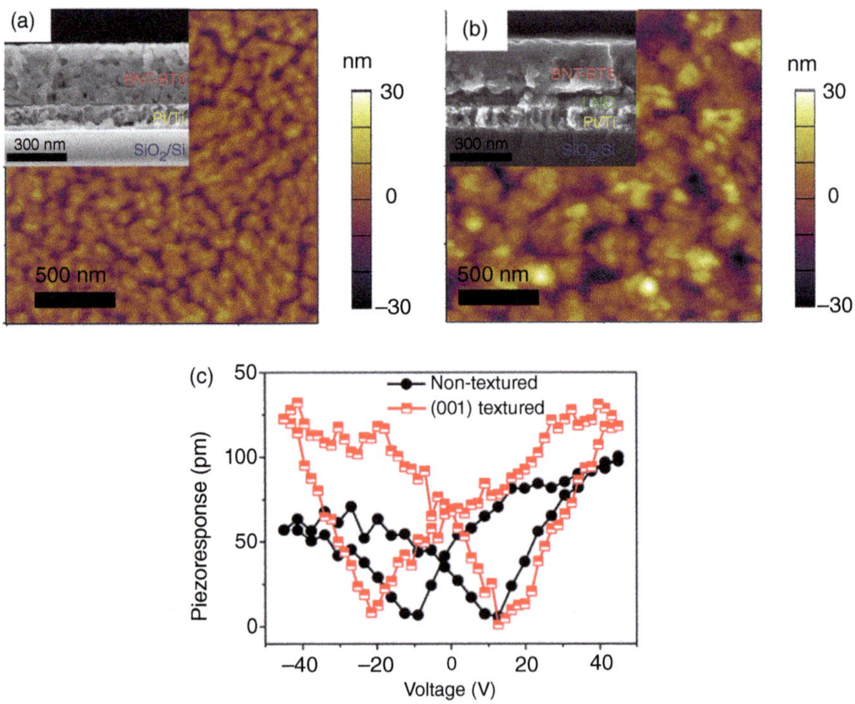

Figure 4.20 Surface atomic force microscopy (AFM) and cross-sectional scanning electron microscopy (SEM) images of the (a) non-textured and (b) textured BNT–BT thin films, and (c) their "butterfly" loops measured by using piezoresponse force microscopy (PFM). Source: (a, b) Zhou et al. [95]; (c) Reprinted with permission from Zhou et al. [95]. Copyright 2018, John Wiley & Sons, Inc.

4.5 Single Crystals

Improved piezoelectricity can be achieved in single crystals oriented to optimize piezoelectric coefficients and for which the mechanical constraint of misaligned neighboring grains is absent. Regarding lead-free piezoelectrics, quite a large numbers of studies have been conducted about the synthesis and characterization of BNT-based single crystals [116–122].

As the first report [116], single crystals of $(Na_{1/2}Bi_{1/2})TiO_3$–$BaTiO_3$ were successfully grown by the flux method using Bi_2O_3 flux by slow cooling from 1350 to 800 °C at a rate of 3.5°C/h. Electric-filed-induced strains as high as 0.25% and 0.85% were obtained in rhombohedral-phase and tetragonal phase crystals, respectively. Later, large-sized single crystals (40 mm in diameter, 10 mm height) of $0.94(Na_{1/2}Bi_{1/2})TiO_3$–$0.06BaTiO_3$, as shown in Figure 4.21, were successfully grown by a precisely controlled top seeded solution growth method in Shanghai Institute of Ceramics, China [117]. Because of the MPB the composition, the crystal shows exhibit rhombohedral–tetragonal phase coexistence at room temperature. High piezoelectric properties d_{33} of 360 pC/N and k_t of 62% were obtained in the (001) oriented crystals [117]. An ultrahigh electromechanical response with large amplitude longitudinal piezoelectric coefficients as high as d^*_{33} = 2500 pm/V was found in $[001]_{pc}$ oriented 0.944BNT–0.056BT single crystals near a depolarization temperature of T_d = 130 °C [118]. A further higher converse piezoelectric coefficient d^*_{33} up to 4600 pm/V was measured by another group in 0935BNT–0.065BT single crystals [119]. It was found that incorporation of a small amount of KNN into BNT–BT lead to great enhancement of piezoelectricity [120, 121]. Ternary lead-free BNT–BT–KNN single crystals were grown also by a precisely controlled top-seeded-solution-growth method used for the fabrication of BNT–BT crystals. An unprecedented piezoelectric coefficient $d_{33} \approx 840$ pC/N was obtained in ternary

Figure 4.21 Lead-free piezoelectric single crystal with a composition $0.94(Na_{1/2}Bi_{1/2})TiO_3$–$0.06BaTiO_3$. Source: Chen et al. [117].

BNT–6BT–KNN lead-free single crystals at room temperature with a large field induced strain of ≈0.9% [121].

BNT-based lead-free single crystals will receive increasing interest not only due to high piezoelectricity but also because they will provide idea specimens for further studies of the phase and multiscale domain structures.

4.6 High-Power Application

The high-power piezoelectric materials have been used for a wide range of applications, such as ultrasonic motors, transformers, and transducers. The high-power application requires a high-level vibration velocity (v_o) determined by the figure of merit (FOM), $k_{ij}\cdot Q_m$, where Q_m is the mechanical quality factor and k_{ij} is the electromechanical coupling coefficient. High Q_m, low tanδ, high E_c, and high k_{ij} values are necessary for high-power applications. Hard PZT, such as PZT-2, PZT-4, and PZT-8, are always used for high power applications. However, Q_m of hard PZT markedly decreases at a vibration velocity of approximately 1.0 m/s and, as a result, the vibration velocity is limited at the high actuating electric field [123] Recently, the BNT-based ceramics were found manifesting a high stable Q_m, which is even superior to the PZT-based ceramics. One of the advantages is that the E_C of the BNT-based ceramics is significantly higher than that of PZT-based ceramics. Generally, high E_C is expected to result in higher domain stability and a lower mechanical loss. However, as aforementioned, the BNT-based ceramics have a shortcoming of low T_d.

The $x(Bi_{1/2}Na_{1/2})TiO_3$–$y(Bi_{1/2}Li_{1/2})TiO_3$–$z(Bi_{1/2}K_{1/2})TiO_3$·(xBNT–yBLT–zBKT) solid solution is one of the most promising BNT-based systems for high-power applications. The system is based on the system BNT–BKT, which has an R–T MPB at around z = 0.18. Hiruma et al. kept y = 0.04 and investigated the high-power properties of different phase structures [124]. Figure 4.22a shows the variation of T_d and Q_m (33) as a function of z. It was found that the R phase side composition has a relatively high Q_m (33) and T_d than that of the T phase side composition.

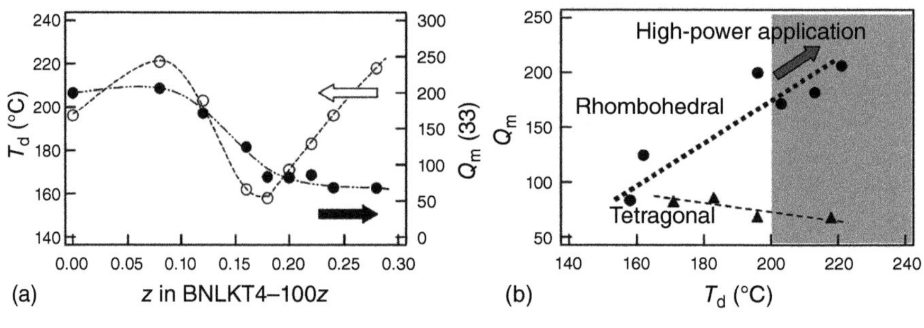

Figure 4.22 (a) Variation of T_d and Q_m (33) as a function of z; (b) the relationships between T_d and Q_m (33) on the rhombohedral and tetragonal sides in (0.96 − z) BNT–0.04BLT–zBKT. Source: Reprinted with permission from Hiruma et al. [124]. Copyright 2008, The Physical Society of Japan.

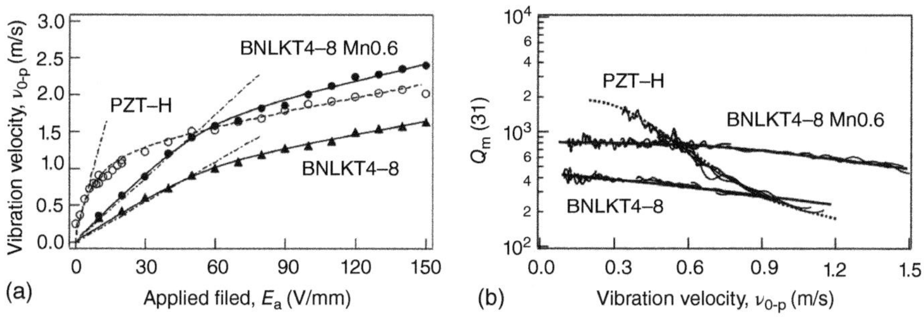

Figure 4.23 (a) Applied field E_a dependences of vibration velocity $v_{0\text{-p}}$ and (b) variations in Q_m as a function of vibration velocity $v_{0\text{-p}}$ for BNLKT4–8, BNLKT4–8$Mn_{0.6}$, and PZT–H. Source: Reprinted with permission from Hiruma et al. [124]. Copyright 2008, The Physical Society of Japan.

Figure 4.22b shows the relationship between T_d and Q_m (33) . It seems that the T_d and Q_m of the R phase have a positive correlation. $MnCO_3$, which is an effective dopant for the lead-based piezoelectrics, was also doped into BNLKT4–8. As the Mn doping increases, the Q_m was enhanced dramatically, while the d_{33} and k_{33} gradually decrease. At 0.6% Mn doping, the Q_m was raised to 400, which was twice as that of the non-doped one, though the k_{33} decreased from 0.51 to 0.49. It is assumed that Mn^{2+} or Mn^{3+} substitutes into B site as an acceptor dopant, and subsequently leads to the formation of oxygen vacancies. The oxygen vacancies pinned on the ferroelectric domain and thus increased the Q_m, accompanied by a slight scarification of k_{33}. Compared with PZT–4H, the BNLKT shows better high-power stability. Figure 4.23a,b depict the vibration velocity $v_{0\text{-p}}$ as a function of applied electric field E_a and the Q_m as the function of vibration velocity, respectively.

It was obvious that the performance of PZT–H decay strongly at high vibration velocity. On the contrary, the Mn-doped BNLKT has an outstanding performance and stability at high vibration velocity. Since the composition deviates from the MPB composition, the depolarization process happens at a relatively high temperature of 200 °C. Hejazi et al. doped 1.5% Mn into BNLKT4–8 and the sintered ceramics show a high planar $Q_m = 970$, low tan$\delta = 0.85\%$, and low $\varepsilon_r = 310$, which are comparable with that of PZT based hard materials [125]. More importantly, as shown in Figure 4.23, the 1.5% Mn-doped BNLKT4–8 manifests the lowest resonant frequency shift, the lowest temperature rise, and the best stability of Q_m as the vibration velocity increases.

$x(Bi_{1/2}Na_{1/2})TiO_3$–$y(Bi_{1/2}K_{1/2})TiO_3$–$zBaTiO_3$ is another suitable BNT-based system for high-power applications. Zhang et al. found that the ceramic with the composition $0.79(Bi_{1/2}Na_{1/2})TiO_3$–$0.14(Bi_{1/2}K_{1/2})TiO_3$–$0.07BaTiO_3$–0.8 wt% MnO_2 has large $Q_m = 1100$, $k_{33} = 0.49$, and low tan$\delta = 0.4\%$ [126]. Also, a high $T_d = 235$ °C is available in this composition, leading to the excellent temperature stability of the piezoelectric performance.

Recently, some researchers integrated the two aforementioned compositions into a solid solution with four components, i.e. $x(Bi_{1/2}Na_{1/2})TiO_3$–$y(Bi_{1/2}K_{1/2})TiO_3$–z

$(Bi_{1/2}Li_{1/2})TiO_3$–$p$$BaTiO_3$ [127]. It is also a system with good velocity stability and a relatively high $Q_m = 800$. Co-doping was effective for adjusting the T_d and other piezoelectric properties. However, the performances are not as good as the aforementioned Mn-doped compositions and more investigations on the compositional optimization are needed to balance different parameters of properties for the high-power applications.

4.7 Summary and Outlook

$(Bi_{1/2}Na_{1/2})TiO_3$ (BNT) was discovered several years later than $(K,Na)NbO_3$ (KNN), but had been extensively studied in an earlier period when attention had not been paid to the latter. This chapter gave a brief review about the recent research progress of BNT-based lead-free piezoelectric ceramics with a particular emphasis on the BNT–$BaTiO_3$ system. The BNT system occupies an important positon as a promising lead-free piezoelectric ceramic because of the following advantages: (i) BNT system has an exceptionally high electric-field-induced strain among the lead-free piezoelectric ceramics including KNN, (ii) the BNT-based ceramics have a better processing repeatability than KNN with narrower processing window, (iii) a PZT-like MPB exist in BNT-rich side with $BaTiO_3$, which is attractive not only for academic studies but also for applications requiring thermal stability of piezoelectricity, and (iv) additionally, the cost of raw materials for BNT is lower than KNN. To push the BNT-based ceramics to applications, more studies should be needed to increase the depolarization temperature and reduce the hysteresis of electric-field-induced strains. Certainly, the phase and domain structures of BNT and its related systems will be interesting topics for a long time.

References

1 Smolenskii, G., Isupov, V., Agranovskaya, A., and Popov, S. (1961). Ferroelectrics with diffuse phase transitions. *Soviet Physics – Solid State* 2 (11): 2584–2594.
2 Buhrer, C.F. (1962). Some properties of bismuth perovskites. *The Journal of Chemical Physics* 36 (3): 798–803.
3 Suchanicz, J. and Kwapulinski, J. (1995). X-ray diffraction study of the phase transitions in $Na_{0.5}Bi_{0.5}TiO_3$. *Ferroelectrics* 165 (1): 249–253.
4 Saito, Y., Takao, H., Tani, T. et al. (2004). Lead-free piezoceramics. *Nature* 432: 84–87.
5 Takenaka, T., Maruyama, K.-I., and Sakata, K. (1991). $(Bi_{1/2}Na_{1/2})TiO_3$–$BaTiO_3$ system for lead-free piezoelectric ceramics. *Japanese Journal of Applied Physics Part 1* 30 (9B): 2236–2239.
6 Takenaka, T. and Nagata, H. (2005). Current status and prospects of lead-free piezoelectric ceramics. *Journal of the European Ceramic Society* 25 (12): 2693–2700.

7 Sasaki, A., Chiba, T., Mamiya, Y., and Otsuki, E. (1999). Dielectric and piezoelectric properties of $(Bi_{0.5}Na_{0.5})TiO_3$–$(Bi_{0.5}K_{0.5})TiO_3$ systems. *Japanese Journal of Applied Physics Part 1* 38: 5564–5567.

8 Zhang, Y.-R., Li, J.-F., Zhang, B.-P., and Peng, C.-E. (2008). Piezoelectric and ferroelectric properties of Bi-compensated $(Bi_{1/2}Na_{1/2})TiO_3$–$(Bi_{1/2}K_{1/2})TiO_3$ lead-free piezoelectric ceramics. *Journal of Applied Physics* 103 (7): 074109.

9 Isupov, V.A. (2005). Ferroelectric $Na_{0.5}Bi_{0.5}TiO_3$ and $K_{0.5}Bi_{0.5}TiO_3$ perovskites and their solid solutions. *Ferroelectrics* 315: 123–147.

10 Yoshii, K., Hiruma, Y., Nagata, H., and Takenaka, T. (2006). Electrical properties and depolarization temperature of $(Bi_{1/2}Na_{1/2})TiO_3$–$(Bi_{1/2}K_{1/2})TiO_3$ lead-free piezoelectric ceramics. *Japanese Journal of Applied Physics Part 1* 45: 4493–4496.

11 Zhang, Y.R., Li, J.F., and Zhang, B.P. (2008). Enhancing electrical properties in NBT–KBT lead-free piezoelectric ceramics by optimizing sintering temperature. *Journal of the American Ceramic Society* 91 (8): 2716–2719.

12 Kounga, A.B., Zhang, S.T., Jo, W. et al. (2008). Morphotropic phase boundary in $(1-x)Bi_{0.5}Na_{0.5}TiO_3$–$xK_{0.5}Na_{0.5}NbO_3$ lead-free piezoceramics. *Applied Physics Letters* 92: 222902.

13 Zhang, S.-T., Kounga, A.B., Aulbach, E. et al. (2007). Giant strain in lead-free piezoceramics $Bi_{0.5}Na_{0.5}TiO_3$–$BaTiO_3$–$K_{0.5}Na_{0.5}NbO_3$ system. *Applied Physics Letters* 91 (11): 112906.

14 Seifert, K.T.P., Jo, W., and Rödel, J. (2010). Temperature-insensitive large strain of $(Bi_{1/2}Na_{1/2})TiO_3$–$(Bi_{1/2}K_{1/2)}TiO_3$–$(K_{0.5}Na_{0.5})NbO_3$ lead-free piezoceramics. *Journal of the American Ceramic Society* 93 (5): 1392–1396.

15 Bai, W., Li, P., Li, L. et al. (2015). Structure evolution and large strain response in BNT–BT lead-free piezoceramics modified with $Bi(Ni_{0.5}Ti_{0.5})O_3$. *Journal of Alloys and Compounds* 649: 772–781.

16 Jo, W., Dittmer, R., Acosta, M. et al. (2012). Giant electric-field-induced strains in lead-free ceramics for actuator applications – status and perspective. *Journal of Electroceramics* 29 (1): 71–93.

17 Hiruma, Y., Imai, Y., Watanabe, Y. et al. (2008). Large electrostrain near the phase transition temperature of $(Bi_{0.5}Na_{0.5})TiO_3$–$SrTiO_3$ ferroelectric ceramics. *Applied Physics Letters* 92 (26): 262904.

18 Daniels, J.E., Jo, W., Rödel, J., and Jones, J.L. (2009). Electric-field-induced phase transformation at a lead-free morphotropic phase boundary: case study in a 93%$(Bi_{0.5}Na_{0.5})TiO_3$–7%$BaTiO_3$ piezoelectric ceramic. *Applied Physics Letters* 95 (3): 032904.

19 Daniels, J.E., Jo, W., Rödel, J. et al. (2010). Electric-field-induced phase-change behavior in $(Bi_{0.5}Na_{0.5})TiO_3$–$BaTiO_3$–$(K_{0.5}Na_{0.5})NbO_3$: a combinatorial investigation. *Acta Materialia* 58 (6): 2103–2111.

20 Jo, W., Schaab, S., Sapper, E. et al. (2011). On the phase identity and its thermal evolution of lead free $(Bi_{1/2}Na_{1/2})TiO_3$–6 mol% $BaTiO_3$. *Journal of Applied Physics* 110 (7): 074106.

21 Jo, W., Granzow, T., Aulbach, E. et al. (2009). Origin of the large strain response in $(K_{0.5}Na_{0.5})NbO_3$-modified $(Bi_{0.5}Na_{0.5})TiO_3$–$BaTiO_3$ lead-free piezoceramics. *Journal of Applied Physics* 105 (9): 094102.

22 Dittmer, R., Jo, W., Rödel, J. et al. (2012). Nanoscale insight into lead-free BNT-BT-*x*KNN. *Advanced Functional Materials* 22 (20): 4208–4215.

23 Paterson, A.R., Nagata, H., Tan, X.L. et al. (2018). Relaxor-ferroelectric transitions: sodium bismuth titanate derivatives. *MRS Bulletin* 43: 600–606.

24 Ma, C. and Tan, X. (2010). Phase diagram of unpoled lead-free $(1-x)(Bi_{1/2}Na_{1/2})TiO_3-xBaTiO_3$ ceramics. *Solid State Communications* 150 (33): 1497–1500.

25 Ma, C., Tan, X., Dul'kin, E., and Roth, M. (2010). Domain structure-dielectric property relationship in lead-free $(1-x)(Bi_{1/2}Na_{1/2})TiO_3-xBaTiO_3$ ceramics. *Journal of Applied Physics* 108 (10): 104105.

26 Aksel, E., Forrester, J.S., Nino, J.C. et al. (2013). Local atomic structure deviation from average structure of $Na_{0.5}Bi_{0.5}TiO_3$: combined X-ray and neutron total scattering study. *Physical Review B* 87 (10): 104113.

27 Rao, B.N. and Ranjan, R. (2012). Electric-field-driven monoclinic-to-rhombohedral transformation in $Na_{1/2}Bi_{1/2}TiO_3$. *Physical Review B* 86 (13): 134103.

28 Rao, B.N., Datta, R., Chandrashekaran, S.S. et al. (2013). Local structural disorder and its influence on the average global structure and polar properties in $Na_{0.5}Bi_{0.5}TiO_3$. *Physical Review B* 88 (22): 224103.

29 Levin, I. and Reaney, I.M. (2012). Nano- and mesoscale structure of $Na_{1/2}Bi_{1/2}TiO_3$: a TEM perspective. *Advanced Functional Materials* 22 (16): 3445–3452.

30 Ma, C., Guo, H., and Tan, X. (2013). A new phase boundary in $(Bi_{1/2}Na_{1/2})TiO_3$–$BaTiO_3$ revealed via a novel method of electron diffraction analysis. *Advanced Functional Materials* 23 (42): 5261–5266.

31 Bokov, A.A. and Ye, Z.-G. (2012). Dielectric relaxation in relaxor ferroelectrics. *Journal of Advanced Dielectrics* 02 (02): 1241010.

32 Bokov, A.A. and Ye, Z.-G. (2006). Recent progress in relaxor ferroelectrics with perovskite structure. *Journal of Materials Science* 41 (1): 31–52.

33 Cowley, R.A., Gvasaliya, S.N., Lushnikov, S.G. et al. (2011). Relaxing with relaxors: a review of relaxor ferroelectrics. *Advances in Physics* 60: 229–327.

34 Li, F., Zhang, S.J., Damjanovic, D. et al. (2018). Local structural heterogeneity and electromechanical responses of ferroelectrics: learning from relaxor ferroelectrics. *Advanced Functional Materials* 28: 1801504.

35 Chen, J., Chan, H.M., and Harmer, M.P. (1989). Ordering structure and dielectric-properties of undoped and LA/Na-doped $Pb(Mg_{1/3}Nb_{2/3})O_3$. *Journal of the American Ceramic Society* 72: 593–598.

36 Jin, H.Z., Zhu, J., Miao, S. et al. (2001). Ordered domains and polar clusters in lead magnesium niobate $Pb(Mg_{1/3}Nb_{2/3})O_3$. *Journal of Applied Physics* 89: 5048–5052.

37 Shvartsman, V.V. and Lupascu, D.C. (2012). Lead-free relaxor ferroelectrics. *Journal of the American Ceramic Society* 95: 1–26.

38 Xu, C.G., Lin, D.M., and Kwok, K.W. (2008). Structure, electrical properties and depolarization temperature of $(Bi_{0.5}Na_{0.5})TiO_3$–$BaTi_{O3}$ lead-free piezoelectric ceramics. *Solid State Sciences* 10: 934–940.

39 Chen, M., Xu, Q., Kim, B.H. et al. (2008). Structure and electrical properties of $(Na_{0.5}Bi_{0.5})(1-x)BaxTiO_3$ piezoelectric ceramics. *Journal of the European Ceramic Society* 28: 843–849.

40 Simons, H., Daniels, J., Jo, W. et al. (2011). Electric-field-induced strain mechanisms in lead-free 94%$(Bi_{1/2}Na_{1/2})TiO_3$–6%$BaTiO_3$. *Applied Physics Letters* 98 (8): 082901.

41 Neagu, A. and Tai, C.-W. (2017). Local disorder in $Na_{0.5}Bi_{0.5}TiO_3$-piezoceramic determined by 3D electron diffuse scattering. *Scientific Reports* 7 (1): 12519.

42 Schmitt, L.A., Kling, J., Hinterstein, M. et al. (2011). Structural investigations on lead-free $Bi_{1/2}Na_{1/2}TiO_3$-based piezoceramics. *Journal of Materials Science* 46 (12): 4368–4376.

43 Luo, C., Bansal, D., Li, J. et al. (2017). Neutron and X-ray scattering study of phonon dispersion and diffuse scattering in $Na_{0.5}Bi_{0.5}TiO_3$–$BaTiO_3$ single crystals near the morphotropic phase boundary. *Physical Review B* 96 (17): 174108.

44 Zhang, H., Deng, G., Studer, A.J. et al. (2014). Neutron diffuse scattering of $(1-x)(Na_{0.5}Bi_{0.5})TiO_3$–$xBaTiO_3$ relaxor ferroelectric single crystals. *Scripta Materialia* 86: 5–8.

45 Cordero, F., Craciun, F., Trequattrini, F. et al. (2010). Phase transitions and phase diagram of the ferroelectric perovskite $(Na_{0.5}Bi_{0.5})_{1-x}Ba_xTiO_3$ by anelastic and dielectric measurements. *Physical Review B* 81 (14): 144124.

46 Jo, W., Daniels, J.E., Jones, J.L. et al. (2011). Evolving morphotropic phase boundary in lead-free $(Bi_{1/2}Na_{1/2})TiO_3$–$BaTiO_3$ piezoceramics. *Journal of Applied Physics* 109 (1): 014110.

47 Wylie-van Eerd, B., Damjanovic, D., Klein, N. et al. (2010). Structural complexity of $(Na_{0.5}Bi_{0.5})TiO_3$–$BaTiO_3$ as revealed by Raman spectroscopy. *Physical Review* 82: 104112.

48 Nagata, H., Yoshida, M., Makiuchi, Y., and Takenaka, T. (2003). Large piezoelectric constant and high Curie temperature of lead-free piezoelectric ceramic ternary system based on bismuth sodium titanate-bismuth potassium titanate-barium titanate near the morphotropic phase boundary. *Japanese Journal of Applied Physics Part 1* 42: 7401–7403.

49 Gorfman, S. and Thomas, P.A. (2010). Evidence for a non-rhombohedral average structure in the lead-free piezoelectric material $Na_{0.5}Bi_{0.5}TiO_3$. *Journal of Applied Crystallography* 43 (6): 1409–1414.

50 Aksel, E., Forrester, J.S., Jones, J.L. et al. (2011). Monoclinic crystal structure of polycrystalline $Na_{0.5}Bi_{0.5}TiO_3$. *Applied Physics Letters* 98 (15): 152901.

51 Jo, W. and Rödel, J. (2011). Electric-field-induced volume change and room temperature phase stability of $(Bi_{1/2}Na_{1/2})TiO_{3-x}$ mol%$BaTiO_3$ piezoceramics. *Applied Physics Letters* 99 (4): 042901.

52 Ma, C., Guo, H., Beckman, S.P., and Tan, X. (2012). Creation and destruction of morphotropic phase boundaries through electrical poling: a case study of lead-free $(Bi_{1/2}Na_{1/2})TiO_3$–$BaTiO_3$ piezoelectrics. *Physical Review Letters* 109 (10): 107602.

53 Zhou, Z., Sun, W., and Li, J.-F. (2016). Sol–gel synthesis of $0.94(Bi_{0.5}Na_{0.5})TiO_3$–$0.06BaTiO_3$ lead-free piezoelectric films: effect of pyrolysis temperature on phase evolution and electrical properties. *Journal of Sol-Gel Science and Technology* 77: 423–429.

54 Dkhil, B. and Kiat, J.M. (2001). Electric-field-induced polarization in the ergodic and nonergodic states of $PbMg_{1/3}Nb_{2/3}O_3$ relaxor. *Journal of Applied Physics* 90 (9): 4676–4681.

55 Chang, W.A., Hong, C.-H., Choi, B.-Y. et al. (2016). A brief review on relaxor ferroelectrics and selected issues in lead-free relaxors. *Journal of the Korean Physical Society* 68: 1481–1494.

56 Hinterstein, M., Knapp, M., Hölzel, M. et al. (2010). Field-induced phase transition in $Bi_{1/2}Na_{1/2}TiO_3$-based lead-free piezoelectric ceramics. *Journal of Applied Crystallography* 43 (6): 1314–1321.

57 Kling, J., Tan, X., Jo, W. et al. (2010). In situ transmission electron microscopy of electric field-triggered reversible domain formation in bi-based lead-free piezoceramics. *Journal of the American Ceramic Society* 93 (9): 2452–2455.

58 Schütz, D., Deluca, M., Krauss, W. et al. (2012). Lone-pair-induced covalency as the cause of temperature- and field-induced instabilities in bismuth sodium titanate. *Advanced Functional Materials* 22 (11): 2285–2294.

59 Hiruma, Y., Nagata, H., and Takenaka, T. (2009). Thermal depoling process and piezoelectric properties of bismuth sodium titanate ceramics. *Journal of Applied Physics* 105: 084112.

60 Do, N.-B., Lee, H.-B., Yoon, C.-H. et al. (2011). Effect of Ta-substitution on the ferroelectric and piezoelectric properties of $Bi_{0.5}(Na_{0.82}K_{0.18})_{0.5}TiO_3$ ceramics. *Transactions on Electrical and Electronic Materials* 12 (2): 64–67.

61 Pham, K.-N., Lee, H.B., Han, H.-S. et al. (2012). Dielectric, ferroelectric, and piezoelectric properties of Nb-substituted $Bi_{1/2}(Na_{0.82}K_{0.18})_{1/2}TiO_3$ lead-free ceramics. *Journal of the Korean Physical Society* 60 (2): 207–211.

62 Hao, J., Xu, Z., Chu, R. et al. (2016). Structure evolution and electrostrictive properties in $(Bi_{0.5}Na_{0.5})_{0.94}Ba_{0.06}TiO_3$–$M_2O_5$ (M = Nb, Ta, Sb) lead-free piezoceramics. *Journal of the European Ceramic Society* 36 (16): 4003–4014.

63 Bai, W., Li, L., Li, W. et al. (2014). Effect of $SrTiO_3$ template on electric properties of textured BNT–BKT ceramics prepared by templated grain growth process. *Journal of Alloys and Compounds* 603: 149–157.

64 Krauss, W., Schütz, D., Mautner, F.A. et al. (2010). Piezoelectric properties and phase transition temperatures of the solid solution of $(1-x)(Bi_{0.5}Na0_{.5})TiO_3$–$xSrTiO_3$. *Journal of the European Ceramic Society* 30 (8): 1827–1832.

65 Hao, J., Shen, B., Zhai, J., and Chen, H. (2014). Effect of $BiMeO_3$ on the phase structure, ferroelectric stability, and properties of lead-free $Bi_{0.5}(Na_{0.80}K_{0.20})_{0.5}TiO_3$ ceramics. *Journal of the American Ceramic Society* 97 (6): 1776–1784.

66 Li, Q., Wang, J., Liu, Z. et al. (2016). Enhanced energy-storage properties of $BaZrO_3$-modified $0.80Bi_{0.5}Na_{0.5}TiO_3$–$0.20Bi_{0.5}K_{0.5}TiO_3$ lead-free ferroelectric ceramics. *Journal of Materials Science* 51 (2): 1153–1160.

67 Hiruma, Y., Nagata, H., and Takenaka, T. (2008). Phase diagrams and electrical properties of $(Bi_{1/2}Na_{1/2})TiO_3$-based solid solutions. *Journal of Applied Physics* 104 (12): 124106.

68 Rahman, J.U., Hussain, A., Maqbool, A. et al. (2014). Dielectric, ferroelectric and field-induced strain response of lead-free $BaZrO_3$-modified $Bi_{0.5}Na_{0.5}TiO_3$ ceramics. *Current Applied Physics* 14 (3): 331–336.

69 Hiruma, Y., Nagata, H., and Takenaka, T. (2009). Detection of morphotropic phase boundary of $(Bi_{1/2}Na_{1/2})TiO_3$–$Ba(Al_{1/2}Sb_{1/2})O_3$ solid-solution ceramics. *Applied Physics Letters* 95 (5): 052903.

70 Riemer, L.M., Lalitha, K.V., Jiang, X. et al. (2017). Stress-induced phase transition in lead-free relaxor ferroelectric composites. *Acta Materialia* 136: 271–280.

71 Janbua, J., Niemchareon, S., Muanghlua, R., and Vittayakorn, N. (2016). High strain response of the $(1-x)(0.94Bi_{0.5}Na_{0.5}TiO_3$–$0.06BaTiO_3)$–$xBaSnO_3$ lead free piezoelectric ceramics system. *Ferroelectrics* 490 (1): 13–22.

72 Nguyen, V.-Q., Han, H.-S., Kim, K.-J. et al. (2012). Strain enhancement in $Bi_{1/2}(Na_{0.82}K_{0.18})_{1/2}TiO_3$ lead-free electromechanical ceramics by co-doping with Li and Ta. *Journal of Alloys and Compounds* 511 (1): 237–241.

73 Maqbool, A., Hussain, A., Ur Rahman, J. et al. (2014). Enhanced electric field-induced strain and ferroelectric behavior of $(Bi_{0.5}Na_{0.5})TiO_3$–$BaTiO_3$–$SrZrO_3$ lead-free ceramics. *Ceramics International* 40 (8, Part A): 11905–11914.

74 Shi, J., Fan, H., Liu, X., and Li, Q. (2014). Giant strain response and structure evolution in $(Bi_{0.5}Na_{0.5})_{0.945-x}(Bi_{0.2}Sr_{0.7}\square_{0.1})_xBa_{0.055}TiO_3$ ceramics. *Journal of the European Ceramic Society* 34 (15): 3675–3683.

75 Liu, X. and Tan, X. (2016). Giant strains in non-textured $(Bi_{1/2}Na_{1/2})TiO_3$-based lead-free ceramics. *Advanced Materials* 28 (3): 574–578.

76 Davies, M., Aksel, E., and Jones, J.L. (2011). Enhanced high-temperature piezoelectric coefficients and thermal stability of Fe- and Mn-substituted $Na_{0.5}Bi_{0.5}TiO_3$ ceramics. *Journal of the American Ceramic Society* 94 (5): 1314–1316.

77 Qin, Y., Zhang, S., Wu, Y. et al. (2017). Impacts of acceptor doping on the piezoelectric properties and domain structure in NBT-based lead-free ceramics. *Journal of the European Ceramic Society* 37 (11): 3493–3500.

78 Aksel, E., Erdem, E., Jakes, P. et al. (2010). Defect structure and materials “hardening” in Fe_2O_3-doped $(Bi_{0.5}Na_{0.5})TiO_3$ ferroelectrics. *Applied Physics Letters* 97 (1): 012903.

79 Li, Z.T., Liu, H., Thong, H.C. et al. Enhanced temperature stability and defect mechanism of BNT-based lead-free piezoceramics investigated by a quenching process. *Advanced Electronic Materials* 5: 1800756.

80 Zhang, S.J., Shrout, T.R., Nagata, H. et al. (2007). Piezoelectric properties in $(K_{0.5}Bi_{0.5})TiO_3$–$(Na_{0.5}Bi_{0.5})TiO_3$–$BaTiO_3$ lead-free ceramics. *IEEE Transactions on Ultrasonics, Ferroelectrics, and Frequency Control* 54 (5): 910–917.

81 Tian, H.Y., Wang, D.Y., Lin, D.M. et al. (2007). Diffusion phase transition and dielectric characteristics of $Bi_{0.5}Na_{0.5}TiO_3$–$Ba(Hf,Ti)O_3$ lead-free ceramics. *Solid State Communications* 142 (1): 10–14.

82 Peng, C., Li, J.-F., and Gong, W. (2005). Preparation and properties of $(Bi_{1/2}Na_{1/2})TiO_3$–$Ba(Ti,Zr)O_3$ lead-free piezoelectric ceramics. *Materials Letters* 59 (12): 1576–1580.

83 Lv, J., Karaki, T., and Adachi, M. (2010). Effect of CuO doping on $(Bi,Na)_{0.83}Ba_{0.17}TiO_3$ piezoelectric ceramics. *Japanese Journal of Applied Physics* 49 (9): 09MD06.

84 Zhang, Q., Zhang, Y., Wang, F. et al. (2009). Enhanced piezoelectric and ferroelectric properties in Mn-doped $Na_{0.5}Bi_{0.5}TiO_3$–$BaTiO_3$ single crystals. *Applied Physics Letters* 95 (10): 102904.

85 Chen, P.-Y., Chen, C.-S., Tu, C.-S. et al. (2016). Effects of texture on microstructure, Raman vibration, and ferroelectric properties in 92.5%$(Bi_{0.5}Na_{0.5})TiO_3$–7.5%$BaTiO_3$ ceramics. *Journal of the European Ceramic Society* 36 (7): 1613–1622.

86 Zhang, J., Pan, Z., Guo, F.-F. et al. (2015). Semiconductor/relaxor 0–3 type composites without thermal depolarization in $Bi_{0.5}Na_{0.5}TiO_3$-based lead-free piezoceramics. *Nature Communications* 6: 6615.

87 Groh, C., Jo, W., and Rödel, J. (2014). Tailoring strain properties of $(0.94-x)Bi_{1/2}Na_{1/2}TiO_3$–$0.06BaTiO_3$–$xK_{0.5}Na_{0.5}NbO_3$ ferroelectric/relaxor composites. *Journal of the American Ceramic Society* 97 (5): 1465–1470.

88 Groh, C., Franzbach, D.J., Jo, W. et al. (2014). Relaxor/ferroelectric composites: a solution in the quest for practically viable lead-free incipient piezoceramics. *Advanced Functional Materials* 24 (3): 356–362.

89 Zhang, H., Groh, C., Zhang, Q. et al. (2015). Large strain in relaxor/ferroelectric composite lead-free piezoceramics. *Advanced Electronic Materials* 1 (6): 1500018.

90 Samara, G.A. (1969). The effects of hydrostatic pressure on ferroelectric properties. *Advances in High Pressure Research* 3: 219.

91 Dittmer, R., Webber, K.G., Aulbach, E. et al. (2013). Electric-field-induced polarization and strain in $0.94(Bi_{1/2}Na_{1/2})TiO_3$–$0.06BaTiO_3$ under uniaxial stress. *Acta Materialia* 61 (4): 1350–1358.

92 Webber, K.G., Zhang, Y., Jo, W. et al. (2010). High temperature stress-induced “double loop-like” phase transitions in Bi-based perovskites. *Journal of Applied Physics* 108 (1): 014101.

93 Schader, F.H., Wang, Z., Hinterstein, M. et al. (2016). Stress-modulated relaxor-to-ferroelectric transition in lead-free $Na_{1/2}Bi_{1/2}TiO_3$–$BaTiO_3$ ferroelectrics. *Physical Review B* 93 (13): 134111.

94 Schader, F.H. (2016). *Mechanical Stability of the Electromechanical Properties and Phase Transitions in Lead-Containing and Lead-Free Ferroelectrics.* Technische Universität Darmstadt.

95 Zhou, Z., Zhang, S.-W., Zhang, Y. et al. (2018). Textured $Bi_{1/2}Na_{1/2}TiO_3$–$BaTiO_3$ lead-free films with enhanced piezoelectric property and depolarization temperature. *Advanced Electronic Materials* 4 (11): 1800351.

96 Harigai, T., Tanaka, Y., Adachi, H., and Fujii, E. (2010). Piezoelectric properties of lead-free (Na,Bi)TiO_3–$BaTiO_3$ (001) epitaxial thin films around the morphotropic phase boundary. *Applied Physics Express* 3: 111501.

97 Adachi, H., Tanaka, Y., Harigai, T. et al. (2011). Large transverse piezoelectricity in strained (Na,Bi)TiO_3–$BaTiO_3$ epitaxial thin films on MgO(110). *Applied Physics Express* 4: 051501.

98 Tanaka, Y., Harigai, T., Ueda, M. et al. (2012). Structural modification and large piezoelectricity of strained (Na,Bi) TiO_3–$BaTiO_3$ epitaxial thin films. *Journal of the American Ceramic Society* 95: 3547–3553.

99 Tanaka, Y., Harigai, T., Adachi, H. et al. (2013). Strain-driven control of piezoelectricity in (Na,Bi)TiO_3–$BaTiO_3$ epitaxial thin films. *Applied Physics Letters* 102: 192901.

100 Tanaka, Y., Hashimoto, K., Harigai, T. et al. (2014). Preparation and piezoelectric properties of self-polarized (Na,Bi)TiO_3–$BaTiO_3$ thin films on Si substrate. *Japanese Journal of Applied Physics* 53: 09PA01.

101 Tanaka, Y., Okamoto, S., Hashimoto, K. et al. (2018). High electromechanical strain and enhanced temperature characteristics in lead-free (Na,Bi)TiO_3–$BaTiO_3$ thin films on Si substrates. *Scientific Reports* 8: 7847.

102 Zhou, Z.H., Xue, J.M., Li, W.Z. et al. (2004). Ferroelectric and electrical behavior of $(Na_{0.5}Bi_{0.5})TiO_3$ thin films. *Applied Physics Letters* 85: 804–806.

103 Cheng, H.W., Zhang, X.J., Zhang, S.T. et al. (2004). Combinatorial studies of $(1\text{–}x)Na_{0.5}Bi_{0.5}TiO_3\text{–}xBaTiO_3$ thin-film chips. *Applied Physics Letters* 85: 2319–2321.

104 Abazari, M., Safari, A., Bharadwaja, S.S.N., and Trolier-McKinstry, S. (2010). Dielectric and piezoelectric properties of lead-free (Bi,Na)TiO_3-based thin films. *Applied Physics Letters* 96: 082903.

105 Hejazi, M.M. and Safari, A. (2011). Temperature-dependent leakage current behavior of epitaxial $Bi_{0.5}Na_{0.5}TiO_3$-based thin films made by pulsed laser deposition. *Journal of Applied Physics* 110: 103710.

106 Scarisoreanu, N., Craciun, F., Ion, V. et al. (2007). Structural and electrical characterization of lead-free ferroelectric $Na_{1/2}Bi_{1/2}TiO_3$–$BaTiO_3$ thin films obtained by PLD and RF-PLD. *Applied Surface Science* 254: 1292–1297.

107 Wang, D.Y., Chan, N.Y., Li, S. et al. (2010). Enhanced ferroelectric and piezoelectric properties in doped lead-free $(Bi_{0.5}Na_{0.5})_{0.94}Ba_{0.06}TiO_3$ thin films. *Applied Physics Letters* 97: 212901.

108 Peng, B.L., Zhang, Q., Li, X. et al. (2015). Giant electric energy density in epitaxial lead-free thin films with coexistence of ferroelectrics and antiferroelectrics. *Advanced Electronic Materials* 1: 1500052.

109 Guo, Y.P., Akai, D., Sawada, K., and Ishida, M. (2008). Dielectric and ferroelectric properties of highly (100)-oriented $(Na_{0.5}Bi_{0.5})_{0.94}Ba_{0.06}TiO_3$ thin films grown on $LaNiO_3/\gamma\text{-}Al_2O_3/Si$ substrates by chemical solution deposition. *Solid State Sciences* 10: 928–933.

110 Cernea, M., Trupina, L., Dragoi, C. et al. (2012). Structural and piezoelectric characteristics of BNT-$BT_{0.05}$ thin films processed by sol–gel technique. *Journal of Alloys and Compounds* 515: 166–170.

111 Cernea, M., Trupina, L., Dragoi, C. et al. (2012). Structural, optical, and electric properties of BNT-$BT_{0.08}$ thin films processed by sol–gel technique. *Journal of Materials Science* 47: 6966–6971.

112 Jeon, Y.H., Patterson, E.A., Cann, D.P. et al. (2013). Large piezoresponse and ferroelectric properties of $(Bi_{0.5}Na_{0.5})TiO_3$–$(Bi_{0.5}K_{0.5})TiO_3$–$Bi(Mg_{0.5}Ti_{0.5})O_3$ thin films prepared by chemical solution deposition. *Journal of the American Ceramic Society* 96: 2172–2178.

113 Zhou, Z., Luo, J., Sun, W., and Li, J.-F. (2017). Temperature and composition dependent phase transitions of lead-free piezoelectric $(Bi_{0.5}Na_{0.5})TiO_3$–$BaTiO_3$ thin films. *Physical Chemistry Chemical Physics* 19: 19992–19997.

114 Lee, H.Y., Sun, W., Luo, J. et al. (2018). Sol–gel processing and nanoscale characterization of $(Bi_{0.5}Na_{0.5})TiO_3$–$SrTiO_3$ lead-free piezoelectric thin films. *Ceramics International* 44: 4114–4120.

115 Lee, H.Y., Luo, J., Zhou, Z. et al. (2018). Macroscopic and local approaches of phase transition in sol–gel synthesized $(Bi_{0.5}Na_{0.5})TiO_3$–$SrTiO_3$ thin films. *Physical Chemistry Chemical Physics* 20: 15236–15243.

116 Chiang, Y.M., Farrey, G.W., and Soukhojak, A.N. (1998). Lead-free high-strain single-crystal piezoelectrics in the alkaline-bismuth-titanate perovskite family. *Applied Physics Letters* 73: 3683–3685.

117 Chen, C., Jiang, X.P., Li, Y.M. et al. (2010). Growth and electrical properties of $Na_{1/2}Bi_{1/2}TiO_3$–$BaTiO_3$ lead-free single crystal with morphotropic phase boundary composition. *Journal of Applied Physics* 108: 124106.

118 Ge, W.W., Luo, C.T., Zhang, Q.H. et al. (2012). Ultrahigh electromechanical response in $(1-x)(Na_{0.5}Bi_{0.5})TiO_3$–$xBaTiO_3$ single-crystals via polarization extension. *Journal of Applied Physics* 111: 093508.

119 Schneider, D., Jo, W., Roedel, J. et al. (2014). Anisotropy of ferroelectric behavior of $(1-x)Bi_{1/2}Na_{1/2}TiO_3$–$xBaTiO_3$ single crystals across the morphotropic phase boundary. *Journal of Applied Physics* 116: 044111.

120 Chen, C., Zhao, X.Y., Wang, Y.J. et al. (2016). Giant strain and electric-field-induced phase transition in lead-free $(Na_{0.5}Bi_{0.5})TiO_3$–$BaTiO_3$–$(K_{0.5}Na_{0.5})NbO_3$ single crystal. *Applied Physics Letters* 108: 022903.

121 Wang, Y.J., Luo, C.T., Wang, S.H. et al. (2020). Large piezoelectricity in ternary lead-free single crystals. *Advanced Electronic Materials* 6: 1900949.

122 Sun, H., Fisher, J.G., Moon, S.H. et al. (2017). Solid-state-growth of lead-free piezoelectric $(Na_{1/2}Bi_{1/2})TiO_3$–$CaTiO_3$ single crystals and their characterization. *Materials Science & Engineering B: Advanced Functional Solid-State Materials* 223: 109–119.

123 Someno, S., Nagata, H., and Takenaka, T. (2014). High-temperature and high-power piezoelectric characteristics of $(Bi_{0.5}Na_{0.5})TiO_3$-based lead-free piezoelectric ceramics. *Journal of the Ceramic Society of Japan* 122 (1426): 406–409.

124 Hiruma, Y., Watanabe, T., Nagata, H., and Takenaka, T. (2008). Piezoelectric properties of $(Bi_{1/2}Na_{1/2})TiO_3$-based solid solution for lead-free high-power applications. *Japanese Journal of Applied Physics* 47 (9S): 7659.

125 Hejazi, M., Taghaddos, E., Gurdal, E. et al. (2014). High power performance of manganese-doped BNT-based Pb-free piezoelectric ceramics. *Journal of the American Ceramic Society* 97 (10): 3192–3196.

126 Zhang, S., Lee, H.J., and Shrout, T.R. (2013). NBT based lead-free piezoelectric materials for high power applications. Google Patents. US Patent 8 501 031 B2, filed 24 September 2009 and issued 6 August 2013

127 Lee, H.J., Ural, S.O., Chen, L. et al. (2012). High power characteristics of lead-free piezoelectric ceramics. *Journal of the American Ceramic Society* 95 (11): 3383–3386.

5

$BaTiO_3$ System

5.1 Brief Introduction of History

$BaTiO_3$ was discovered in 1941 during World War II and soon became an important strategic material for various military technologies [1, 2]. The search for high-performance dielectric materials led to the independent discovery of $BaTiO_3$ in the United States, the United Kingdom, Russia, and Japan [1, 3, 4]. The high dielectric properties of $BaTiO_3$ were associated with its ferroelectricity [5–7]. $BaTiO_3$ was also the first demonstration of ferroelectricity in a simple ternary oxide compound. This was unacceptable at that time because it was believed that ferroelectricity existed only in potassium dihydrogen phosphate and related hydrogen-bonded crystals displaying order–disorder phase transition. Different from those materials, $BaTiO_3$ has a simpler crystal structure and go through a displacive phase transition upon Curie temperature (T_C). Only after several new ferroelectrics were discovered, a material classification based on the displacive phase transition became possible [8].

$BaTiO_3$ has an ABO_3 chemical formula. The perovskite structure of $BaTiO_3$ can be described as a network of corner-linked oxygen octahedra, with Ti^{4+} and Ba^{2+} filling the octahedral hole and the dodecahedral hole, respectively. $BaTiO_3$ exhibits a series of phase transformation as shown in Figure 5.1 [9]. $BaTiO_3$ has a hexagonal structure (H, space group: $P6_3/mmc$) above 1432 °C. The high-temperature hexagonal structure transforms into a cubic phase (C, space group: $Pm3m$) with an ideal perovskite-type structure around 1432 °C. The paraelectric C phase transforms to a ferroelectric tetragonal phase at 132 °C (T, space group: $P4mm$). Upon further cooling, the tetragonal phase transforms to an orthorhombic phase (O, space group: $Amm2$) at 5 °C and a rhombohedral phase (R, space group: $R3m$) at −90 °C. The anomalies in relative permittivity appear during the sequence of phase transitions in $BaTiO_3$ as shown in Figure 2.2.

After World War II, the phase transitions and physical properties of $BaTiO_3$ became hot research topics [10, 11], with some important scientific advances in crystallography. The first detailed crystal structures and changes in symmetry related to the phase transitions were provided by the researchers from the United Kingdom and Japan [12–14]. During the 1940s, the ferroelectric switching, domain

Lead-Free Piezoelectric Materials, First Edition. Jing-Feng Li.

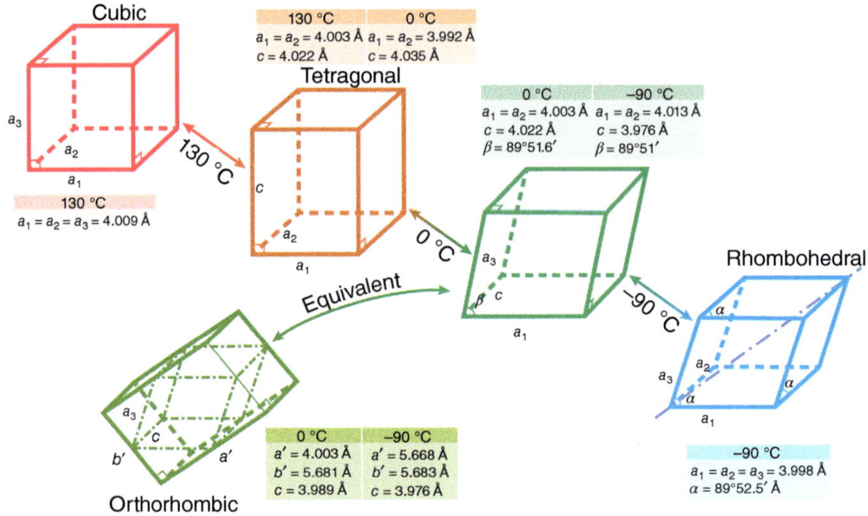

Figure 5.1 Unit cells and parameters of phases and transitions in $BaTiO_3$. Note that the cell-doubled orthorhombic structure is represented as a primitive pseudo-monoclinic cell for ease of comparison with the tetragonal and rhombohedral phases. Source: Jona and Shirane [9]. © 1993, Dover Publications.

structure, and crystal structure evolution of $BaTiO_3$ were revealed, which greatly contributed to the advance in ferroelectricity [11].

Another milestone was the discovery of the poling process. At that time, it was not yet recognized that the polarization direction in a randomly oriented ferroelectric ceramic could be reoriented. Hence, the understanding of the poling process was a breakthrough. This also made the polycrystalline ceramics possible for piezoelectric applications, as they were mechanically more robust and less expensive than single crystals. The poling process was first reported to be used for manufacturing transducers [15]. Following this discovery, the electrostriction, piezoelectricity, and the evolution of domains under external excitations were investigated in $BaTiO_3$ [16–18].

The first commercialized products based on $BaTiO_3$ became available around 1950, including phonograph pickups, transducers, accelerometers, and capacitors [19, 20]. Meanwhile, several theories were developed to explain the spectacular experimental phenomena [21]. For example, the phenomenological theory was developed to reconcile the relationships among the structural, dielectric, electromechanical, and thermal properties of $BaTiO_3$ [22, 23], which was a particular form of the statistical thermodynamic theories of phase transitions. In 1954, bulk single crystals with high quality were successfully synthesized [24], which allowed further characterization on $BaTiO_3$ in detail and consequently to be correlated with the existing theoretical descriptions [25]. The period witnessed a progressive improvement in the dielectric and piezoelectric properties of new perovskite-type ferroelectric materials [26]. Some studies focused on the phase diagram and synthesis optimization of polycrystalline $BaTiO_3$ [27, 28], whereas others expanded the research scope to encompass more complex $BaTiO_3$-based solid solutions such as the quaternary system between $BaTiO_3$ and $BaZrO_3/CaTiO_3/CaZrO_3$, and the quaternary system among $BaTiO_3$, $BaSnO_3$, $SrSnO_3$, and $CaSnO_3$ [29, 30]. Even with the development of these complex solid solutions, the poor temperature stability of the piezoelectric properties was not improved.

In the following decades, the research interest mainly focused on the dielectric properties and related applications on capacitors. Between the 1960s and 1970s, new solid-state physics concepts were developed to explain ferroelectricity in the perovskite oxides, including lattice dynamic approaches and the soft mode theory [31–33]. By the time around the 1970s–1990s, the community started to diversify and work on $Pb(Zr,Ti)O_3$ and $BaTiO_3$-based materials, which were marked by the appearance of new fields associated with electro-optics and thermistors. In fact, $BaTiO_3$-based materials were indispensable for applications on thermistors, due to their positive temperature coefficient of resistance [32]. During the 1980s and 1990s, the integration of ferroelectrics and miniaturization of devices began [32], as a result, the physical theory and chemistry of piezoelectric materials continued to evolve rapidly. It was until 2009 when high-performance $(Ba,Ca)(Zr,Ti)O_3$ piezoelectric ceramics were synthesized, the scientific interest in the $BaTiO_3$ system was reignited [34].

5.2 $BaTiO_3$-Based Ceramics and Single Crystals

5.2.1 Ceramics

For several decades, $BaTiO_3$ ceramics have been considered as "a mediocre class of piezoelectric materials" with a piezoelectric constant d_{33} of 120–190 pC/N, which were significantly inferior to lead zirconate titanate (PZT) [35]. Nevertheless, high d_{33} values were continuously reported in the past decades. The first high d_{33} up to 350 pC/N was obtained in the $BaTiO_3$ ceramics sintered by the microwave sintering using the hydrothermally synthesized fine powders (100 nm) [36–38]. By using the same powders, further high d_{33} values of 416 and 485 pC/N were obtained by spark plasma sintering (SPS) [39] and hot-press sintering [40]. Karaki et al. applied two-step sintering to fabricate fine-grained $BaTiO_3$ ceramics from the same hydrothermally synthesized 100 nm powders and obtained large d_{33} of 460 pC/N [41, 42]. By further optimizing the conditions for the two-step sintering, Huan et al. obtained a higher d_{33} of 519 pC/N [43]. An extremely high d_{33} up to 788 pC/N was achieved in textured $BaTiO_3$ ceramics fabricated by the templated grain growth method also using the hydrothermally synthesized fine $BaTiO_3$ powders [44]. Besides using advanced sintering techniques, high d_{33} values were also obtained in the $BaTiO_3$ ceramics fabricated by the conventional sintering process. For example, Ma et al. attained a maximum d_{33} of 355 pC/N by optimizing the sintering temperature in the $BaTiO_3$ ceramics using hydrothermally synthesized fine $BaTiO_3$ powders [45]. High d_{33} up to 419 pC/N was obtained even in the $BaTiO_3$ ceramics derived from the conventional solid-state reaction route starting from ordinary $BaCO_3$ and TiO_2 powders [46, 47]. Figure 5.2 shows the reported highest d_{33} in the past for the $BaTiO_3$ ceramics.

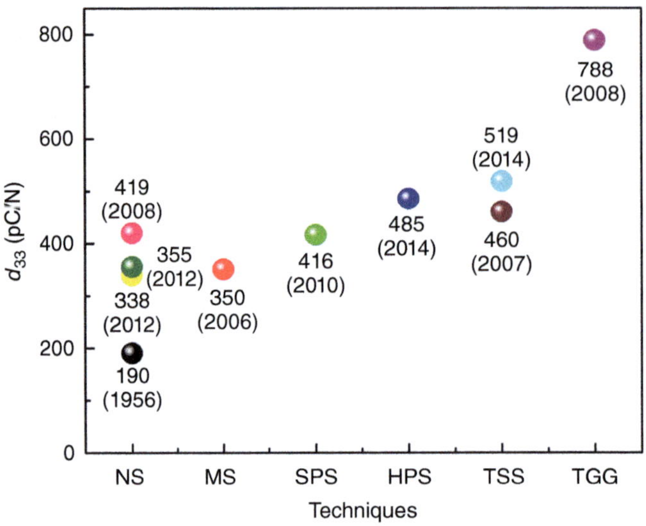

Figure 5.2 The d_{33} record reported in the past for the pure $BaTiO_3$ ceramics processed by different sintering techniques. NS, normal sintering; MS, microwave sintering; SPS, spark plasma sintering; HPS, hot-pressed sintering; TSS, two-step sintering; TGG, templated grain growth.

The underlying mechanism for the increased piezoelectricity in the pure $BaTiO_3$ ceramics is interesting. The aforementioned works revealed that the enhancement is closely related to the nano-domain structures. The enhanced piezoelectric properties of 350 pC/N in $BaTiO_3$ ceramics prepared by the microwave sintering were attributed to the nano-domain structure with an average domain width less than 50 nm [36–38]. In the SPS-processed $BaTiO_3$ ceramics with high d_{33} values, the domain width was also identified <50 nm by transmission electron microscopy (TEM), whereas such domains were not observed in the $BaTiO_3$ ceramics with much lower d_{33} (216 pC/N), even though the same SPS process was employed but using 500 nm powders [39]. The surprisingly high d_{33} up to 788 pC/N was obtained in [110]-oriented $BaTiO_3$ ceramics [44] that had much finer domain sizes in addition to the orientation effect. It is generally accepted that the domain sizes can be reduced by refining the grain size. Actually, the domain width was found proportional to the (grain size)$^{1/2}$ in the unpoled $BaTiO_3$ ceramics [48]. Therefore, the two-step sintering was applied to suppress the grain growth, and hence the resultant piezoelectricity enhancement was also attributed to the nano-domain structure.

Ma et al. revealed the importance of the nano-domain structure for the enhanced piezoelectricity in the pure $BaTiO_3$ ceramics [45]. They obtained a high d_{33} up to 355 pC/N even in the $BaTiO_3$ ceramics prepared by the normal sintering. The grains were as large as 50–60 μm, but the domain structure was very special. As shown in Figure 5.3, in an individual grain, there existed several stripes of micron-sized domains, which consisted of nano-domain with a width of 10–20 nm [45]. Certainly, such nano-domain structures should be responsible for the high piezoelectricity, but

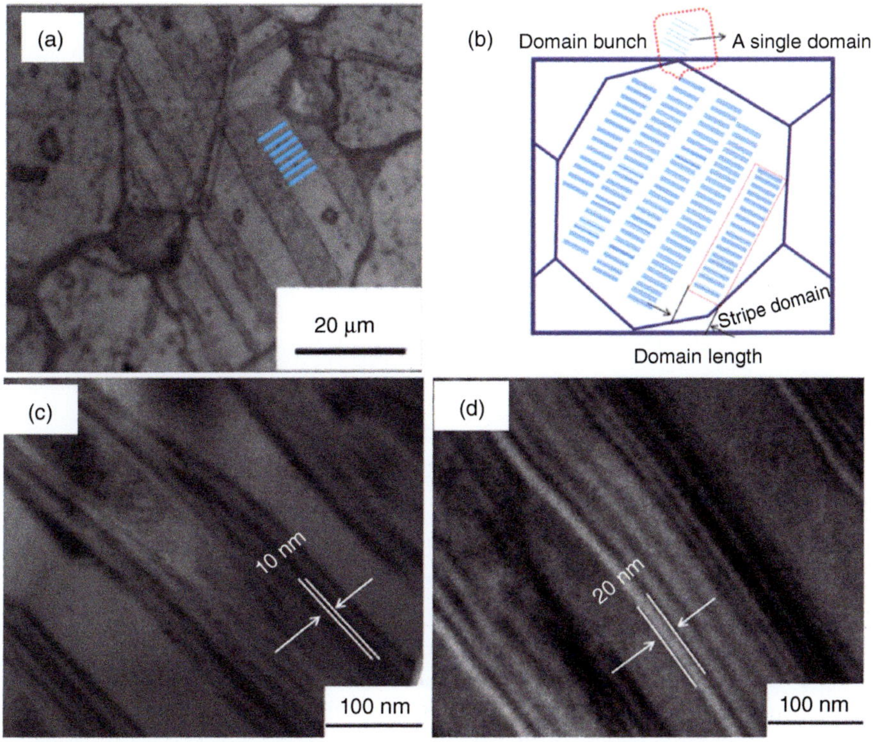

Figure 5.3 Domain structures observed in normally sintered $BaTiO_3$ ceramics. (a) Optical microscopy microstructure, (b) schematic image of domains inside a grain, (c, d) TEM images of nano-domains observed in the poled samples. Source: Ma et al. [45]. © (2012), with permission from Elsevier.

why could they be formed even in the ceramics with coarse grains. It seems inconsistent with the aforementioned grain size effect on the domain structure. They pointed out that the phase transition may be another factor for high-performance $BaTiO_3$ ceramics [45]. The formation of nano-domains might result from the coexistence of orthorhombic and tetragonal phases in the ceramics since the transition temperature in $BaTiO_3$ is close to room temperature. Zhang et al. revealed an interesting result that the d_{33} increased clearly when measured at around 10 °C lower than room temperature [40]; this is because the tetragonal-to-orthorhombic transition temperature is around 5 °C. Later, more discussions will be given in this chapter about the piezoelectricity enhancement mechanisms based on phase structure engineering in chemically modified $BaTiO_3$-based ceramics.

In addition to the previous works focused on the control of microstructure and domain structures, many studies have been conducted to improve the electrical properties of the $BaTiO_3$ ceramics by chemical doping. Although the dopant concentration may be very low, they can considerably alter the functional properties by modifying the extrinsic contribution. For $BaTiO_3$, chemical doping has been extensively investigated to develop positive temperature coefficient of resistivity

(PTCR) materials whose electrical conductivity abruptly changes from a conductive to a non-conductivity state with increasing temperature. There are some studies focused on piezoelectricity enhancement by chemical doping [49–52]. Kimura et al. obtained a high d_{33} of 260 pC/N in Li-doped $BaTiO_3$ ceramics [49]. Although most dopants also reduce T_C, the T_C in Li-doped $BaTiO_3$ was 130 °C, which is almost unchanged as compared with the pristine one. Li-doping was also found to be effective to lower the sintering temperature, so that LiF was often used as a sintering aid and also chemical dopant. A higher d_{33} of 270 pC/N was reported in the $BaTiO_3$ sintered with 0.04 at% LiF [50]. The d_{33} values of the $BaTiO_3$ ceramics sintered with LiF were raised to 340 pC/N when an excess of 0.025 at% Ba was added, and even to 361 pC/N when 0.02 at % Ca^{2+} excess was added [51]. A high d_{33} up to 454 pC/N was obtained in the $Ba(Ti_{0.96}Sn_{0.04})O_3$–$xY_2O_3$ ceramics [52]. However, the LiF and/or other element-doped $BaTiO_3$ ceramics showed lower T_C than Li-doped ones.

5.2.2 Single Crystal

Remeika and Jackson successfully invented a method for preparing large and clear $BaTiO_3$ single crystals, having desirable ferroelectric characteristics [24]. The process started from $BaTiO_3$ powder and KF as a flux heated at a temperature between 1150°C and 1200°C for eight hours in a furnace, which was then cooled at a slow constant rate. They found that the Fe-doped $BaTiO_3$ crystals showed much lower dielectric loss and the T_C was about 105–110 °C [24].

Later, Saifi et al. developed the laser-heated pedestal growth (LHPG) technique to grow $BaTiO_3$ single-crystal fibers from sintered ceramic rods with single-crystal $BaTiO_3$ seed [53]. The grown single-crystal fibers (with a diameter of 100 μm) were a ferroelectric tetragonal phase. However, the previous studies revealed that the growth from stoichiometric $BaTiO_3$ melt always results in its hexagonal non-ferroelectric phase.

The LHPG technique was successfully applied to grow large-sized single crystals of the ferroelectric $Ba(Ti_{1-x}Zr_x)O_3$ ($x = 0.05$–0.20) by Yu et al. [54, 55]. Figure 5.4a shows the pictures of as-grown $Ba(Ti_{1-x}Zr_x)O_3$ crystals with $x = 0.05$ and 0.08 [55]. The remnant polarization and strain behavior along different crystal orientations were studied. Higher strain and lower polarization were obtained along the pseudocubic [001] direction compared with those observed along the [110, 111] directions in both orthorhombic ($x = 0.05$) and rhombohedral ($x = 0.08$) phase [56]. The results for the [001]-oriented orthorhombic crystals with $x = 0.05$ indicate the highest strain level (similar to 0.25%) with almost hysteresis-free behavior as shown in Figure 5.4b [56]. In the non-polar (001) direction, the strain exhibits the highest value among the three measured directions, due to its orthorhombic symmetry at room temperature for $x = 0.05$ crystals. This shows that in lead-free crystals, the crystal orientation determines the piezoelectric behavior as in the lead-containing ferroelectric relaxor based materials.

The top-seeded solution growth (TSSG) technique was successfully applied to the commercialization of optical-graded $BaTiO_3$ single crystals in the 1990s [57]. Park et al. obtained high piezoelectric properties with an electromechanical coupling

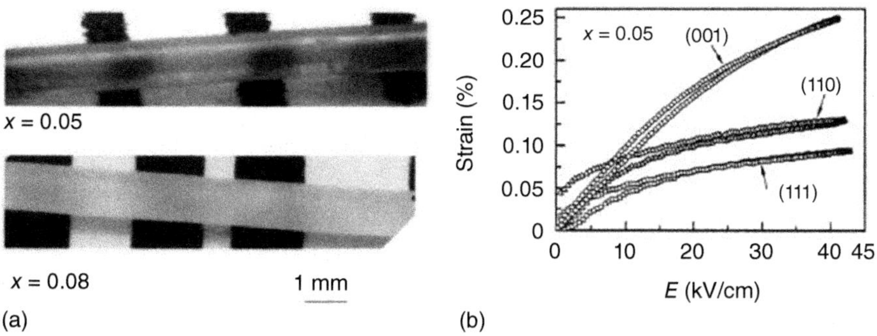

Figure 5.4 (a) Photos of as-grown $Ba(Ti_{1-x}Zr_x)O_3$ crystals with $x = 0.05$ and 0.08. (b) The unipolar strain versus electric field (E) for the $Ba(Ti_{1-x}Zr_x)O_3$ single crystals with $x = 0.05$ along various crystal orientations. Source: (a) Yu et al. [55]; (b) Yu et al. [55, 56].

coefficient (k_{33}) ~ 85% and d_{33} ~ 500 pC/N by the introduction of crystallographic engineering in the commercialized $BaTiO_3$ single crystals [57]. Their work demonstrated that $BaTiO_3$ single crystals are promising candidates for high-performance lead-free piezoelectrics at low temperatures. Wada et al. further investigated the size effect of engineered domains of the $BaTiO_3$ single crystals, which found that the piezoelectric properties such as d_{31} and k_{31} increased significantly with decreasing domain sizes [58]. This work gave guidance for the piezoelectricity enhancement in the $BaTiO_3$-based ceramics.

In 2004, Ren achieved a giant electric-field-induced strain in $BaTiO_3$ (doped with 0.02 at% Fe) crystal by using the reversible domain switching mechanism [59]. As shown in Figure 5.5A, the strain reached 0.7%, despite a large hysteresis. Such a giant strain was believed to be associated with reversible domain switching in aged ferroelectrics due to the symmetry-conforming property of point defects. As shown in Figure 5.5B, a cubic $BaTiO_3$ crystal has cubic-symmetric point-defect distributions. When cooled down below T_C, it transforms into a multi-domain tetragonal ferroelectric phase (Figure 5.5B). The cubic-symmetric point-defect distributions remained unchanged because of the diffusionless transition, but it can be corrected into a polar tetragonal one when it is aged at a temperature high enough but below the T_C for sufficient duration. Now, when an electric field is applied on it, the domain switching occurs abruptly (without diffusion) with spontaneous polarization following the field direction, and in the meantime, the 90° switching of domains generates a large strain. Because the defect symmetry and defect dipole moment cannot be rotated in such a diffusionless process, the firm defect symmetry and the associated defect dipole moment provide a restoring force or a reverse internal field, which leads to domain backswitching when the applied electric field is removed.

5.3 $BaTiO_3$-Based Solid Solution Ceramics

In 2009, a high d_{33} up to 620 pC/N was reported in $BaTiO_3$-based ceramics with a binary solid solution composition of $0.7Ba(Ti_{0.8}Zr_{0.2})O_3$–$0.3(Ba_{0.7}Ca_{0.3})TiO_3$, which

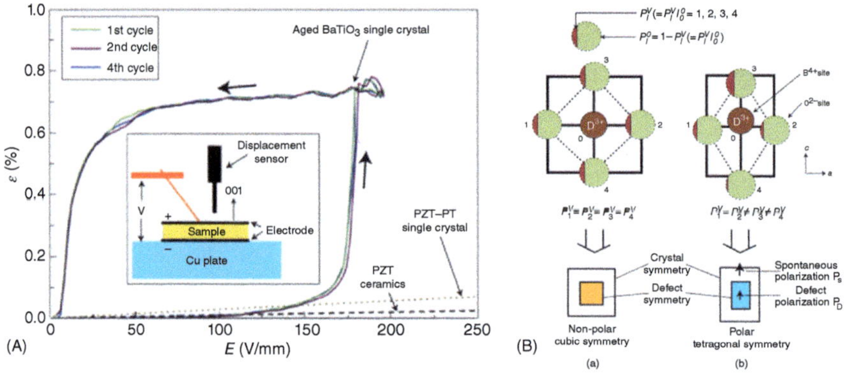

Figure 5.5 (A) Large recoverable electric-field-induced strain in an aged $BaTiO_3$ single crystal along the [001] pseudo-cubic direction, (B) symmetry-conforming property of point defects in centrosymmetric cubic paraelectric state and polar tetragonal ferroelectric state [59]. Source: Reprinted by permission from Ren [59]. © 2004, Springer Nature.

is located at a tricritical triple point of a cubic paraelectric phase (C), ferroelectric rhombohedral (R), and tetragonal (T) phases [34]. This work stimulated the studies about the $BaTiO_3$-based solid solution ceramics with high piezoelectricity. The major systems are $(Ba,Ca)(Zr,Ti)O_3$, $(Ba,Ca)(Sn,Ti)O_3$, and $(Ba,Ca)(Hf,Ti)O_3$ systems [35, 60].

5.3.1 $(Ba,Ca)(Ti,Zr)O_3$

Liu and Ren found the phase diagram in the $Ba(Zr_{0.2}Ti_{0.8})O_3$–$(Ba_{0.7}Ca_{0.3})TiO_3$ (BZT–BCT) system as shown in Figure 5.6A [34]. There is a phase boundary separating tetragonal and rhombohedral phases, which is called the morphotropic phase boundary (MPB) in the PZT system. It should be noted that this boundary bends from the BCT-rich side to the BZT-rich side, but that in PZT is almost normal to the compositional axis. In fact, as indicated by the dielectric permittivity curves in Figure 5.6B, an additional peak was detected at a lower temperature across the MPB [34]. There is a triple point where the MPB crosses the transition line from cubic to non-cubic phases. The high piezoelectricity of 620 pC/N was obtained at room temperature in the middle composition (BZT–50BCT) as shown in Figure 5.6C [34]. Although more studies are needed to attain a full understanding of the mechanism, it is believed that the enhanced piezoelectricity originates from a multiphase coexistence effect. However, the T_C of the BZT–BCT ceramics is low, around 75 °C as indicated by the permittivity curves. Nevertheless, the BZT–BCT ceramics showed high piezoelectric coefficients than other lead-free systems and even the PZT family at the time when this paper was published, as compared in Figure 5.6D [34].

Since 2009, this lead-free ceramic system has been extensively studied [61–70]. A lot of research has been conducted to further investigate the phase structures [61–65]. Keeble et al. investigated the BZT–BCT phase diagram using high-resolution synchrotron X-ray powder diffraction [61]. They found an intermediate orthorhombic phase, isostructural to that present in the parent phase, $BaTiO_3$,

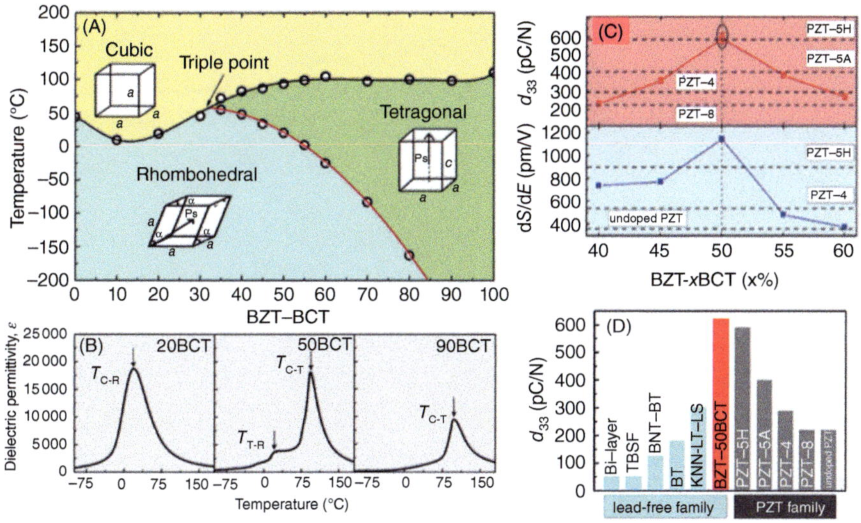

Figure 5.6 (A) Phase diagram of $Ba(Zr_{0.2}Ti_{0.8})O_3$–$(Ba_{0.7}Ca_{0.3})TiO_3$ (BZT–BCT) system, (B) dielectric permittivity curves of selected compositions as a function of temperature. (C) Compositional dependence of piezoelectric charge coefficient and piezoelectric strain coefficients. (D) Comparison of d_{33} among lead-free and PZT families. Source: Reprinted with permission from Liu and Ren [34]. © (2016) by the American Physical Society.

and identified a tetragonal-to-orthorhombic transition instead of the previously proposed tetragonal-to-rhombohedral transition, as shown in Figure 5.7a. More recent investigations also suggested the existence of an orthorhombic (O) phase in BZT–BCT between the T and R phases [62, 63]. Brajesh et al. conducted a detailed temperature-dependent dielectric and lattice thermal expansion study, which revealed a weak dielectric relaxation that is characteristic of a relaxor ferroelectric material [62]. Their structural analysis indicated that a ferroelectric state at room temperature is composed of three-phase coexistences, tetragonal ($P4mm$) + orthorhombic ($Amm2$) + rhombohedral ($R3m$) [62]. A recent study from Ren's group also supported the existence of an intermediate orthorhombic (O) phase between tetragonal (T) and rhombohedral (R) ones, as shown in Figure 5.7b [63]. $BaTiO_3$ experiences a series of phase transitions from the tetragonal phase to the orthorhombic phase and then to rhombohedral upon cooling [9].

Research on processing includes normal sintering, microwave sintering, and texturing techniques [66–70]. Chandrakala et al. investigated the effect of sintering temperature on structural, dielectric, piezoelectric, and ferroelectric properties of sol–gel derived BZT–BCT ceramics [66]. The $0.5Ba(Zr_{0.2}Ti_{0.8})O_3$–$0.5(Ba_{0.7}Ca_{0.3})TiO_3$ ceramics sintered at 1550 °C for two hours showed a high piezoelectric charge coefficient d_{33} of 617 pC/N and a large planar-mode electromechanical coupling factor (k_p) of 54% [66]. Wu et al. found that the dwell time during sintering also largely affected the piezoelectric properties of the $(Ba_{0.85}Ca_{0.15})(Ti_{0.90}Zr_{0.10})O_3$ ceramics [67]. The d_{33} values reached a maximum at 492 pC/N when sintered for five hours, while a longer sintering time led to a decrease in d_{33} but did not affect other dielectric and

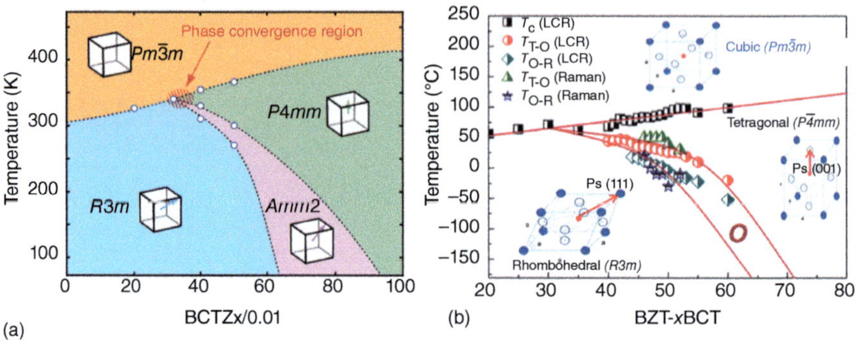

Figure 5.7 Revised phase diagrams of the $Ba(Zr_{0.2}Ti_{0.8})O_3-(Ba_{0.7}Ca_{0.3})TiO_3$ ceramics with an orthorhombic phase between tetragonal and rhombohedral phases. Source: (a) Keeble et al. [61] and (b) Zhang et al. [63].

ferroelectric properties, suggested that the poled state might be more sensitive to the microstructure. Microwave sintering of BZT–BCT ceramics was studied, by which dense microstructures with better electrical properties could be obtained at lower temperatures with shorter dwelling time compared to that of the conventional sintering [68, 69]. (001)-textured $(Ba_{0.85}Ca_{0.15})(Zr_{0.1}Ti_{0.9})O_3$ (BCZT) lead-free piezoelectric ceramics were fabricated by the templated grain growth method using plate-like $BaTiO_3$ as template [70]. However, the d_{33} obtained in the textured ceramics was lower than that reported in the non-textured ceramics.

5.3.2 $(Ba,Ca)(Ti,Sn)O_3$

In addition to $(Ba,Ca)(Ti,Zr)O_3$, another important system is $(Ba,Ca)(Ti,Sn)O_3$ (abbreviated BCTS), in which large d_{33} values of 360–680 pC/N were reported [71–82]. Xue et al. found a similar phase diagram, as shown in Figure 5.8, in the pseudo-binary system of $Ba(Sn_{0.12}Ti_{0.88})O_3$–$x(Ba_{0.7}Ca_{0.3})O_3$ (BTS–xBCT), which is also characterized by a phase boundary starting from a tricritical triple point of a paraelectric cubic phase, ferroelectric rhombohedral phase, and ferroelectric tetragonal phases [71]. The optimal composition BTS–30BCT exhibits a high d_{33} of 530 pC/N at room temperature. It should be noted that the optimal composition also deviates a little from the tricritical triple point, analogous to the case of the BZT–BCT system.

As compared with the BZT–BCT system, the investigated compositions in the BCTS system were focused on the range with less than 10% substitution amounts of Ca and Sn for Ba and Ti, respectively [72–82]. As shown in Figure 5.9, either Ca or Sn doping significantly affects the phase transition temperatures of $BaTiO_3$ [83, 84]. As shown in Figure 5.9a, the Ca doping can effectively shift both tetragonal-to-orthorhombic and orthorhombic-to-rhombohedral phase transition temperatures downward, but has limited impact on the T_C corresponding to the cubic-to-tetragonal phase transition temperature. Therefore, the tetragonal phase can exist in a wider temperature range. On the other hand, the Sn doping shifts both

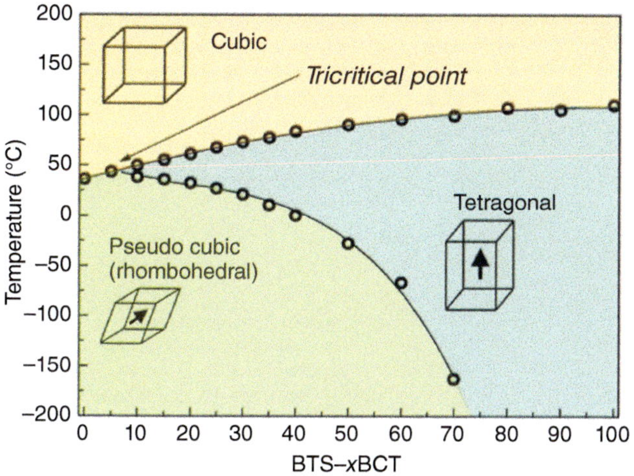

Figure 5.8 Phase diagram of $Ba(Sn_{0.12}Ti_{0.88})O_3-x(Ba_{0.7}Ca_{0.3})O_3$ ceramics. Source: Reprinted from Xue et al. 2011 [71] with the permission of AIP Publishing.

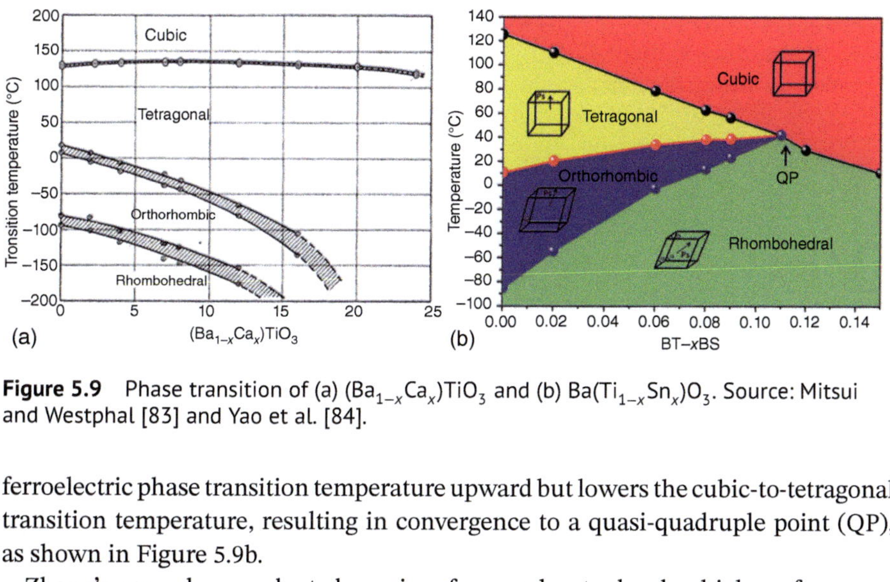

Figure 5.9 Phase transition of (a) $(Ba_{1-x}Ca_x)TiO_3$ and (b) $Ba(Ti_{1-x}Sn_x)O_3$. Source: Mitsui and Westphal [83] and Yao et al. [84].

ferroelectric phase transition temperature upward but lowers the cubic-to-tetragonal transition temperature, resulting in convergence to a quasi-quadruple point (QP), as shown in Figure 5.9b.

Zhang's group has conducted a series of researches to develop high-performance lead-free piezoelectric ceramics based on the BCTS system [72–80]. They conducted detailed studies on the phase transitions in the $(Ba_{0.95}Ca_{0.05})(Ti_{1-x}Sn_x)O_3$ ceramics and found that the phase structure evolved from an inceptive orthorhombic (O) phase at $0 \leq x \leq 0.04$ to two-phase coexistence of pseudocubic–orthorhombic (PC–O) at $0.06 \leq x \leq 0.10$ and further to a multiple-phases coexistence of rhombohedral–pseudocubic–orthorhombic (R–PC–O) at $x = 0.11$ [72]. High d_{33} values over 520 pC/N were obtained in a wide compositional range of $0.07 \leq x \leq 0.11$ with two peak values over 670 pC/N. The phase diagram and the d_{33} and k_p, as well as the mechanical quality factor (Q_m) as a function of x, are shown in Figure 5.10.

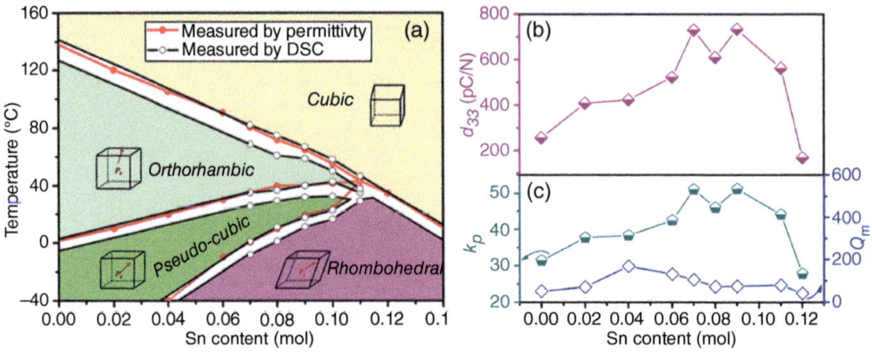

Figure 5.10 (a) Phase diagram of $(Ba_{0.95}Ca_{0.05})(Ti_{1-x}Sn_x)O_3$ ceramics. (b, c) Piezoelectric coefficient d_{33} and planar mode electromechanical coupling coefficient k_p and the mechanical quality factor Q_m as a function of x [72]. Source: Reprinted from Zhu et al. [72] with permission of AIP Publishing.

Zhou et al. obtained a higher d_{33} of 683 pC/N and an ultrahigh converse piezoelectric coefficient ($dS/dE = 1257$ pm/V) in the CuO-doped $(Ba_{0.95}Ca_{0.05})(Ti_{0.90}Sn_{0.10})O_3$ ceramics that were sintered at 1250 °C, which was 200 °C lower than that for the same composition without CuO [73]. Research on the sintering of $BaTiO_3$-based ceramics, including the binary solid solution, will be introduced later in this chapter.

5.3.3 $(Ba,Ca)(Ti,Hf)O_3$

$(Ba_{1-x}Ca_x)(Hf_yTi_{1-y})O_3$ (BCHT) is another $BaTiO_3$-based solid solution, which shows a similar phase diagram to the aforementioned two systems. In fact, Hf and Zr are two neighboring elements with very close characteristics. An early study conducted by Payne and Tennery gave a partial solid-state phase diagram for $BaTiO_3$-rich end of the $BaTiO_3$–$BaHfO_3$ system, as shown in Figure 5.11 [85]. Substitution of Hf^{4+} into Ti^{4+} in $BaTiO_3$ increases the two ferroelectric phase transition temperatures but decreases the T_C, resulting in a convergence of phase transitions at 15 mol% $BaHfO_3$, which is similar to the phase diagrams shown in Figures 5.9b and 5.10a.

The piezoelectric properties of the BCHT ceramics have been investigated in association with the phase structure [86–91]. For example, Zhou et al. obtained a high d_{33} of 550 pC/N at room temperature in the $Ba(Ti_{0.8}Hf_{0.2})O_3(Ba_{0.7}Ca_{0.3})TiO_3$ ceramics [86]. Wang et al. further investigated the phase diagram and piezoelectric properties of the BCHT ceramics with three composition groups in the $BaTiO_3$–$CaTiO_3$–$BaHfO_3$ ternary system, $(1-x)Ba(Hf_yTi_{1-y})O_3$–$x(Ba_{1-z}Ca_z)TiO_3$ (abbreviated as BH_yT–xBC_zT with $x = 0$–1.0, $y = 0.16$, $z = 0.20$; $x = 0$–1.0, $y = 0.16$, $z = 0.30$; $x = 0$–1.0, $y = 0.20$, $z = 0.30$, respectively) [87]. The enhanced piezoelectric properties are achieved in the orthorhombic-to-tetragonal (O–T) phase boundary compositions of the three groups, which are $BH_{0.16}T$–$0.58BC_{0.20}T$, $BH_{0.16}T$–$0.48BC_{0.30}T$, and $BH_{0.20}T$–$0.53BC_{0.30}T$, respectively. d_{33} of 448 pC/N is obtained in $BH_{0.20}T$–$0.53BC_{0.30}T$, which is higher than those of the other two

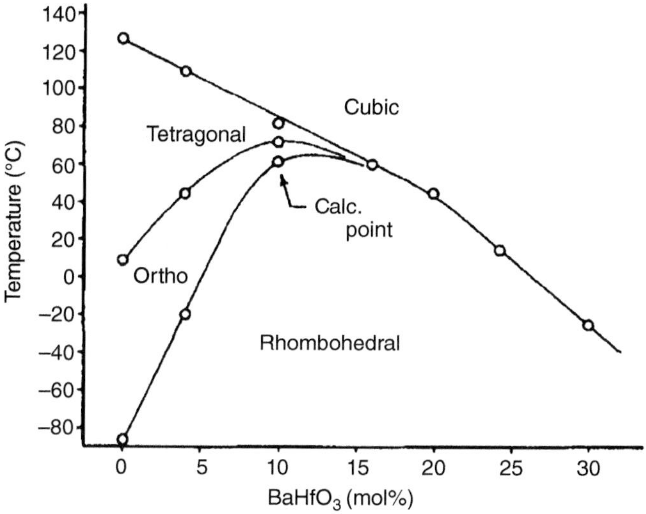

Figure 5.11 Phase diagram of the $BaTiO_3$–$BaHfO_3$ system at the $BaTiO_3$-rich end. Source: Reprinted with permission from Payne and Tennery [85]. © 1965, John Wiley & Sons, Inc.

phase boundary compositions. This might be due to the fact that the O-phase zone in $BH_{0.20}T$–$xBC_{0.30}T$ with more Hf substitutions is narrower than those in $BH_{0.16}T$–$xBC_{0.20}T$ and $BH_{0.16}T$–$xBC_{0.30}T$ [87]. Zhao et al. investigated the composition dependence of the phase structures and electrical properties of the $(Ba_{1-y}Ca_y)(Ti_{1-x}Hf_x)O_3$ ceramics. It was found that rhombohedral-to-orthorhombic and orthorhombic-to-tetragonal (R–O/O–T) phase transitions converged in the ceramics with $x = 0.10$ and $y = 0.15$, resulting in an enhanced piezoelectricity of $d_{33} \sim 540$ pC/N and a large strain of ~0.21% [88].

However, the number of researches about the BCHT ceramics is less and the reported piezoelectric properties also are not high if compared with the rest of the systems.

5.4 Piezoelectricity Enhancement

5.4.1 Phase Engineering

Regardless of the lead-containing or lead-free systems, phase engineering is often an effective way to enhance the piezoelectricity. In the lead-containing PZT system, the piezoelectric properties can be maximized at the MPB compositions where tetragonal and rhombohedral phases coexist. In the lead-free system, the phase-coexisting effects have been well utilized to enhance the piezoelectricity in the KNN systems as described in Chapter 3, but the phase coexistence was realized mostly by moving phase transition boundary to room temperature via chemical modification. Despite the difference in transition temperature, both KNN and $BaTiO_3$ show the same sequence of tetragonal-to-orthorhombic and

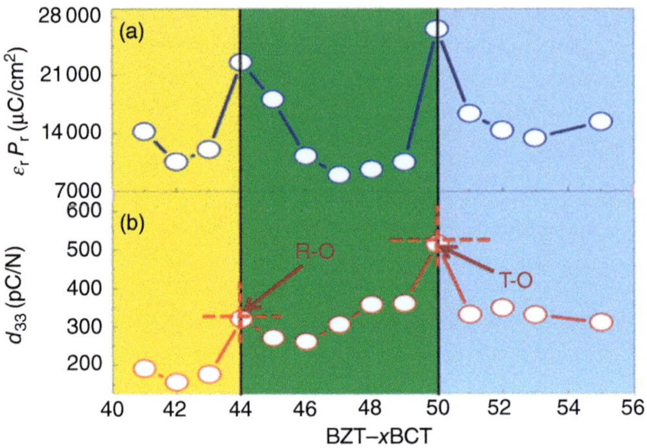

Figure 5.12 The changes of (a) $\varepsilon_r P_r$ and (b) d_{33} measured at room temperature as a function of BCT content. Source: Reprinted from Zhang et al. 2014 [63] with the permission of AIP Publishing.

orthorhombic-to-rhombohedral ferroelectric phase transitions. It should be noted that the tetragonal-to-orthorhombic phase transition in $BaTiO_3$ is very close to room temperature (~5 °C), which should be one of the reasons why high d_{33} values could be obtained even in the pure $BaTiO_3$ ceramics, as mentioned earlier.

Many studies revealed the importance of phase coexistence on the enhanced piezoelectricity in the binary solid-solution ceramics [63, 84, 92–94]. As shown in Figure 5.6, the maximum was obtained at the $Ba(Zr_{0.2}Ti_{0.8})O_3$–50%$(Ba_{0.7}Ca_{0.3})TiO_3$ (BZT–BCT) composition, which is located at the boundary between the tetragonal and rhombohedral phases. Zhang et al. conducted a detailed investigation about the phase transitions in association with the piezoelectricity in the BZT–BCT ceramics using Raman spectroscopy, temperature-dependent permittivity curves, and X-ray diffraction [63]. They found two peaks for ferroelectric (the product of relative dielectric constant ε_r and remanent polarization P_r) and piezoelectric properties at the 56BZT–44BCT and 50BZT–50BCT compositions, which correspond to R–O/O–T phase boundaries, respectively, as shown in Figure 5.12.

In the $(Ba,Ca)(Ti,Sn)O_3$ system, multiple phases converge to a point called quasi-quadruple point as shown in Figures 5.9b and 5.10a. It should be noted that such a converged point also exists in the revised phase diagram of the $(Ba,Ca)(Ti,Zr)O_3$ system, as shown in Figure 5.7, although the old version contained a triple point. Yao et al. measured the d_{33} of the $BaTiO_3$–$x$$BaSnO_3$ ceramics at different temperatures and obtained a contour map of the d_{33} in temperature-composition dimensions as shown in Figure 5.13 [84]. The d_{33} measured at 40 °C reached 697 pC/N at the composition of $BaTiO_3$–0.11$BaSnO_3$ as shown in Figure 5.12b, which is located at the quasi-quadruple point. As shown in Figure 5.12b, if measured at 28 °C, the d_{33} values decreased greatly due to the deviation from the multiple-phases coexisting region. Their calculation based on the Landau–Devonshire model showed that four-phase coexisting leads to minimized

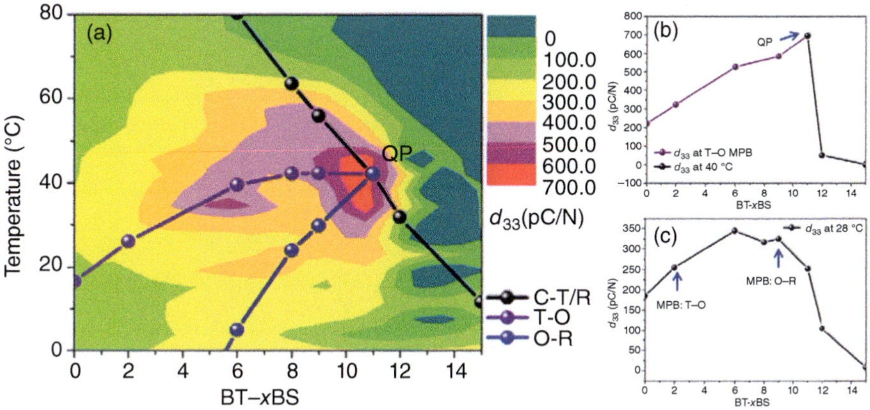

Figure 5.13 (a) Contour map of piezoelectric coefficient d_{33} in temperature-composition dimensions. (b) The d_{33}-composition curve along T–O MPB and at 40 °C. (c) d_{33}-composition curve at 28 °C. Source: Reprinted with permission from Yao et al. [84]. © 2012, IOP Publishing.

energy barriers for both polarization rotation and extension, which might be the origin of the giant dielectric and piezoelectric properties [84].

5.4.2 Domain Engineering

Domain engineering is an effective strategy to maximize the piezoelectric and dielectric properties by creating a stable and desirable domain structure. It has been applied to single crystals [58, 95, 96] but has not been widely used for property tailoring of polycrystalline materials. The domain miniaturization and hierarchical domain morphology found in the convergence region and near the phase transition boundaries has generally been considered as a major contributor to piezoelectric properties in BZT–BCT ceramics [97–103]. As shown in Figure 5.14, the distinct feature of MPB composition is that miniaturized nanodomains with the average size of 20–100 nm are developed on these lamellas forming a "domain hierarchy," as compared with the two-side compositions. Further convergent beam electron diffraction (CBED) measurement shows that rhombohedral and tetragonal crystal symmetries coexist among the miniaturized domains [97]. The evolution of such nanodomain structures as a function of the applied electric field was studied by using *in situ* TEM. It was found that the multidomain state transformed into a unique single-domain state at a moderate poling field. The single-domain state was suggested to be the major reason for the outstanding piezoelectricity.

A high degree of domain switching was corroborated, which leads to a metastable single-domain state [100–102]. It was found that the macroscopic properties such as maximum polarization and strain correlate with the coercive fields and the fields required to trigger the single-domain state. BZT–50BCT ceramics have higher domain wall mobility than $BaTiO_3$. Thus, it was suggested that high switching dynamics may also contribute to the large piezoelectricity in BZT–50BCT ceramics [100]. Switching was the main contributor to the strains for all the T-phase

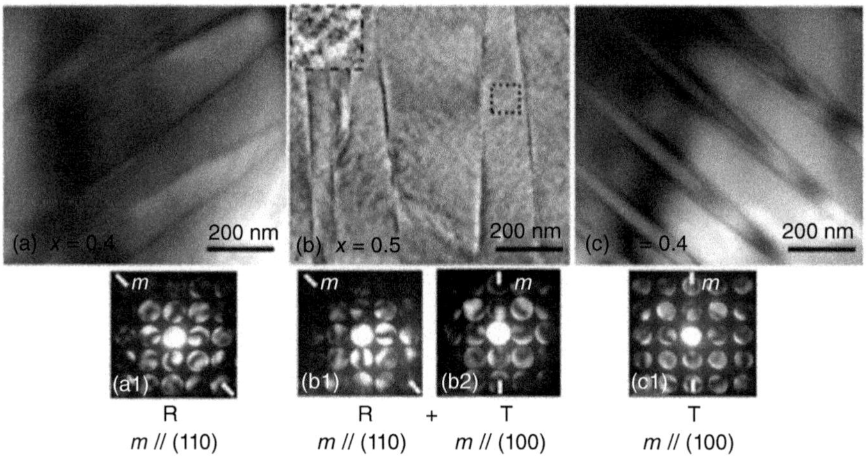

Figure 5.14 TEM observation of domain structures in BZT−xBCT (x = 0.4, 0.5, and 0.6). The bright-field image of BZT−xBCT from $[001]_{pc}$ incidence (a) x = 0.4, (b) x = 0.5, and (c) x = 0.6. The convergent beam electron diffraction (CBED) pattern obtained from one domain in (a), indicating R symmetry. (b1, b2) The CBED pattern obtained from one domain lamella in (b), indicating the coexistence of R and T symmetry in the MPB region. (c1) The CBED pattern obtained within one domain in (c), indicating T symmetry. Source: Gao et al. [97].

compositions [104, 105]. In the vicinity of the O–T boundary, higher macroscopic strain occurs due to higher extrinsic and intrinsic contributions [105]. The higher extrinsic contribution was ascribed to the enhanced domain wall motion and alignment under either weak or high electric field [105].

It is known that grain sizes also affect the domain structure and hence the piezoelectricity. An optimized grain size generally leads to increased small-signal (d_{33}) and large-signal (d_{33}^*) electromechanical properties, enhanced dielectric constant, and decreased coercive field [106–108]. That is because there exists a maximized domain wall density that is favorable for switching with minimal back switching [108–110]. If the grains are too large, the corresponding microstructure often suppresses piezoelectric and dielectric properties as a consequence of excessive back switching [110]. In contrast, excessively small grains also lead to suppression of ferroelectricity due to reduced lattice distortion [111–113]. It was revealed that a strong increase in the functional properties in $BaTiO_3$ occurs at grain sizes between 1 and 2.3 μm and domain sizes below 100 nm [37, 108, 109, 114]. In contrast, a critical grain size between 10 and 100 nm induces a paraelectric state in $BaTiO_3$ at room temperature [111–113]. Several studies on $BaTiO_3$ have determined that a fine domain structure through engineered synthesis routes can greatly enhance the piezoelectric and dielectric properties [36, 39, 42, 46].

It was also speculated that the weak relaxor-like properties found in the BZT–50BCT ceramic may contribute to the high piezoelectric activity since polar nano-regions may favor the electric-field-induced phase transformation and domain wall motion [115].

5.4.3 Texturing

Texturing is an effective way to enhance the piezoelectric properties, dielectric properties, and temperature stability [116–119]. In contrast to other approaches, texturing does not lower the temperature range at which the ferroelectric state remains stable [120]. Templated grain growth and reactive templated grain growth are the most commonly used texturing techniques [121]. In lead-free perovskite ferroelectrics, texturing degrees up to 90% were achieved in various systems. A high texturing degree enhanced d_{33} by 1.4–2.1 times with respect to the non-textured polycrystalline materials [121]. The maximum piezoelectric and dielectric properties achievable in textured polycrystalline materials are limited by the properties of the corresponding single crystals [122].

$BaTiO_3$ ceramics with [001]-, [110]-, and [111]-texture orientation were found demonstrating improved electromechanical properties than the non-textured polycrystalline materials [46, 123, 124]. Enhanced piezoelectricity up to 274 pC/N was obtained in [001]-oriented $BaTiO_3$ with a 97% texturing degree by using [001]-oriented $BaTiO_3$ plate-like particles [124]. [111]-oriented $BaTiO_3$ with a 100% texturing degree led to an increased d_{33} by 30% and reduced strain hysteresis [123]. Texturing is also applicable in the materials with engineered grain and domain sizes, although it increases the processing complexity and costs [125]. An ultrahigh d_{33} up to 788 pC/N was obtained in [110]-oriented $BaTiO_3$ with an average grain size of 75 μm and an average domain size of 800 nm [44, 126, 127]. Texturing maximizes the piezoelectric and dielectric properties since it increases the fraction of domains oriented along the applied electric field [120, 121].

5.5 Key Issues of Sintering Processes

As compared with other piezoelectric ceramics, the $BaTiO_3$-based ceramics usually require high sintering temperatures above 1400 °C [128]. Therefore, for multilayered devices with metallic electrodes, it is necessary to reduce the sintering temperatures. The reduction of sintering temperature can be realized by physical modifications such as using starting materials with smaller particle sizes or improving the preparation process, and chemical modifications such as adding sintering aids. Usually, the fluxes must lower the sintering temperature markedly, and at the same time cause no significant deterioration of the dielectric or piezoelectric properties of the host materials. Lots of studies have been conducted to reduce the sintering temperature of $BaTiO_3$ ceramics as dielectric or piezoelectric materials [128–134].

Dense $BaTiO_3$ ceramics with a relative density >98% could be prepared at 1210 °C by the conventional sintering technique using hydrothermally synthesized powders with an average particle size around 100 nm [39]. For improving the preparation process, dense $BaTiO_3$ ceramics with 98% relative density were obtained by microwave sintering at 1100 °C using powders derived from the modified solid-state reaction [135]. Compared with the strategy of refining the particle sizes of the starting powders, it is more effective in reducing the sintering temperature by incorporating sintering aids into $BaTiO_3$ ceramics. Nevertheless, the addition of sintering aids may lead to the formation of a secondary phase or the change of phase structure and hence deteriorates the dielectric or piezoelectric properties. Unfortunately, a trade-off between the reduction of sintering temperature and maintaining good dielectric or piezoelectric properties has been often reported. An ideal sintering aid should effectively lower the sintering temperature while maintaining the original dielectric or piezoelectric properties. The typical sintering aids can be divided into two categories: Li-containing sintering additives (such as LiF, Li_2CO_3, Li_2O, and their compounds) and glass compositions.

5.5.1 Li-containing Sintering Additives

Adding LiF or other Li-containing compounds like Li_2O or Li_2CO_3 is an effective way to lower the sintering temperature of $BaTiO_3$ ceramics, which has been proved by many reports [50, 136–143]. For example, an addition of 0.5–3 wt% LiF allowed densification of $BaTiO_3$ with a relative density over 98% at temperatures as much as 500 °C lower than that without additives [136]; more than 98% of the theoretical density was reached at a sintering temperature of 900 °C by addition of LiF-containing sintering additives, such as 1.5–2.0 wt% LiF–$SrCO_3$, LiF–$BaCO_3$, and LiF–$CaCO_3$ [137]; the densification of the sample with 3 wt% LiF was nearly completed as sintered at 850 °C and the relative density of the sample containing 1 wt% LiF and 1 mol% $BaCO_3$ sintered at 900 °C could reach 95% [138]. The sintering temperature was reduced to 1000 °C by adding 6 mol% LiF in $BaTiO_3$ ceramic, which has a high relative density of 95.5% [50]. Besides LiF and LiF-based flux additions, the other two Li-containing additives Li_2O and Li_2CO_3 also have been used to lower the sintering temperature of $BaTiO_3$ ceramics. It was reported that the sintering temperature of $BaTiO_3$ can be reduced to 1050 °C by adding 0.5 mol% Li_2O or 1.0 wt% Li_2CO_3. Meanwhile, better piezoelectric property with d_{33} = 200 pC/N was obtained [139, 140].

The sintering mechanisms also attracted interest. LiF has a low melting point of 846 °C. It is a consensus that the densification of $BaTiO_3$ ceramics with LiF or LiF-based flux additions at low temperatures is attributed to the liquid phase sintering. A later study showed the interaction of LiF with $BaTiO_3$ matrix [141]. It showed the emergence of a liquid phase as early as 740 °C and the formation of a Li_2TiO_3 compound at grain corners [141]. The nature of the liquid phase formed during the liquid phase sintering in 20 wt% LiF and 20 wt% LiF–10 wt% $BaCO_3$ doped $BaTiO_3$ was investigated [142]. A $LiTiO_2$ intermediate phase was found in the mixtures quenched from 750 °C for both systems due to the solubility of Ti in the melt is less than that of Ba [143]. $LiTiO_2$ is the origin of the liquid phase sintering in the LiF-containing $BaTiO_3$ ceramics. The low-temperature densification behavior exhibited by LiF-containing $BaTiO_3$ in the presence of excess $BaCO_3$ is clearly due to two eutectic points at around 610 °C and 700 °C. Li_2CO_3 with a low melting point about 723 °C may form a liquid phase directly and promote the densification in the low-temperature sintering process [140]. So far, few reports refer to the detailed sintering mechanism of Li_2CO_3 in $BaTiO_3$ ceramics. Li_2O has a high melting point above 1500 °C, making it impossible to form a liquid phase directly in the

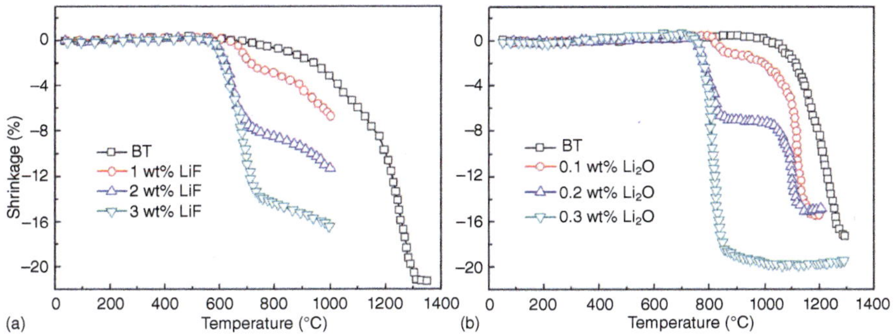

Figure 5.15 Sintering curves of $BaTiO_3$ with LiF (a) and $BaTiO_3$ with Li_2O (b) compared with $BaTiO_3$. Sources: Liu et al. [138] and Valent et al. [128].

low-temperature sintering process. $BaCO_3$ was observed in the $BaTiO_3$–0.3 wt% Li_2O sintered at 750 °C for five hours, suggesting that the sintering process involved an intermediate liquid phase $BaCO_3$ with a melting point of 811 °C [128]. Figure 5.15 shows the sintering curves of $BaTiO_3$, $BaTiO_3$–LiF, and $BaTiO_3$–Li_2O [128, 138]. Both the LiF- and Li_2O-containing $BaTiO_3$ show reaction liquid sintering rather than self-melting.

Besides $BaTiO_3$ ceramics, low-temperature sintering of the $(Ba_{0.85}Ca_{0.15})(Ti_{0.90}Zr_{0.10})O_3$ and $(Ba_{0.95}Ca_{0.05})(Ti_{0.90}Sn_{0.10})O_3$ ceramics have also been investigated because of their attractive high performance of 620 and ~630 pC/N as shown in Table 5.1. The addition of LiF and CuO can effectively lower the sintering temperature of the $(Ba_{0.85}Ca_{0.15})(Ti_{0.90}Zr_{0.10})O_3$ ceramic from 1450–1500 °C to 1350–1400 °C but the piezoelectric properties decrease [73, 77–79, 144–146]. The sintering temperature of the $(Ba_{0.95}Ca_{0.05})(Ti_{0.90}Sn_{0.10})O_3$ ceramic can be reduced to 1250–1300 °C from 1480 °C by adding Li_2CO_3, Li_2O, LiF, and CuO [77–79]. $(Ba_{0.95}Ca_{0.05})(Ti_{0.90}Sn_{0.10})O_3$ ceramic has a higher sintering temperature than $BaTiO_3$ ceramics, which shrink at 1050 and 950 °C, respectively. It was found that the sintering shrinkage stated at 750 °C in 4 mol% Li_2CO_3 and 4 mol% Li_2O added $(Ba_{0.95}Ca_{0.05})(Ti_{0.90}Sn_{0.10})O_3$, which is attributed to the formation of liquid phase. However, as shown in Table 5.1, the piezoelectric properties of $(Ba_{0.95}Ca_{0.05})(Ti_{0.90}Sn_{0.10})O_3$ ceramics were deteriorated by adding Li_2CO_3 or Li_2O or LiF. The exciting result is that the addition of CuO lowers the sintering temperature to 1250 °C and also enhances the piezoelectric properties to 683 pC/N in the $(Ba_{0.95}Ca_{0.05})(Ti_{0.90}Sn_{0.10})O_3$ ceramic [73].

5.5.2 Glass Compositions

Mixing glass frit into $BaTiO_3$ is an effective method to enhance densification. As a result, numerous types of glasses have been developed as sintering additives for $BaTiO_3$ [133, 147–154]. Early in 1982, Burn reported a series of studies on fluxed sintering of $BaTiO_3$, in which the fluxes containing B_2O_3, CdO, Bi_2O_3, and/or PbO were systematically investigated [149]. The sintering temperature of $BaTiO_3$ can be reduced to 900 °C or lower by adding ZnO–B_2O_3–SiO_2, B_2O_3–Bi_2O_3–SiO_2–ZnO,

Table 5.1 Piezoelectric property of the $(Ba_{0.85}Ca_{0.15})(Ti_{0.90}Zr_{0.10})O_3$ and $(Ba_{0.95}Ca_{0.05})(Ti_{0.90}Sn_{0.10})O_3$ piezoelectric ceramics with different sintering additives [73, 77–79, 144–146].

Ceramic systems	Sintering temperature (°C)	d_{33} (pC/N)	k_p (%)	Phase structure
$(Ba_{0.85}Ca_{0.15})(Ti_{0.90}Zr_{0.10})O_3$	1450–1500	620	—	R–T
$(Ba_{0.85}Ca_{0.15})(Ti_{0.90}Zr_{0.10})O_3$–6 mol% LiF	1350	380	39.5	R–T
$(Ba_{0.85}Ca_{0.15})(Ti_{0.90}Zr_{0.10})O_3$–0.04 wt% CuO	1350	510	45.0	R–T
$(Ba_{0.85}Ca_{0.15})(Ti_{0.90}Zr_{0.10})O_3$–0.5 mol% CuO	1400	403	44.6	—
$(Ba_{0.95}Ca_{0.05})(Ti_{0.90}Sn_{0.10})O_3$	1480	~630	—	R–PC–O
$(Ba_{0.95}Ca_{0.05})(Ti_{0.90}Sn_{0.10})O_3$–3 mol% Li_2CO_3	1300	485	39.0	R–T
$(Ba_{0.95}Ca_{0.05})(Ti_{0.90}Sn_{0.10})O_3$–4 mol% Li_2O	1300	457	25.2	R–T
$(Ba_{0.95}Ca_{0.05})(Ti_{0.90}Sn_{0.10})O_3$–5 mol% LiF	1250	510	40.9	R–T
$(Ba_{0.95}Ca_{0.05})(Ti_{0.90}Sn_{0.10})O_3$–2 mol% CuO	1250	683	55.0	R–O–T

$ZnO–B_2O_3–Li_2O–Nb_2O_5–Co_2O_3$, or BaO : CuO(0.8 : 1)–$Bi_2O_3$–$B_2O_3$ glass [133, 150–154]. However, glass-mixed $BaTiO_3$ did not yield the desired dielectric or piezoelectric properties due to the presence of the grain-boundary phase. It is necessary to tailor the glass composition to promote densification with limited grain growth and to reduce the volume fraction of second phases as well as porosity. The secondary glass phase in the $BaTiO_3$ matrix is fatal for its piezoelectric properties. Another important factor is to control diffusion into $BaTiO_3$ lattice. If the cations present in the glass substitute into the $BaTiO_3$ lattice, the glass plays a dual role in flux and modifier. The composition and the interaction between the flux and matrix are very important in characterizing, understanding, and controlling the dielectric and piezoelectric properties [155].

5.6 Mechanical Property

The mechanical property of ferroelectric materials was an interesting research topic receiving attention for a long time [156]. For ferroelectric materials, a substantial alternating electric field can induce domain switching, and the accompanied strains could lead to fatigue cracking and subsequently degrade the dielectric, ferroelectric, and mechanical properties of the materials. The mechanically related problems in ferroelectric ceramics arise from internal stresses generated by phase transitions and the reorientation of ferroelectric domains that are subjected to external loads. Consequently, the electric fields might affect the mechanical properties of the material. Accordingly, understanding the fracture behavior of the ferroelectric ceramics is the key to improve reliability and to assess the lifetime of the dielectric and piezoelectric devices. Since $BaTiO_3$ is a classic ferroelectric material with the typical brittle feature of ceramics, its single crystals and polycrystalline ceramics were used as the model

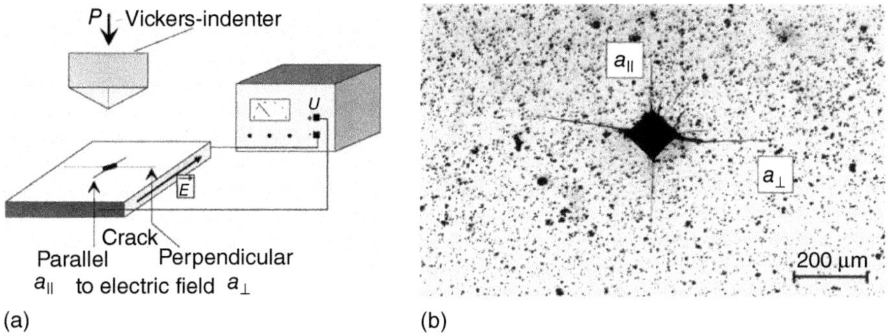

Figure 5.16 (a) Experimental setup of the Vickers indentation method for studying the crack growth under an applied electric field. (b) A typical Vickers indentation with longer crack normal to the applied electric field. Source: (a) Reprinted from Schneider and Heyer [158] with permission from Elsevier; (b) Schneider and Heyer [158].

material for the relevant studies to understand the fracture behavior and fatigue strength of ferroelectric brittle materials under alternating electric fields [157–160].

By using compact tension (CT) specimens standardized for fracture mechanics studies, Meschke et al. investigated the fracture resistance curve of $BaTiO_3$ ceramics with different grain sizes [157]. They found that the crack resistance increase with increasing crack length, resulting in an increase of fracture toughness from 0.5–0.7 MPa $\sqrt{m}$ to 0.7–1.2 MPa $\sqrt{m}$, depending on the grain sizes. The main toughening mechanism was attributed to the ferroelastic domain switching, leading to the development of a process zone around the crack. A simple and convenient Vickers indentation method was used to study the crack growth of ferroelectric ceramics and single crystals by several groups [158, 159]. For example, Schneider and Heyer investigated the crack growth in the $BaTiO_3$ ceramic under applied electric fields up to four times the coercive field by using the Vickers indentation technique and found that cracks perpendicular to the electric field were always longer than cracks parallel to it, which can be explained by the stress-induced ferroelastic domain switching mechanism [158]. The Vickers indentation method also enables the observation of domain switching and electrically induced fatigue cracking, as shown in Figure 5.16. Such a study conducted in $BaTiO_3$ single crystals by using atomic force microscopy (AFM) revealed the interaction between the crack tip and domain switching [160].

The mechanical properties of the $BaTiO_3$-based ceramics with high piezoelectric properties were also investigated by some researchers [156, 161–163]. For example, the crack growth resistance behavior of the lead-free $0.615Ba(Zr_{0.2}Ti_{0.8})O_3$–0.385 $(Ba_{0.7}Ca_{0.3})TiO_3$ (BZT–38.5BCT) ceramic was characterized using compact-tension specimens from 25 to 60 °C. Compared with similar measurements of soft PZT, the fracture toughness of the BZT–38.5BCT is 37% lower at room temperature and 45% lower at 60 °C. It was found that the maximum fracture toughness decrease at elevated temperature, which was related to the temperature-dependent elastic and ferroelastic properties [161]. Srinivas et al. investigated the mechanical properties of $0.5Ba(Zr_{0.2}Ti_{0.8})O_3$–$0.5(Ba_{0.7}Ca_{0.3})TiO_3$ ceramic, and found that Young's modulus, hardness, and fracture toughness are comparable with those of PZT [162]. Multilayer actuators are composed of inactive electrodes and active ferroelectric layers to reduce the driving voltage as compared with monolithic actuators. There is a substantial crack driving force caused by the strain incompatibility between active and inactive layers [164]. Hence, the information about the residual stress and fracture toughness as a function of poling direction and subcritical crack growth is important.

The mechanical properties include electrical fatigue behavior for piezoelectric materials. There have been a few studies investigating the fatigue of BZT–50BCT ceramics. For bipolar fatigue, BZT–50BCT ceramic shows reduction in remanent polarization, maximum polarization, relative permittivity as a function of bias-field and d_{33} with increasing cycles, asymmetry of the strain hysteresis, and increase in dielectric losses as a function of bias-field [165, 166]. For unipolar fatigue, BZT–50BCT ceramics show a decrease of 30% in the maximum polarization and of 9% in the maximum strain after 10^3 unipolar cycles. The bipolar polarization, bipolar strain loops, and small-signal properties as a function of bias-field became asymmetric after unipolar cycling. An internal positive bias-field was developed after 5×10^6 cycles. After $\sim10^7$ unipolar cycles, BZT–50BCT ceramic shows decrease in the remanent polarization about 12% and d_{33}^* about 13% [166]. Fatigued ceramics have a reduction in the electromechanical properties owing to a reduction in irreversible domain wall movement since intrinsic contributions remain unaffected [166]. The properties of the fatigued ceramics recovered almost totally after thermal annealing at 400 °C [165, 166]. It was proposed that charge agglomeration and concomitant domain wall pinning account for the unipolar fatigue of all the materials, although the orientation of defect dipoles under cycling may also have contributions [165, 166]. Overall, it can be stated that BZT–BCT ceramics seem quite resistant to fatigue. It has high bipolar fatigue resistance as compared with lead-containing and other lead-free materials. Its unipolar fatigue resistance remains also high as compared with BNT-based compositions, although the degradation of properties under unipolar cycling is slightly higher than that in KNN-based materials and commercial soft PZT [165, 166].

5.7 Summary and Outlook

$BaTiO_3$, whose ferroelectric nature was discovered in the middle of the 1940s, has an ABO_3 perovskite structure and undergoes sequential phase transitions from R, O, T, C, to H phase with increasing temperature. $BaTiO_3$ ceramics suffer from the low piezoelectric properties ($d_{33} \leq 190$ pC/N), high sintering temperature ($T_s > 1300$ °C), and low T_C ($T_C \sim 130$ °C) as lead-free piezoelectric materials. Nevertheless, the studies by chemical modification and advanced sintering processing raised the d_{33} to a high level, and the extremely high piezoelectricity achieved in the $(Ba,Ca)(Ti,Zr)O_3$ ceramic reignited the scientific interest in $BaTiO_3$ system. After that, similar $(Ba,Ca)(Ti,Sn)O_3$ and $(Ba,Ca)(Ti,Hf)O_3$ systems were developed. In addition, the relevant studies about the phase structure and domain engineering in $BaTiO_3$-based ceramics enriched the knowledge about the mechanisms for enhanced piezoelectricity. However, if compared with KNN-based and BNT-based lead-free piezoelectrics, the $BaTiO_3$-based system has a clear disadvantage of low T_C, which limits its applications, especially as actuators. Additionally, the sintering temperature of the $BaTiO_3$-based ceramics is also higher than those two systems. Adding sintering aids is an effective way to lower sintering temperature but at the expense of d_{33} and T_C. It seems to be a great challenge to realize a combination of high d_{33} and high T_C as well as low sintering temperature in a material. In general, $BaTiO_3$-based piezoelectric ceramics may be more suitable for applications such as sensors and transducers at low temperatures. It may be an important research direction to increase T_C of the $BaTiO_3$-based ceramics.

References

1 Coursey, P.R. and Brand, K.G. (1946). Dielectric constants of some titanates. *Nature* 157: 297–298.

2 Randall, C.A., Newnham, R.E., and Cross, L.E. (2004). *History of the First Ferroelectric Oxide, $BaTiO_3$*, 1–11. University Park, PA: Materials Research Institute, The Pennsylvania State University.

3 Acosta, M., Novak, N., Rojas, V. et al. (2017). $BaTiO_3$-based piezoelectrics: fundamentals, current status, and perspectives. *Applied Physics Reviews* 4: 041305.

4 Ogawa, T. and Waku, S. (1947). On barium titanate ceramics. *Busseiron Kenkyu* 6 (1): 27.

5 Von Hippel, A., Breckenridge, R.G., Chesley, F.G. et al. (1946). High dielectric constant ceramics. *Industrial and Engineering Chemistry* 38 (11): 1097–1109.

6 Wul, B. and Goldman, J.M. (1946). Ferroelectric switching in $BaTiO_3$ ceramics. *Comptes rendus de l'Académie des sciences de l'URSS* 51: 21.

7 Vul, B.M. and Vereshchagin, L.F. (1945). The dielectric permittivity as a function of pressure. *Doklady Akademii Nauk SSSR* 48: 662–665.

8 Nakamura, E., Mitsui, T., and Furuichi, J. (1963). A note on the classification of ferroelectrics. *Journal of the Physical Society of Japan* 18 (10): 1477–1481.

9 Jona, F. and Shirane, G. (1993). *Ferroelectric Crystals*. Dover Publications.

10 Wainer, E. (1946). High titania dielectrics. *Transactions of the Electrochemical Society* 89 (1): 331–356.

11 Von Hippel, A. (1950). Ferroelectricity, domain structure, and phase transitions of barium titanate. *Reviews of Modern Physics* 22 (3): 221.

12 Megaw, H.D. (1945). Crystal structure of barium titanate. *Nature* 155: 484–485.

13 Kay, H.F. and Vousden, P. (1949). XCV. Symmetry changes in barium titanate at low temperatures and their relation to its ferroelectric properties. *The London, Edinburgh, and Dublin Philosophical Magazine and Journal of Science* 40 (309): 1019–1040.

14 Miyake, S. and Ueda, R. (1946). On polymorphic change of $BaTiO_3$. *Journal of the Physical Society of Japan* 1 (1): 32–33.

15 Gray, R.B. (1949). Transducer and method of making the same. US Patent No. 2 486 560, filed 20 September 1946 and issued 1 November 1949.

16 Cherry, W.L. Jr., and Adler, R. (1947). Piezoelectric effect in polycrystalline barium titanate. *Physical Review* 72 (10): 981–982.

17 Roberts, S. (1947). Dielectric and piezoelectric properties of barium titanate. *Physical Review* 71 (12): 890.

18 Mason, W.P. (1948). Electrostrictive effect in barium titanate ceramics. *Physical Review* 74 (9): 1134–1147.

19 Roberts, S. (1950). Dielectric properties of lead zirconate and barium-lead zirconate. *Journal of the American Ceramic Society* 33 (2): 63–66.

20 Dranetz, A.E., Howatt, G.N., and Crownover, J.W. (1949). Barium titanates as circuit elements: I, II. *Tele-Tech* 8 (4): 29–31.

21 Mason, W.P. and Matthias, B.T. (1948). Theoretical model for explaining the ferroelectric effect in barium titanate. *Physical Review* 74 (11): 1622–1636.

22 Devonshire, A.F. (1949). Theory of barium titanate – Part I. *The London, Edinburgh, and Dublin Philosophical Magazine and Journal of Science* 40: 1040–1063.

23 Devonshire, A.F. (1951). CIX. Theory of barium titanate – Part II. *The London, Edinburgh, and Dublin Philosophical Magazine and Journal of Science* 42 (333): 1065–1079.

24 Remeika, J.P. and Jackson, W.M. (1954). A method for growing barium titanate single crystals. *Journal of the American Chemical Society* 76 (3): 940–941.

25 Shirane, G., Danner, H., and Pepinsky, R. (1957). Neutron diffraction study of orthorhombic $BaTiO_3$. *Physical Review* 105 (3): 856–860.

26 Cross, L.E. and Newnham, R.E. (1987). History of ferroelectrics. *Ceramics and Civilization* 3: 289–305.

27 Rase, D.E. and Roy, R. (1955). Phase equilibria in the system BaO–TiO_2. *Journal of the American Ceramic Society* 38 (3): 102–113.

28 Egerton, L. and Koonce, S.E. (1955). Effect of firing cycle on structure and some dielectric and piezoelectric properties of barium titanate ceramics. *Journal of the American Ceramic Society* 38 (11): 412–418.

29 McQuarrie, M. and Behnke, F.W. (1954). Structural and dielectric studies in the system $(Ba,Ca)(Ti,Zr)O_3$. *Journal of the American Ceramic Society* 37 (11): 539–543.

30 Coffeen, W.W. (1953). Dielectric bodies in the quaternary system $BaTiO_3$–$BaSnO_3$–$SrSnO_3$–$CaSnO_3$. *Journal of the American Ceramic Society* 36 (7): 215–221.

31 Fousek, J. (1991). Ferroelectricity: remarks on historical aspects and present trends. *Ferroelectrics* 113 (1): 3–20.

32 Newnham, R.E. and Cross, L.E. (2005). Ferroelectricity: the foundation of a field from form to function. *Materials Research Society Bulletin* 30 (11): 845–848.

33 Blinc, R. (1987). The soft mode concept and the history of ferroelectricity. *Ferroelectrics* 74 (1): 301–303.

34 Liu, W.F. and Ren, X.B. (2009). Large piezoelectric effect in Pb-free ceramics. *Physical Review Letters* 103 (25): 257602.

35 Gao, J.H., Ke, X.Q., Acosta, M. et al. (2018). High piezoelectricity by multiphase coexisting point: barium titanate derivatives. *MRS Bulletin* 43: 595–599.

36 Takahashi, H., Numamoto, Y., Tani, J. et al. (2006). Lead-free barium titanate ceramics with large piezoelectric constant fabricated by microwave sintering. *Japanese Journal of Applied Physics* 45 (1): 30–32.

37 Takahashi, H., Numamoto, Y., Tani, J. et al. (2006). Piezoelectric properties of $BaTiO_3$ ceramics with high performance fabricated by microwave sintering. *Japanese Journal of Applied Physics* 45 (9S): 7405–7408.

38 Takahashi, H., Numamoto, Y., Tani, J. et al. (2007). Domain properties of high-performance barium titanate ceramics. *Japanese Journal of Applied Physics* 46 (10S): 7044–7047.

39 Shen, Z.Y. and Li, J.F. (2010). Enhancement of piezoelectric constant d_{33} in $BaTiO_3$ ceramics due to nano-domain structure. *Journal of the Ceramic Society of Japan* 118 (1382): 940–943.

40 Zhang, J.L., Ji, P.F., Wu, Y.Q. et al. (2014). Strong piezoelectricity exhibited by large-grained $BaTiO_3$ ceramics. *Applied Physics Letters* 104 (22): 222909.

41 Karaki, T., Yan, K., and Adachi, M. (2007). Barium titanate piezoelectric ceramics manufactured by two-step sintering. *Japanese Journal of Applied Physics* 46 (10B): 7035–7038.

42 Karaki, T., Yan, K., and Adachi, M. (2008). Subgrain microstructure in high-performance $BaTiO_3$ piezoelectric ceramics. *Applied Physics Express* 1 (11): 111402.

43 Huan, Y., Wang, X.H., Fang, J. et al. (2014). Grain size effect on piezoelectric and ferroelectric properties of $BaTiO_3$ ceramics. *Journal of the European Ceramic Society* 34 (5): 1445–1448.

44 Wada, S., Takeda, K., Muraishi, T. et al. (2007). Preparation of [110] grain oriented barium titanate ceramics by templated grain growth method and their piezoelectric properties. *Japanese Journal of Applied Physics* 46: 7039–7043.

45 Ma, N., Zhang, B.P., Yang, W.G. et al. (2012). Phase structure and nano-domain in high performance of $BaTiO_3$ piezoelectric ceramics. *Journal of the European Ceramic Society* 32 (5): 1059–1066.

46 Shao, S.F., Zhang, J.L., Zhang, Z. et al. (2008). High piezoelectric properties and domain configuration in $BaTiO_3$ ceramics obtained through the solid-state reaction route. *Journal of Physics D: Applied Physics* 41 (12): 125408.

47 Zheng, P., Zhang, J.L., Tan, Y.Q. et al. (2012). Grain-size effects on dielectric and piezoelectric properties of poled $BaTiO_3$ ceramics. *Acta Materialia* 60 (13–14): 5022–5030.

48 Hoshina, T., Kigoshi, Y., Hatta, S. et al. (2009). Domain contribution to dielectric properties of fine-grained $BaTiO_3$ ceramics. *Japanese Journal of Applied Physics* 48 (9): 09KC01.

49 Kimura, T., Dong, Q., Yin, S. et al. Synthesis and piezoelectric properties of Li-doped $BaTiO_3$ by a solvothermal approach. *Journal of the European Ceramic Society* 33: 1009–1015.

50 Yang, W.G., Zhang, B.P., Ma, N. et al. (2012). High piezoelectric properties of $BaTiO_3$–xLiF ceramics sintered at low temperatures. *Journal of the European Ceramic Society* 32 (4): 899–904.

51 Zhu, L.F., Zhang, B.P., Yang, W.G. et al. (2013). High piezoelectric properties of (Ba,Ca)TiO_3–0.04LiF ceramics sintered at a low temperature. *Journal of Electroceramics* 30: 24–29.

52 Chen, M.L., Xu, Z.J., Chu, R.Q. et al. Y_2O_3-modified $Ba(Ti_{0.96}Sn_{0.04})O_3$ ceramics with improved piezoelectricity and raised Curie temperature. *Materials Research Bulletin* 59: 305–310.

53 Saifi, M., Dubois, B., Vogel, E.M. et al. (1986). Growth of tetragonal $BaTiO_3$ single crystal fibers. *Journal of Materials Research* 1: 452–456.

54 Yu, Z., Guo, R.Y., and Bhalla, A.S. (2000). Dielectric behavior of $Ba(Ti_{1-x}Zr_x)O_3$ single crystals. *Journal of Applied Physics* 88: 410–415.

55 Yu, Z., Guo, R.Y., and Bhalla, A. (2001). Growth of $Ba(Ti_{1-x}Zr_x)O_3$ single crystals by the laser heated pedestal growth technique. *Journal of Crystal Growth* 233: 460–465.

56 Yu, Z., Guo, R.Y., and Bhalla, A.S. (2000). Orientation dependence of the ferroelectric and piezoelectric behavior of $Ba(Ti_{1-x}Zr_x)O_3$ single crystals. *Applied Physics Letters* 77: 1535–1537.

57 Park, S.E., Wada, S., Cross, L.E. et al. (1999). Crystallographically engineered $BaTiO_3$ single crystals for high-performance piezoelectrics. *Journal of Applied Physics* 86: 2746–2750.

58 Wada, S., Yako, K., Kakemoto, H. et al. (2005). Enhanced piezoelectric properties of barium titanate single crystals with different engineered-domain sizes. *Journal of Applied Physics* 98 (1): 014109.

59 Ren, X.B. (2004). Large electric-field-induced strain in ferroelectric crystals by point-defect-mediated reversible domain switching. *Nature Materials* 3: 91–94.

60 Zheng, T., Wu, J.G., Xiao, D.Q., and Zhu, J.G. (2018). Recent development in lead-free perovskite piezoelectric bulk. *Progress in Materials Science* 98: 552–624.

61 Keeble, D.S., Benabdallah, F., Thomas, P.A. et al. (2013). Revised structural phase diagram of $(Ba_{0.7}Ca_{0.3}TiO_3)$–$(BaZr_{0.2}Ti_{0.8}O_3)$. *Applied Physics Letters* 102: 092903.

62 Brajesh, K., Tanwar, K., Abebe, M. et al. (2015). Relaxor ferroelectricity and electric-field-driven structural transformation in the giant lead-free piezoelectric (Ba,Ca)(Ti,Zr)O_3. *Physical Review* 92: 224112.

63 Zhang, L., Zhang, M., Wang, L. et al. (2014). Phase transitions and the piezoelectricity around morphotropic phase boundary in $Ba(Zr_{0.2}Ti_{0.8})O_3$–$x(Ba_{0.7}Ca_{0.3})TiO_3$ lead-free solid solution. *Applied Physics Letters* 105: 162908.

64 Haugen, A.B., Forrester, J.S., Damjanovic, D. et al. (2013). Structure and phase transitions in $0.5(Ba_{0.7}Ca_{0.3}TiO_3)$–$0.5(BaZr_{0.2}Ti_{0.8}O_3)$ from −100 °C to 150 °C. *Journal of Applied Physics* 113: 014103.

65 Damjanovic, D., Biancoli, A., Batooli, L. et al. (2012). Elastic, dielectric, and piezoelectric anomalies and Raman spectroscopy of $0.5Ba(Ti_{0.8}Zr_{0.2})O_3$–$0.5(Ba_{0.7}Ca_{0.3})TiO_3$. *Applied Physics Letters* 100: 192907.

66 Chandrakala, E., Praveen, J.P., Hazra, B.K. et al. (2016). Effect of sintering temperature on structural, dielectric, piezoelectric and ferroelectric properties of sol–gel derived BZT–BCT ceramics. *Ceramics International* 42: 4964–4977.

67 Wu, J.G., Xiao, D.Q., Wu, W.J. et al. (2011). Effect of dwell time during sintering on piezoelectric properties of $(Ba_{0.85}Ca_{0.15})(Ti_{0.90}Zr_{0.10})O_3$ lead-free ceramics. *Journal of Alloys and Compounds* 509: L359–L361.

68 Orlik, K., Lorgouilloux, Y., Marchet, P. et al. (2020). Influence of microwave sintering on electrical properties of BCTZ lead free piezoelectric ceramics. *Journal of the European Ceramic Society* 40: 1212–1216.

69 Sun, Z.X., Pu, Y.P., Dong, Z.J. et al. (2014). Impact of fast microwave sintering on the grain growth, dielectric relaxation and piezoelectric properties on $Ba_{0.18}Ca_{0.02}Ti_{0.09}Zr_{0.10}O_3$ lead-free ceramics prepared by different methods. *Materials Science and Engineering B* 185: 114–122.

70 Bai, W.F., Chen, D.Q., Li, P. et al. (2016). Enhanced electromechanical properties in (001)-textured $(Ba_{0.85}Ca_{0.15})(Zr_{0.1}Ti_{0.9})O_3$ lead-free piezoceramics. *Ceramics International* 42: 3429–3436.

71 Xue, D.Z., Zhou, Y.M., Bao, H.X. et al. (2011). Large piezoelectric effect in Pb-free $Ba(Ti,Sn)O_3$–$x(Ba,Ca)TiO_3$ ceramics. *Applied Physics Letters* 99 (12): 122901.

72 Zhu, L.F., Zhang, B.P., Zhao, X.K. et al. (2013). Phase transition and high piezoelectricity in $(Ba,Ca)(Ti_{1-x}Sn_x)O_3$ lead-free ceramics. *Applied Physics Letters* 103 (7): 072905.

73 Zhou, P.F., Zhang, B.P., Zhao, L. et al. (2013). High piezoelectricity due to multiphase coexistence in low-temperature sintered $(Ba,Ca)(Ti,Sn)O_3$–CuO_x ceramics. *Applied Physics Letters* 103: 172904.

74 Zhu, L.F., Zhang, B.P., Zhao, X.K. et al. (2013). Enhanced piezoelectric properties of $(Ba_{1-x}Ca_x)(Ti_{0.92}Sn_{0.08})O_3$ lead-free ceramics. *Journal of the American Ceramic Society* 96: 241–245.

75 Zhu, L.F., Zhang, B.P., Zhao, L. et al. (2014). High piezoelectricity of $BaTiO_3$–$CaTiO_3$–$BaSnO_3$ lead-free ceramics. *Journal of Materials Chemistry C* 2: 4764–4771.

76 Zhu, L.F., Zhang, B.P., Zhao, L. et al. (2016). Large piezoelectric effect of $(Ba,Ca)TiO_3$–$xBa(Sn,Ti)O_3$ lead-free ceramics. *Journal of the European Ceramic Society* 36: 1017–1024.

77 Zhao, L., Zhang, B.P., Zhou, P.F. et al. (2016). Piezoelectric and ferroelectric properties of $(Ba,Ca)(Ti,Sn)O_3$ lead-free ceramics sintered with Li_2O additives: analysis of point defects and phase structures. *Ceramics International* 42: 1086–1093.

78 Zhao, L., Zhang, B.P., Zhou, P.F. et al. (2015). Effect of Li_2O addition on sintering and piezoelectric properties of $(Ba,Ca)(Ti,Sn)O_3$ lead-free piezoceramics. *Journal of the European Ceramic Society* 35 (2): 533–540.

79 Zhao, L., Zhang, B.P., Zhou, P.F. et al. (2014). Phase structure and property evaluation of (Ba,Ca)(Ti,Sn)O_3 sintered with Li_2CO_3 addition at low temperature. *Journal of the American Ceramic Society* 97 (7): 2164–2169.

80 Zhao, L., Zhang, B.P., Wang, W.Y. et al. (2016). Effect of B_2O_3 on phase structure and electrical properties of CuO-modified (Ba,Ca)(Ti,Sn)O_3 lead-free piezoceramics sintered at a low-temperature. *Ceramics International* 42: 7366–7372.

81 Wu, B., Ma, J., Wu, W.J. et al. (2017). Structure and electrical properties of MnO doped $(Ba_{0.96}Ca_{0.04})(Ti_{0.92}Sn_{0.08})O_3$ lead free ceramics. *Journal of Materials Science: Materials in Electronics* 28: 2358–2365.

82 Chen, Z.H., Li, Z.W., Ding, J.N. et al. Piezoelectric and ferroelectric properties of $Ba_{0.9}Ca_{0.1}Ti_{0.9}Sn_{0.1}O_3$ lead-free ceramics with La_2O_3 addition. *Journal of Alloys and Compounds* 704: 193–196.

83 Mitsui, T. and Westphal, W.B. (1961). Dielectric and X-ray studies of $Ca_xBa_{1-x}TiO_3$ and $Ca_xSr_{1-x}TiO_3$. *Physical Review* 124: 1354–1359.

84 Yao, Y.G., Zhou, C., Lv, D.C. et al. (2012). Large piezoelectricity and dielectric permittivity in $BaTiO_3$–$x$$BaSnO_3$ system: the role of phase coexisting. *Europhysics Letters* 98: 27008.

85 Payne, W.H. and Tennery, V.J. (1965). Dielectric and structural investigations of system $BaTiO_3$–$BaHfO_3$. *Journal of the American Ceramic Society* 48: 413–417.

86 Zhou, C., Liu, W.F., Xue, D.Z. et al. (2012). Triple-point-type morphotropic phase boundary based large piezoelectric Pb-free material: $Ba(Ti_{0.8}Hf_{0.2})O_3$–$(Ba_{0.7}Ca_{0.3})TiO_3$. *Applied Physics Letters* 100: 222910.

87 Wang, D.L., Jiang, Z.H., Yang, B. et al. (2014). Phase diagram and enhanced piezoelectric response of lead-free $BaTiO_3$–$CaTiO_3$–$BaHfO_3$ system. *Journal of the American Ceramic Society* 97: 3244–3251.

88 Zhao, C.L., Wu, W.J., Wang, H. et al. (2016). Site engineering and polarization characteristics in $(Ba_{1-y}Ca_y)(Ti_{1-x}Hf_x)O_3$ lead-free ceramics. *Journal of Applied Physics* 119: 024108.

89 Zhao, C.L., Wang, H., Xiong, J. et al. (2016). Composition-driven phase boundary and electrical properties in $(Ba_{0.94}Ca_{0.06})(Ti_{1-x}M_x)O_3$ (M = Sn, Hf, Zr) lead-free ceramics. *Dalton Transactions* 45: 6466–6480.

90 Wang, D.L., Jiang, Z.H., Yang, B. et al. (2014). Phase transition behavior and high piezoelectric properties in lead-free $BaTiO_3$–$CaTiO_3$–$BaHfO_3$ ceramics. *Journal of Materials Science* 49: 62–69.

91 Dai, Z.H., Li, D.Y., Xie, J.L. et al. (2019). Phase diagram determination of large piezoelectric response in BHT–BCT ceramics. *Functional Materials Letters* 12: 1950070.

92 Zhang, Y., Glaum, J., Groh, C. et al. (2014). Correlation between piezoelectric properties and phase coexistence in (Ba,Ca)(Ti,Zr)O_3 ceramics. *Journal of the American Ceramic Society* 97 (9): 2885–2891.

93 Wang, W., Li, W.L., Xu, D. et al. (2014). Phase transitions in $(1-x)BaZr_{0.2}Ti_{0.8}O_3$–$x$$Ba_{0.7}Ca_{0.3}TiO_3$ powders and ceramic pellets. *Ceramics International* 40 (3): 3933–3937.

94 Acosta, M., Novak, N., Jo, W. et al. (2014). Relationship between electromechanical properties and phase diagram in the $Ba(Zr_{0.2}Ti_{0.8})O_3$–$x(Ba_{0.7}Ca_{0.3})TiO_3$ lead-free piezoceramic. *Acta Materialia* 80: 48–55.

95 Wada, S. and Tsurumi, T. (2004). Enhanced piezoelectricity of barium titanate single crystals with engineered domain configuration. *British Ceramic Transactions* 103 (2): 93–96.

96 Lin, D., Zhang, S.J., Li, Z.R. et al. (2011). Domain size engineering in tetragonal $Pb(In_{1/2}Nb_{1/2})O_3$–$Pb(Mg_{1/3}Nb_{2/3})O_3$–$PbTiO_3$ crystals. *Journal of Applied Physics* 110 (8): 084110.

97 Gao, J.H., Xue, D.H., Wang, Y. et al. (2011). Microstructure basis for strong piezoelectricity in Pb-free $Ba(Zr_{0.2}Ti_{0.8})O_3$–$(Ba_{0.7}Ca_{0.3})TiO_3$ ceramics. *Applied Physics Letters* 99: 092901.

98 Zakhozheva, M., Schmitt, L.A., Acosta, M. et al. (2014). In situ electric field induced domain evolution in $Ba(Zr_{0.2}Ti_{0.8})O_3$–$0.3(Ba_{0.7}Ca_{0.3})TiO_3$ ferroelectrics. *Applied Physics Letters* 105 (11): 112904.

99 Zakhozheva, M., Schmitt, L.A., Acosta, M. et al. (2015). Wide compositional range in situ electric field investigations on lead-free $Ba(Zr_{0.2}Ti_{0.8})O_3$–$x(Ba_{0.7}Ca_{0.3})TiO_3$ piezoceramic. *Physical Review Applied* 3 (6): 064018.

100 Turygin, A.P., Neradovskiy, M.M., Naumova, N.A. et al. (2015). Domain structures and local switching in lead-free piezoceramics $Ba_{0.85}Ca_{0.15}Ti_{0.90}Zr_{0.10}O_3$. *Journal of Applied Physics* 118 (7): 072002.

101 Guo, H., Zhou, C., Ren, X. et al. (2014). Unique single-domain state in a polycrystalline ferroelectric ceramic. *Physical Review B* 89 (10): 100104(R).

102 Lu, S.B., Xu, Z.K., Su, S. et al. (2014). Temperature driven nano-domain evolution in lead-free $Ba(Zr_{0.2}Ti_{0.8})O_3$–$50(Ba_{0.7}Ca_{0.3})TiO_3$ piezoceramics. *Applied Physics Letters* 105 (3): 032903.

103 Acosta, M., Novak, N., Rossetti, G.A. et al. (2015). Mechanisms of electromechanical response in $(1-x)Ba(Zr_{0.2}Ti_{0.8})O_3$–$x(Ba_{0.7}Ca_{0.3})TiO_3$ ceramics. *Applied Physics Letters* 107 (14): 142906.

104 Kalyani, A.K., Khatua, D.K., Loukya, B. et al. (2015). Metastable monoclinic and orthorhombic phases and electric field induced irreversible phase transformation at room temperature in the lead-free classical ferroelectric $BaTiO_3$. *Physical Review B* 91 (10): 104104.

105 Sendova, M. and Hosterman, B.D. (2014). Raman spectroscopic study of the size-dependent order parameter of barium titanate. *Journal of Applied Physics* 115 (21): 214104.

106 Randall, C.A., Kim, N., Kucera, J.P. et al. (1998). Intrinsic and extrinsic size effects in fine-grained morphotropic-phase-boundary lead zirconate titanate ceramics. *Journal of the American Ceramic Society* 81 (3): 677–688.

107 Hao, J.G., Bai, W.F., Li, W. et al. (2012). Correlation between the microstructure and electrical properties in high-performance $(Ba_{0.85}Ca_{0.15})(Zr_{0.1}Ti_{0.9})O_3$ lead-free piezoelectric ceramics. *Journal of the American Ceramic Society* 95 (6): 1998–2006.

108 Tan, Y.Q., Zhang, J.L., Wu, Y.Q. et al. (2015). Unfolding grain size effects in barium titanate ferroelectric ceramics. *Scientific Reports* 5 (1): 9953.

109 Ghosh, D., Sakata, A., Carter, J. et al. (2014). Domain wall displacement is the origin of superior permittivity and piezoelectricity in $BaTiO_3$ at intermediate grain sizes. *Advanced Functional Materials* 24 (7): 885–896.

110 Tan, Y.Q., Zhang, J.L., Wang, C.L. et al. (2015). Enhancement of electric field-induced strain in $BaTiO_3$ ceramics through grain size optimization. *Physica Status Solidi A: Applications and Material Science* 212 (2): 433–438.

111 Uchino, K., Sadanaga, E., and Hirose, T. (1989). Dependence of the crystal structure on particle size in barium titanate. *Journal of the American Ceramic Society* 72 (8): 1555–1558.

112 Tsunekawa, S., Ito, S., Mori, T. et al. (2000). Critical size and anomalous lattice expansion in nanocrystalline $BaTiO_3$ particles. *Physical Review B* 62 (5): 3065–3070.

113 Zhao, Z., Buscaglia, V., Viviani, M. et al. (2004). Grain-size effects on the ferroelectric behavior of dense nanocrystalline $BaTiO_3$ ceramics. *Physical Review B* 70 (2): 024107.

114 Takahashi, H. (2012). Development of lead-free $BaTiO_3$ ceramics possessing enhanced piezoelectric properties. *Electronics and Communications in Japan* 95 (4): 20–26.

115 Dong, L., Stone, D.S., and Lakes, R.S. (2012). Enhanced dielectric and piezoelectric properties of $x$$BaZrO_3$–(1−$x$)$BaTiO_3$ ceramics. *Journal of Applied Physics* 111 (8): 084107.

116 Jones, J.L., Iverson, B.J., and Bowman, K.J. (2007). Texture and anisotropy of polycrystalline piezoelectrics. *Journal of the American Ceramic Society* 90 (8): 2297–2314.

117 Zhou, J.E., Yan, Y.K., Priya, S. et al. (2017). Computational study of textured ferroelectric polycrystals: dielectric and piezoelectric properties of template-matrix composites. *Journal of Applied Physics* 121 (2): 024101.

118 Liu, L. (2015). Progress on the fabrication of lead-free textured piezoelectric ceramics: perspectives over 25 years. *Journal of Materials Science: Materials in Electronics* 26 (7): 4425–4437.

119 Messing, G.L., Poterala, S., Chang, Y.F. et al. (2017). Texture-engineered ceramics – property enhancements through crystallographic tailoring. *Journal of Materials Research* 32 (17): 3219–3241.

120 Roedel, J., Jo, W., Seifert, K.T.P. et al. (2009). Perspective on the development of lead-free piezoceramics. *Journal of the American Ceramic Society* 92 (6): 1153–1177.

121 Messing, G.L., Trolier-McKinstry, S., Sabolsky, E.M. et al. (2004). Templated grain growth of textured piezoelectric ceramics. *Critical Reviews in Solid State and Materials Sciences* 29 (2): 45–96.

122 Leontsev, S.O. and Eitel, R.E. (2010). Progress in engineering high strain lead-free piezoelectric ceramics. *Science and Technology of Advanced Materials* 11 (4): 044302.

123 Kobayashi, E., Ueno, S., Nakashima, K. et al. (2015). Fabrication of (111)-oriented tetragonal $BaTiO_3$ ceramics by an electrophoretic deposition in a high magnetic field. *Transactions of the Materials Research Society of Japan* 40 (3): 223–226.

124 Vriami, D., Damjanovic, D., Vleugels, J. et al. (2015). Textured $BaTiO_3$ by templated grain growth and electrophoretic deposition. *Journal of Materials Science* 50 (24): 7896–7907.

125 Zhao, Z.H., Li, X.L., Dai, Y.J. et al. (2016). Texture development in $Ba_{0.85}Ca_{0.15}Ti_{0.90}Zr_{0.10}O_3$ lead-free ceramics prepared by reactive template grain growth with different Ba and Ca sources. *Ceramics International* 42: 18756.

126 Wada, S. and Pulpan, P. (2010). Domain wall engineering in lead-free piezoelectric materials. *Key Engineering Materials* 421: 13–16.

127 Wada, S., Takeda, K., Muraishi, T. et al. (2008). Domain wall engineering in lead-free piezoelectric grain-oriented ceramics. *Ferroelectrics* 373 (1): 11–21.

128 Valant, M., Suvorov, D., Pullar, R.C. et al. (2006). A mechanism for low-temperature sintering. *Journal of the European Ceramic Society* 26: 2777–2783.

129 Choi, S.Y. and Kang, S.J.L. (2004). Sintering kinetics by structural transition at grain boundaries in barium titanate. *Acta Materialia* 52: 2937–2943.

130 Kim, H.T. and Han, Y.H. (2004). Sintering of nanocrystalline $BaTiO_3$. *Ceramics International* 30: 1719–1723.

131 Polotai, A., Breece, K., Dickey, E. et al. (2005). A novel approach to sintering nanocrystalline barium titanate ceramics. *Journal of the American Ceramic Society* 88: 3008–3012.

132 Wang, X.H., Deng, X.Y., Bai, H.L. et al. (2006). Two-step sintering of ceramics with constant grain-size, II: $BaTiO_3$ and Ni–Cu–Zn ferrite. *Journal of the American Ceramic Society* 89: 438–443.

133 Naghib-Zadeh, H., Glitzky, C., Oesterle, W. et al. (2011). Low temperature sintering of barium titanate based ceramics with high dielectric constant for LTCC applications. *Journal of the American Ceramic Society* 31 (4): 589–596.

134 Gong, H.L., Wang, X.H., Zhang, S.P. et al. (2015). Sintering behavior and reliability characteristics of $BaTiO_3$-based ceramics prepared by different methods. *Journal of Materials Science* 50: 3523–3530.

135 Sonia, Patel, R.K., Kumar, P. et al. (2012). Low temperature synthesis and dielectric, ferroelectric and piezoelectric study of microwave sintered $BaTiO_3$ ceramics. *Ceramics International* 38 (2): 1585–1589.

136 Walker, B.E. (1976). Densification and strength of $BaTiO_3$ with LiF and MgO additives. *American Ceramic Society Bulletin* 55 (3): 274–276.

137 Naghib-zadeh, H., Glitzky, C., Dörfel, I. et al. (2010). Low temperature sintering of barium titanate ceramics assisted by addition of lithium fluoride-containing sintering additives. *Journal of the European Ceramic Society* 30 (1): 81–86.

138 Liu, G., Jiang, Y., and Button, T.W. (2011). Low temperature sintering and dielectric properties of $BaTiO_3$ ceramics incorporating nano-sized powders. *Ferroelectrics* 421: 72–81.

139 Ma, N., Zhang, B.P., and Yang, W.G. (2012). Low-temperature sintering of Li_2O-doped $BaTiO_3$ lead-free piezoelectric ceramics. *Journal of Electroceramics* 28 (4): 275–280.

140 Ma, N., Zhang, B.P., Li, H.T. et al. (2011). Low temperature sintering of Li_2CO_3 doped $BaTiO_3$ lead-free piezoelectric ceramics. *Rare Metal Materials and Engineering* 40: 395–398.

141 Guha, J.P. and Anderson, H.U. (1986). Reaction during sintering of barium titanate with lithium fluoride. *Journal of the American Ceramic Society* 69 (8): C193–C194.

142 Wang, S.-F., Yang, T.C.K., Huebner, W. et al. (2000). Liquid-phase sintering and chemical inhomogeneity in the $BaTiO_3$–$BaCO_3$–LiF system. *Journal of Materials Research* 15 (2): 407–416.

143 Lin, J.N. and Wu, T.B. (1989). Wetting reaction between lithium fluoride and barium titanate. *Journal of the American Ceramic Society* 72 (9): 1709–1712.

144 Tan, C.K.I., Yao, K., and Ma, J. (2013). Effects of LiF on the structure and properties of $Ba_{0.85}Ca_{0.15}Zr_{0.1}Ti_{0.9}O_3$ lead-free piezoelectric ceramics. *International Journal of Applied Ceramic Technology* 10 (4): 701–706.

145 Cui, Y.R., Liu, X.Y., Jiang, M.H. et al. (2012). Lead-free $(Ba_{0.85}Ca_{0.15})(Ti_{0.9}Zr_{0.1})O_3$–$CeO_2$ ceramics with high piezoelectric coefficient obtained by low-temperature sintering. *Ceramics International* 38 (6): 4761–4764.

146 Chen, T., Zhang, T., Wang, G.C. et al. (2012). Effect of CuO on the microstructure and electrical properties of $Ba_{0.85}Ca_{0.15}Ti_{0.90}Zr_{0.10}O_3$ piezoceramics. *Journal of Materials Science* 47 (11): 4612–4619.

147 Matsuo, Y., Fujimura, M., Sasaki, H. et al. (1968). Semiconducting $BaTiO_3$ with additions of Al_2O_3, SiO_2 and TiO_2. *American Ceramic Society Bulletin* 47 (3): 292.

148 Matsuo, Y. and Sasaki, H. (1971). Exaggerated grain growth in liquid-phase sintering of $BaTiO_3$. *Journal of the American Ceramic Society* 54 (9): 471.

149 Burn, I. (1982). Flux-sintered $BaTiO_3$ dielectrics. *Journal of Materials Science* 17 (5): 1398–1408.

150 Hsiang, H., Hsi, C.S., Huang, C.C. et al. (2008). Sintering behavior and dielectric properties of $BaTiO_3$ ceramics with glass addition for internal capacitor of LTCC. *Journal of Alloys and Compounds* 459 (1–2): 307–310.

151 Ling, W.W., Zhang, H.W., Li, Y.X. et al. (2010). Effect of B_2O_3–Bi_2O_3–SiO_2–ZnO glass on the dielectric and magnetic properties of ferroelectric/ferromagnetic composite for low temperature cofired ceramic technology. *Journal of Applied Physics* 107 (9): 09D911.

152 Hsiang, H.I., Hsi, C.S., Huang, C.C. et al. (2009). Low temperature sintering and dielectric properties of $BaTiO_3$ with glass addition. *Materials Chemistry and Physics* 113 (2–3): 658–663.

153 Gong, Y.X., Deng, W., and Zhang, W. (2015). Effect of a BaO–CuO–Bi_2O_3–B_2O_3 glass flux, and its processing on the dielectric properties of $BaTiO_3$. *Ceramics International* 41 (1): 671–680.

154 Chen, M.Y., Juuti, J., and Hsi, C.S. (2015). Dielectric $BaTiO_3$–BBSZ glass ceramic composition with ultra-low sintering temperature. *Journal of the European Ceramic Society* 35 (1): 139–144.

155 Prakash, D., Sharma, B.P., Mohan, T.R.R. et al. (2000). Flux additions in barium titanate: overview and prospects. *Journal of Solid State Chemistry* 155 (1): 86–95.

156 Webber, K.G., Voegler, M., Khansur, N.H. et al. (2017). Review of the mechanical and fracture behavior of perovskite lead-free ferroelectrics for actuator applications. *Smart Materials and Structures* 26 (6): 063001.

157 Meschke, F., Kolleck, A., and Schneider, G.A. (1997). R-curve behaviour of $BaTiO_3$ due to stress-induced ferroelastic domain switching. *Journal of the European Ceramic Society* 17: 1143–1149.

158 Schneider, G.A. and Heyer, V. (1999). Influence of the electric field on Vickers indentation crack growth in $BaTiO_3$. *Journal of the European Ceramic Society* 19: 1299–1306.

159 Kolleck, A., Schneider, G.A., and Meschke, F.A. (2000). R-curve behavior of $BaTiO_3$- and PZT ceramics under the influence of an electric field applied parallel to the crack front. *Acta Materialia* 48: 4099–4113.

160 Wang, F., Su, Y.J., He, J.Y. et al. (2006). In situ AFM observation of domain switching and electrically induced fatigue cracking in $BaTiO_3$ single crystal. *Scripta Materialia* 54: 201–205.

161 Vogler, M., Acosta, M., Brandt, D.R.J. et al. (2015). Temperature-dependent R-curve behavior of the lead-free ferroelectric $0.615Ba(Zr_{0.2}Ti_{0.8})O_3$–0.385 $(Ba_{0.7}Ca_{0.3})TiO_3$ ceramic. *Engineering Fracture Mechanics* 144: 68–77.

162 Srinivas, A., Krishnaiah, R.V., Niranjani, V.L. et al. (2015). Piezoelectric and mechanical properties in lead free $0.5Ba(Zr_{0.2}Ti_{0.8})O_3$–$0.5(Ba_{0.7}Ca_{0.3})TiO_3$ electroceramics. *Ceramics International* 41 (2): 1980–1985.

163 Brandt, D.R.J., Acosta, M., Koruza, J. et al. (2014). Mechanical constitutive behavior and exceptional blocking force of lead-free BZT–*x*BCT piezoceramics. *Journal of Applied Physics* 115 (20): 204107.

164 Lucato, S.D.S.E., Lupascu, D.C., Kamlah, M. et al. (2001). Constraint-induced crack initiation at electrode edges in piezoelectric ceramics. *Acta Materialia* 49 (14): 2751–2759.

165 Zhang, Y., Glaum, J., Ehmke, M.C. et al. (2016). Unipolar fatigue behavior of BCTZ lead-free piezoelectric ceramics. *Journal of the American Ceramic Society* 99 (4): 1287–1293.

166 Rojas, V., Koruza, J., Patterson, E.A. et al. (2017). Influence of composition on the unipolar electric fatigue of $Ba(Zr_{0.2}Ti_{0.8})O_3$–$(Ba_{0.7}Ca_{0.3})TiO_3$ lead-free piezoceramics. *Journal of the American Ceramic Society* 100 (10): 4699–4709.

6

$BiFeO_3$ System

6.1 Introduction

Bismuth ferrite, i.e. $BiFeO_3$, has been known since 1950s [1]. $BiFeO_3$ has been studied intensively as the only single-phase multiferroic at room temperature, owing to its high Curie temperature (T_c = 1103 K) [2] and Néel temperature (T_N = 673 K) [3]. It has another attractive merit in the research fields of ferroics, i.e. lead-free and environmental-friendly feature, making it a potential substitute for commercial lead-based $Pb(Zr,Ti)O_3$ materials. It is further noteworthy that the domain wall conductivity found in $BiFeO_3$ makes it more attractive for nanoelectronic devices [4]. There are, however, problems that hinder it entering actual applications. Four are major ones, i.e. processing difficulty of high phase-purity ceramics, high leakage, large coercive field and difficulty of poling, and tendency of fatigue. To solve the previous problems, there have been many routines and efforts, including modifications of the processing procedure and the chemical compositions (by doping and forming solid solutions) in bulk materials, single crystals, and thin films. These may push it to actual applications eventually after the extensive investigations, taking advantages of the outstanding characteristics of $BiFeO_3$.

In this chapter, we first give a brief introduction to multiferroic materials, and then introduce the multiferroicity and its origins in $BiFeO_3$. This is followed by a discussion of its phase diagrams in both ceramics and films, with an emphasis on the main issues of processing (i.e. avoiding impurities during the fabrication). Electrical properties, including macroscopic dielectric permittivity and leakage current, and microscopic domain wall conductivity are included. We then give a conclusion on $BiFeO_3$-based substitution and solid solution studies. Some initial ideas of prototypical devices and future applications have been given in Section 6.8 of this chapter.

6.2 Brief Introduction to Multiferroic Materials

A multiferroic material couples more than one ferroic order parameter (e.g. ferroelectricity, ferromagnetism, and ferroelasticity) and thus allows multiple degrees of

Lead-Free Piezoelectric Materials, First Edition. Jing-Feng Li.

freedom in device design [5–8]. This unique class of functional materials offers a wide range of applications. The most exciting and widely discussed application is for novel memory media that utilize the coupling between ferroelectric polarization and magnetization [9–14]. This enables the writing, reading, and storing of data in multiple ways. In addition, only a few multiferroics, in either natural existence or artificially synthesized, have been discovered thus far. Therefore, such a phenomenon, which seems to be not so common, contains rich and fascinating fundamental science.

Nickel iodine boracite ($Ni_3B_2O_{13}I$) was discovered as the first multiferroic material in 1966 [15]. Some early history of the development of multiferroics dates back to the 1980s, when several multiferroics were synthesized in laboratories. This was achieved by replacing the d_0 cations with magnetic d_n cations on the B site of a perovskite ferroelectric structure (for example, $Pb_2(Co,W)O_6$ and Pb_2FeTaO_6) [16]. However, these synthesized materials generally have very low Néel and Curie temperatures.

In recent years, interest in studying perovskite-based multiferroics has been driven toward finding materials with high Néel and Curie temperatures and large magnetoelectric effects. One example is the study of hexagonal (space group $P6_3cm$) rare-earth manganese oxides ($RMnO_3$ with R = Ho, Er, Tm, Yb, Lu, and Y) [17], which have been investigated in considerable detail and are known as Type-I multiferroics. Type-I multiferroics are regular ferroelectrics which are also (anti)ferromagnetic. A special single-phase Type-I multiferroic is $BiFeO_3$, with Curie (1103 K) and Néel temperature (673 K) both well above room temperature. Its remnant polarization is as large as $100\,\mu C/cm^2$ [18]. The functional capabilities of $BiFeO_3$ have attracted attention for potential magnetoelectric devices and high-temperature piezoelectrics [19]. Additionally, $BiFeO_3$ displays conductive ferroelectric domain walls, the existence of [4] which, on one hand, has impeded the investigation of bulk $BiFeO_3$ as it inhibits electrical poling [20], but on the other hand has triggered significant research effort into novel domain-wall based electronic applications [21, 22]. With intensive interest, there have been a vast number of investigations on $BiFeO_3$. However, there are many remaining issues, such as the processing and poling of pure bulk $BiFeO_3$, the understanding of electromechanical and ferroelectric–antiferromagnetic coupling, and the influence of domain wall conductivity on the microscopic and macroscopic properties.

In contrast, Type II multiferroics are materials in which one of the ferroic orders (generally ferroelectricity) is a consequence of the other (generally antiferromagnetism). One typical example of this group is the orthorhombic (space group *Pbnm*) $RMnO_3$ series of perovskite manganites, where R = (Gd, Tb, Dy) [23].

In addition, nanostructures, composites, and hybrid inorganic–organic multiferroics have been intensively investigated as these approaches potentially promise desired magnetoelectric or multiferroic behavior, taking advantage of nanoengineering or combining a number of properties that usually are not found together in a single-phase material [11, 14, 24].

6.3 Multiferroicity of $BiFeO_3$

The conventional mechanism of ferroelectricity is cation off-centering and requires formally empty d-orbitals, while the formation of magnetic moments is usually related to partially filled d-orbitals. Therefore, in a single-phase material, the coexistence of ferroelectricity and magnetism is rare, but is possible through the following mechanism: ferroelectricity results from one cation while magnetism results from another cation. This is indeed the origin of the multiferroicity of $BiFeO_3$. $BiFeO_3$ has the perovskite (ABO_3) structure, with Bi^{3+} and Fe^{3+} cations on the A and B site, respectively. $BiFeO_3$ is a multiferroic well above room temperature with ferroelectricity up to 1103 K (result from stereochemical activity of the lone pair of larger Bi cation) and canted G-type antiferromagnetic ordering at 673 K (from smaller magnetic Fe cation). At room temperature, bulk $BiFeO_3$ has a rhombohedral $R3c$ structure (unit cell parameters $a = 3.96$ Å and $\alpha = 89.4°$), while in thin $BiFeO_3$ films there are richer phenomenon, including structure and phase transitions. We will focus on the intrinsic physics of $BiFeO_3$, while in Section 6.4.3 we will include the thin film introduction.

6.3.1 Ferroelectricity

The ferroelectricity of $BiFeO_3$ is due to the spontaneous polarization formed by bismuth (Bi) on the A-site [19], as shown in Figure 6.1. It shows that the polarization is along the (111) directions due to Bi displacements.

The remnant polarization along the polar (111) direction of $BiFeO_3$ crystals is approximately 100 μC/cm^2, which is the largest among known perovskite ferroelectrics [18]. There are eight possible polar directions along the four body diagonals of the pseudo-cubic unit cell of rhombohedral $BiFeO_3$. Therefore, these poles can form eight different domains, and the separation (ferroelectric domain walls) of these domains falls into three types. The domain wall types are labeled according to the angle between the polarization vectors on each side of the wall. These domain wall types are schematically depicted in Figure 6.2 [19]. Figure 6.2a presents a 71° wall, indicating that the polar vectors on each side of such a wall form an angle of approximately 71°. For this type of domain wall, the indices of the polarization vectors on both sides have only one reversed component (e.g. one domain has a [111] orientation and the adjacent one is [$11\bar{1}$]). When the polarization vector indices have two reversed components, it is a 109° wall (Figure 6.2b), and three reversed components correspond to a 180° wall (Figure 6.2c). Typically, 71° domain walls occur on $\{101\}_{pc}$ planes, and 109° walls on $\{100\}_{pc}$ planes (where pc denotes for pseudo-cubic structure) and 180° walls should occur on planes containing the (111) polar vector, as have been reported previously [4, 25]. The domain shape has been reported to range from being wedge-like, lamellar, and irregular in bulk ceramics, observed with atomic force microscopy [26].

6.3.2 Antiferromagnetism and Weak Ferromagnetism

Antiferromagnetism is associated with the B-site iron cation (Fe) [19]. It has been reported that $BiFeO_3$ is G-type antiferromagnetic below a Néel temperature of 673 K

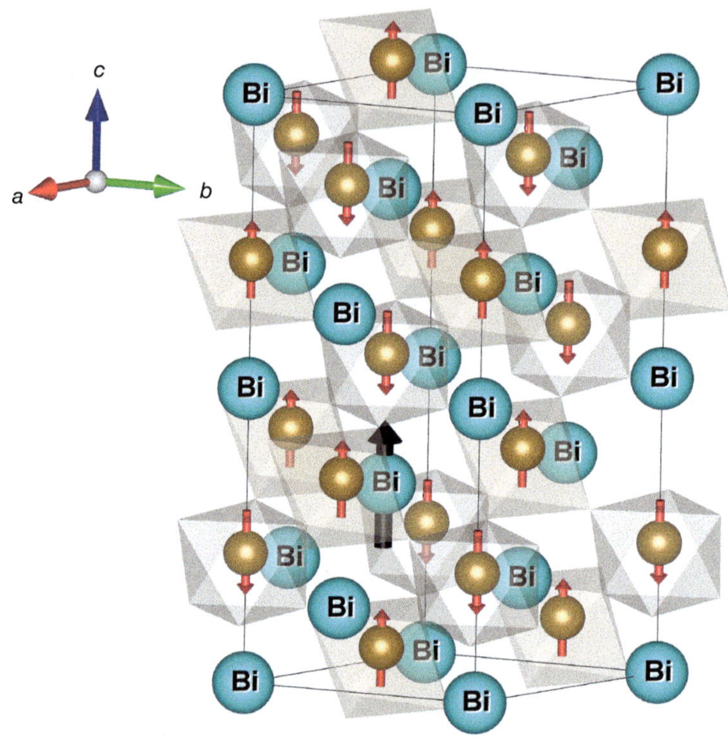

Figure 6.1 The *R*3*c* structure of $BiFeO_3$ represented with the hexagonal unit cell (black lines) and the pseudocubic unit cell (red lines) with the polarization from Bi cations along the [111] direction as marked by a black arrow. Oxygen octahedra surrounding the Fe atoms are also shown.

with magnetic moments of the nearest neighbor Fe aligned antiparallel to each other [3].

Superimposing on the antiferromagnetism below the Néel temperature, T_N, there exists a modulated incommensurate cycloidal spin structure of the antiferromagnetically ordered sublattice with a long periodicity of ~620 Å [27]. The modulation can propagate along three symmetry-equivalent wave vectors $\mathbf{k}_1 = \delta[1\bar{1}0]$, $\mathbf{k}_2 = \delta[10\bar{1}]$, and $\mathbf{k}_3 = \delta[0\bar{1}1]$ (sometimes use τ_1, τ_2, and τ_3), as shown in Figure 6.3, where $\delta \sim 0.0045$ reciprocal lattice units. It has been suggested by first principles calculations of the magneto-crystalline anisotropy energy that the individual spins show a preferred orientation perpendicular to the rhombohedral axis (also the polarization direction) [28]. The cycloidal spins, in ~620 Å periodicity, define the easy magnetic plane, the normal of which aligns with the polarization. In each of the three possible magnetic domains, the spins rotate in the plane defined by the polarization vector, [111], and the propagation vector, **k**.

Furthermore, it can be deduced that the polarization and magnetic plane coupling would also result in a coupling between the antiferromagnetic state and the ferroelastic strain state.

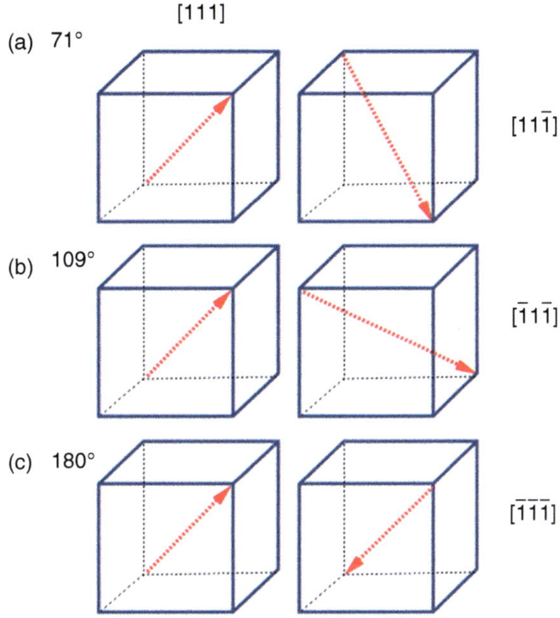

Figure 6.2 Schematic diagram of domain and domain wall types in $BiFeO_3$: (a) 71° with one component of the polarization switched; (b) 109° with two components of the polarization switched, and (c) 180° with all three components of the polarization switched.

In fact, the spins are not perfectly antiparallel, as there is a weak canting moment caused by the local magnetoelectric coupling to the polarization as discussed earlier. This is shown in Figure 6.3 at the left bottom. Normally, the weak local canting moment is completely canceled by the averaging-out effect of the cycloid at room temperature, ruling out a linear magnetoelectric coupling between the polarization and magnetization in $BiFeO_3$ (note that this does not necessarily rule out higher-order magnetoelectric coupling). However, density functional theory calculations predicted, a Dzyaloshinskii–Moriya type weak ferromagnetism as a result of the canting of antiferromagnetic moments in $BiFeO_3$ (i.e. a weak magnetization of ~0.05 μB per unit cell from ~1° canting), also confirmed by low temperature measurements [28]. This reflects an underlying competition between antiferromagnetic and ferromagnetic interactions.

6.3.3 Magnetoelectric Coupling

6.3.3.1 Antiferromagnetic Switching on Electric Field

Polarization switching by 71° or 109° can give rise to the reorientation of the easy magnetization plane, as a demonstration of electric control of magnetic domains in $BiFeO_3$ thin films [29, 30], single crystals [31, 32], and ceramics [33] at room temperature. One exemplified schematic diagram of electrically-induced antiferromagnetic switching is shown in Figure 6.4. This electrically induced switching of the magnetic easy plane is of fundamental importance, because it opens the possibility of magnetoelectric devices based on the voltage control of magnetization (see Section 6.8.2).

However, it is worth noting that the magnetic order switching by electric field is not necessarily always accompanied by polarization switching. This means that even

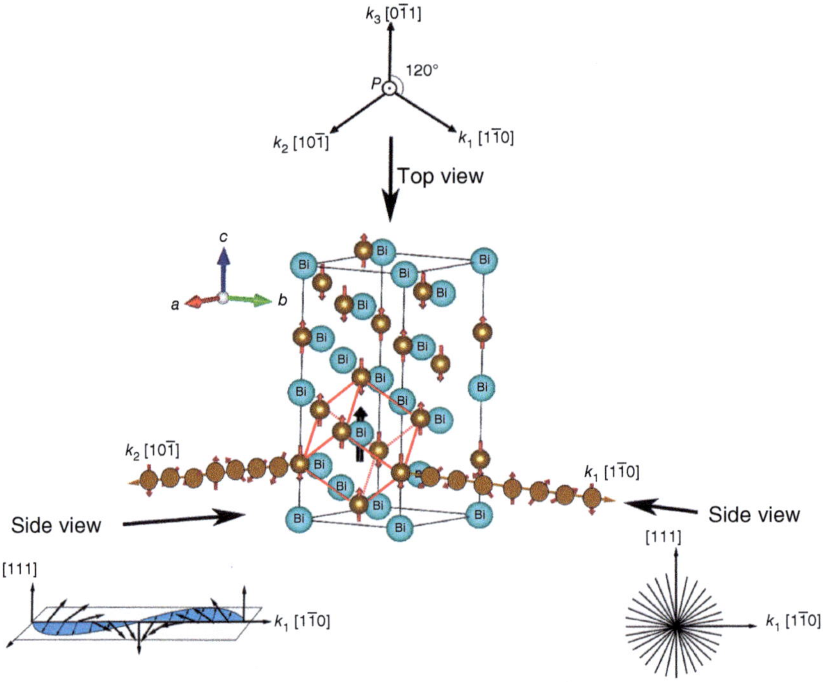

Figure 6.3 The *R*3*c* structure of $BiFeO_3$ represented with the hexagonal unit cell (black lines) and the pseudocubic unit cell (red lines) with the magnetic moment of Fe atoms displayed. Antiparallel spins of the nearest Fe atom in $BiFeO_3$ illustrates the antiferromagnetism, while there is also a modulated incommensurate cycloidal spin structure with a long periodicity of ~620 Å.

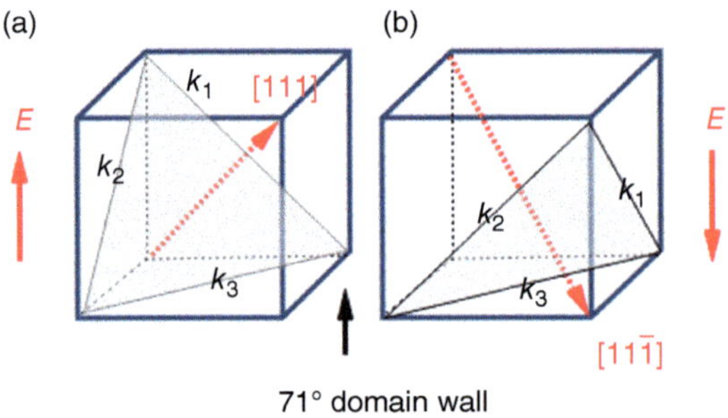

Figure 6.4 Polarization (***P***, dashed arrows) along the ⟨111⟩ axis and three possible propagation directions of the spin cycloid, $\boldsymbol{k}_1$, $\boldsymbol{k}_2$, and $\boldsymbol{k}_3$, which are normal to the ferroelectric polarization vector ***P***. Polarization switching with electric field induces a flop of magnetic spin cycloid (magnetic easy plane).

without polarization switching, there still can be magnetic order switching under electric field. This may happen between different magnetic spin cycloids as found previously [34].

6.3.3.2 Ferroelectricity on Magnetic Field

Macroscopically, up to magnetic fields of several Tesla the magnetically induced polarization is found to be proportional to the square of the magnetic field as shown in Figure 6.5, aligning a higher-order magnetoelectric coupling [35]. Above a certain

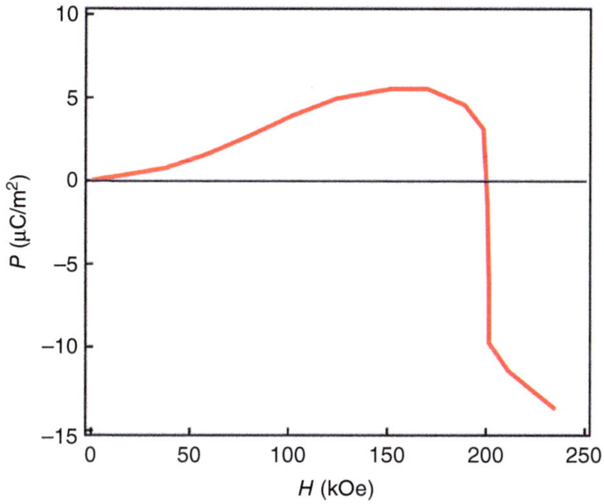

Figure 6.5 Magnetoelectric (ME) effect in BiFeO$_3$: at low fields, P is proportional to H^2 (quadratic ME coupling). Above 200 kOe, P is linearly dependent on H instead, indicating the destruction of cycloid (since the linear ME is forbidden in the presence of a cycloid). Source: Reproduced with permission from Zvezdina et al. [35]. Copyright 2006, Elsevier.

critical field, however, the magnetoelectric polarization markedly changes, indicating a change in the spin configuration (Figure 6.5). Another critical field appears to be at a much higher value above which the magnetoelectric polarization changes sign and becomes linearly dependent on magnetic field (Figure 6.5). Since the linear magnetoelectric effect is forbidden by the cycloid spin configuration, its onset shows that the cycloid has been destroyed by the high magnetic field. Note that, in any case, the magnetically induced polarization is very small ($\sim 10^5$ times smaller than the ferroelectric polarization). These experimental results may indicate the polarization switching on application of magnetic field occurs in a similar way as have demonstrated earlier for the magnetic easy plane switching in unit cells.

6.4 Phase Diagram of BiFeO$_3$

6.4.1 High Curie Temperature and Processing Issues

The phase diagram of Bi_2O_3 and Fe_2O_3 is critical to understand BiFeO$_3$. Figure 6.6 presents the phase diagram reported previously for Bi_2O_3 and Fe_2O_3 [36, 37], showing the main equilibrium phases and their transition temperatures. On the left-hand Bi-rich side, there exists the cubic $Bi_{25}FeO_{39}$ (space group $I23$) phase. In the middle with an equal mole ratio of these two oxides, the phase is the rhombohedral perovskite BiFeO$_3$ (space group $R3c$). On the right-hand Fe-rich side, the phase is the orthorhombic $Bi_2Fe_4O_9$ (space group $Pbam$).

Polycrystalline rhombohedral BiFeO$_3$ (α phase in the figure) can usually be prepared from stoichiometric starting oxides, i.e. Bi_2O_3 and Fe_2O_3, using solid-state

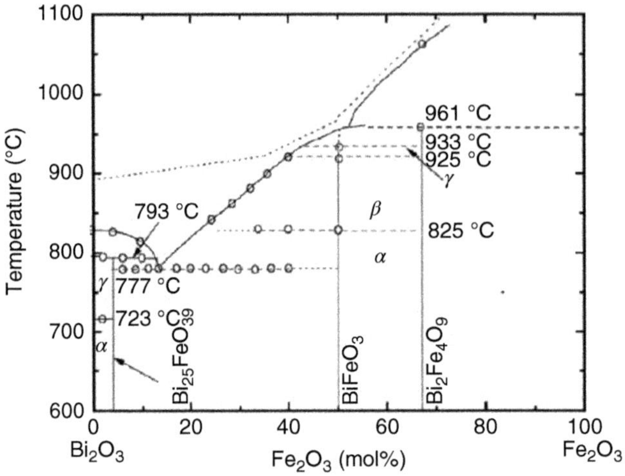

Figure 6.6 Compositional phase diagram of Bi_2O_3 and Fe_2O_3. Source: Reprinted with permission from Palai et al. [36]. Copyright 2008, American Physical Society.

methods. The reaction between the starting materials is as follows:

$$Bi_2O_3 + Fe_2O_3 \rightarrow 2BiFeO_3 \tag{6.1}$$

The first-order transition from the α phase to β phase at 825 °C, with a sudden volume contraction and a peak in the dielectric constant [36], is reckoned as the ferroelectric–paraelectric transition. This Curie temperature of $BiFeO_3$ is significantly higher than commercial lead-based $Pb(Zr,Ti)O_3$ (~160–350 °C) [38] and lead-free $BaTiO_3$ (~110–130 °C) [39]. It is actually the known ferroelectrics with the highest Curie temperature, making it a promising candidate for high temperature sensors and actuators. Even forming solid solutions (for example, with $BaTiO_3$) to mitigate three other disadvantages, i.e. the difficulty of processing, the high coercive field and the large leakage current, the Curie temperature is still higher than the existing widely used materials (~400 K). The exact phase of β-$BiFeO_3$ or the transition from α- to β-phase $BiFeO_3$ has been argued for a long time with different beliefs. The conclusion, from *in situ* neutron diffraction, synchrotron X-ray diffraction, differential thermal analysis, and optical viewgraphs, is that it undergoes an intermediate monoclinic phase [40, 41] or a coexistence of monoclinic and orthorhombic phases [42], and finally goes to a coexistence of rhombohedral phase and orthorhombic phase [36].

Above the temperature of 933 °C as shown in the figure, $BiFeO_3$ is in its γ phase belonging to the cubic $Pm\bar{3}m$ space group. So the overall phase transformation path as a function of temperature is rhombohedral–orthorhombic–cubic. However, cubic $BiFeO_3$ is unstable at high temperatures, decomposing easily back into the starting oxides or transforming into impurity phases (i.e. iron-rich phase $Bi_2Fe_4O_9$ and bismuth-rich phase $Bi_{25}FeO_{39}$).

The thermodynamic instability of $BiFeO_3$ has been clarified from both an experimental and theoretical point of view [43]. One concern is the Bi_2O_3 loss during sintering or annealing at elevated temperature, which may result in the formation of large iron-rich $Bi_2Fe_4O_9$ crystals represented by the reaction [44–46]:

$$8BiFeO_3(s) \rightarrow 2Bi_2Fe_4O_9(s) + 4BiO(g) + O_2(g) \quad (6.2)$$

Another concern is the transformation of $BiFeO_3$ to these impurities as described by the following equilibrium equation:

$$BiFeO_3 \leftrightarrow 1/49Bi_{25}FeO_{39} + 12/49Bi_2Fe_4O_9 \quad (6.3)$$

The Gibbs free energy (Δ_rG^0) for this equilibrium reaction has been calculated. It was found that the Δ_rG^0 is below 0 at the temperature range of 447–767 °C, and thus the $BiFeO_3$ is not stable against secondary phases under such conditions. At temperatures of 447 and 767 °C, $\Delta_rG^0 = 0$, the equilibrium condition is reached, and the equilibrium reaction implies a coexistence of the three phases. In reality, the coexistence of these three phases has been reported for a much broader temperature range, which may be explained by the small driving force (Δ_rG^0) of the forward and backward reactions [43]. The study of thermodynamic instability of $BiFeO_3$ enables optimizations of the sintering temperature range of $BiFeO_3$ using the solid-state synthesis.

The $Bi_{25}FeO_{39}$ and $Bi_2Fe_4O_9$ impurity phases are easy to appear if a third compound exists in the Bi_2O_3–Fe_2O_3 system, resulting in a coexistence of the perovskite $BiFeO_3$ and these secondary phases [47]. Prior works studied systematically the influence of a third component to the stabilization of undesired $Bi_{25}FeO_{39}$-like and $Bi_2Fe_4O_9$-like phases, finding that the former component, even added in a very small quantity, for example, <1 wt% SiO_2, can result in a large amount (several tens of vol%) of impurities. This can be explained by the solubility of the third component in $Bi_{25}FeO_{39}$ and/or $Bi_2Fe_4O_9$ phases (Gibbs phase rule).

In addition to the thermodynamics of the Bi_2O_3–Fe_2O_3 system, the reaction kinetics between the starting oxides has been investigated. Based on experimental results, it is assumed that Bi_2O_3 diffuses toward Fe_2O_3 during the solid reaction thereby forming $BiFeO_3$ [20]. This process is likely to be accompanied by the easy formation of $Bi_2Fe_4O_9$ secondary phase in the Fe-rich counterpart core area and $Bi_{25}FeO_{39}$ around the edges, due to an incomplete reaction between the starting oxides.

According to the thermodynamics and reaction kinetics of $BiFeO_3$ discussed above, impure parasitic phases are prone to appear very easily. Minimizing the secondary phases requires careful optimization of processing parameters, including particle size of the starting materials, clean environment during the sintering (or calcining), sintering (or calcining) temperature, reacting time, and oxygen pressure during sintering.

6.4.2 Influence of Pressure on Phase Diagram

A schematic pressure–temperature phase diagram has been given by Catalan and Scott based on: (i) the existing experimental data and (ii) the pressure/doping effects and the generally known behavior of the phase transitions in other orthoferrites (see Ref. [19] and references therein). The diagram illustrates that at room temperature,

$BiFeO_3$ goes through a sequence of transitions from the ground-state rhombohedral *R3c* to a coexistence of rhombohedral and orthorhombic phases (the latter is also sometimes defined as a monoclinic phase), then to an orthorhombic *Pnma* phase, and finally to a cubic $Pm\bar{3}m$ phase. In general, with increasing temperature, the pressures of the above transition sequence reduce gradually, before these transitions are compressed to a triple point. Though intensive work and careful exploration are still required to complete the view, the sketch of this given diagram provides an instructive guide for future investigations in this direction.

6.4.3 Thin Film and Strain Effect on Phase Structure

Thin-film growth, especially with epitaxial characteristics, provides an extremely versatile method for modifying the crystal structure of $BiFeO_3$. In those systems, the space group of the material can easily deviate from its ground structure, caused by an epitaxial strain and mismatch to the substrate on which the thin film stands. This is known as "strain engineering." While the bulk structure of $BiFeO_3$ is rhombohedral (in space group *R3c*), when it is grown in thin films, epitaxial strain imposed by the substrate may cause the compound to take on a structure of reduced symmetry. Reports of tetragonal-like [48], rhombohedral [49], orthorhombic [50], monoclinic M_A [51], monoclinic M_B [52], and triclinic [53] $BiFeO_3$ films can be found. This wide range of possible crystal structures highlight the strong effect that the substrate can have on the film's structural properties, and we shall see that these modifications may have substantial effects on the physical properties as well.

Figure 6.7 shows a series of phases exhibited in the $BiFeO_3$ films depending on the substrate strain [14, 54]. These various crystal symmetries, summarized by Sando et al. (see also a number of references in the review), are shown relative to the primitive pseudocubic perovskite unit cell. Epitaxial $BiFeO_3$ film is bulk-like rhombohedral phase at zero strain. On $(111)_{pc}$-oriented cubic substrates (for example, $SrTiO_3$), bulk phase (rhombohedral *R3c*) $BiFeO_3$ is typically obtained. In the compressively strained regime, $BiFeO_3$ goes from monoclinic M_A phase to monoclinic M_C phase, and for very strong compressive strain, a highly distorted tetragonal-like monoclinic phase is obtained (Figure 6.7, achieving very large c/a ratio ~ 1.3). This is often obtained on a $(001)_{pc}$ orientated substrate (for example, $YAlO_3$), and experimental data show that ultrathin films (<10 nm) of $BiFeO_3$ on $YAlO_3$ $(001)_{pc}$ may be purely tetragonal. In these monoclinic phases, the unit cell is doubled in volume and rotated by 45° with respect to the pseudocubic cell. In the tensile strain direction, monoclinic M_B phase is normally obtained before it finally goes to a highly-distorted orthorhombic phase.

The induced strain in $BiFeO_3$ by the single crystal substrate is related to the lattice mismatch between these two phases. It is worth noting that though the lattice mismatch is essential to tune the phase transition, the substrate should in general have close lattice matching to obtain epitaxial $BiFeO_3$ films. There are a number of single crystals that can provide such close lattice matching for epitaxial $BiFeO_3$. Examples of commonly used substrates to provide compressive and tensile strains are listed in Table 6.1:

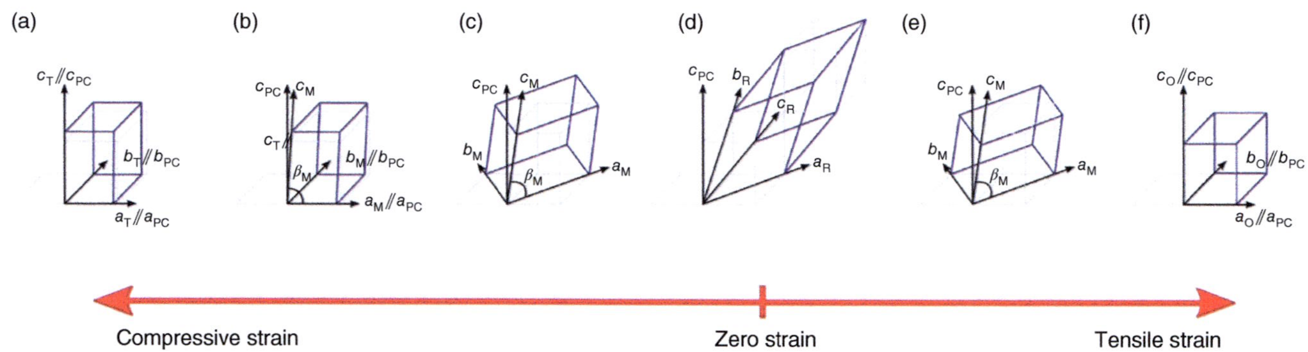

Figure 6.7 Phase transition according to substrate strain induced on the $BiFeO_3$ films. (a) Tetragonal *P4mm*. (b) Monoclinic M_C (*Cm*). (c) Monoclinic M_A (*Cc*). (d) Rhombohedral *R3c*. (e) Monoclinic M_B (*Cc*). (f) Orthorhombic *Ima2*. Source: Reprinted with permission from Sando et al. [54]. Copyright 2014, IOP Publishing. The subscripts PC, T, M, R, and O in the figure denote for pseudo-cubic, tetragonal, monoclinic, rhombohedral, and orthorhombic structures, respectively.

Table 6.1 Lattice parameters of the substrates commonly used for epitaxial $BiFeO_3$ films growth [14, 55–60].

Substrate materials	Substrate phase	Lattice parameters (Å)	Misfit strain with $BiFeO_3$ (%)	Strain types
$NdCaAlO_4$	$(001)_T$	$3.685(a = b)/12.12(c)$	−7.0	Compressive
$YAlO_3$	$(110)_O$	$3.69(a = b = c)$	−6.9	
$LaSrAlO_4$	$(001)_T$	$3.755(a = b)/12.6(c)$	−5.2	
$LaAlO_3$	$(001)_R$	$3.790(a = b = c)$	−4.4	
$NdGaO_3$	$(110)_O$	$3.861(a)/3.854(b)$	−2.6	
$SrTiO_3$	$(001)_C$	$3.905(a = b = c)$	−1.5	
$SrRuO_3$	$(001)_C$	$3.923(a = b = c)$	−1.0	
$DyScO_3$	$(110)_O$	$3.946(a)/3.952(b)$	−0.3	
$TbScO_3$	$(110)_O$	$3.960(a)/3.959(b)$	+0.1	Tensile
$GdScO_3$	$(110)_O$	$3.970(a)/3.966(b)$	+0.2	
$KTaO_3$	$(001)_C$	$3.989(a = b = c)$	+0.6	
$SmScO_3$	$(110)_O$	$3.991(a)/3.983(b)$	+0.7	
NdScO	(110)o	$4.014(a)/4.002(b)$	+1.2	
MgO	$(001)_C$	$4.216(a = b = c)$	+6.3	

The calculation of the misfit strain is made relative to the lattice parameters of bulk $BiFeO_3$ (rhombohedral, $a = b = c = 3.965$ Å).

Another important factor to influence the material symmetry, other than the lattice mismatch, is the thickness of the films. It greatly alters the symmetry of the material, which is known as the strain relaxation of the film. Thinner films can more easily maintain the structure of the substrate, while thicker films are more prone to revert to its bulk structures. This is experimentally observed for the phase structures of $BiFeO_3$ thin films. For example, $BiFeO_3$ films fabricated on $SrRuO_3/SrTiO_3$ substrates with a thickness smaller than 50 nm exhibited a tetragonal phase with a tetragonality of ~1.036, but a decreased tetragonality to 1.015 was found for the film thickness increasing to 400 nm [61]. Three phases (i.e. orthorhombic-like, tetragonal-like, and rhombohedral-like phase) coexist in the (110) $BiFeO_3$ films as the film thickness increases from 10 to 190 nm, as confirmed by high resolution synchrotron X-ray diffraction [62].

Also worth noting is that since $BiFeO_3$ is a multiferroic, it is foreseeable that besides the ferroelectric phase transitions, the magnetic order (including the case of maintaining and losing cycloidal order) may be also influenced significantly by temperature and strain in thin films. Figure 6.8 (taken from Ref. [54]) gives the temperature–strain dependent phase diagram of $BiFeO_3$ films as constructed from a number of theoretical and experimental results. Around the zero-strain region as shown in the diagram, the magnetic order maintains the intrinsic cycloidal feature in general. In strained films with broken cycloidal order, a ferromagnetic

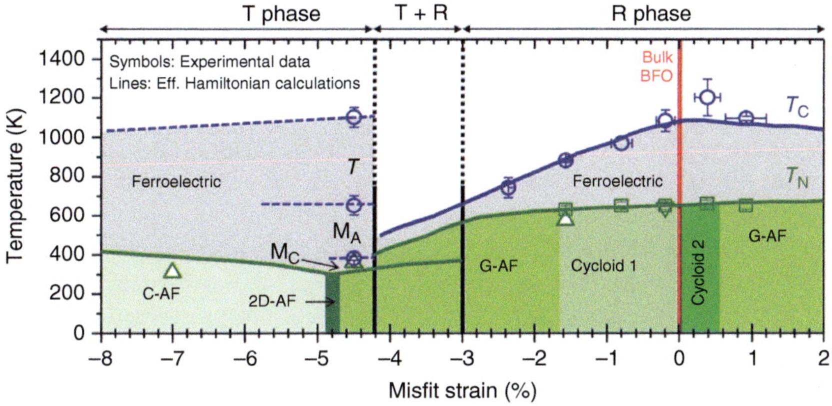

Figure 6.8 Temperature–strain phase diagram for $BiFeO_3$ thin films. Note that the dashed lines indicate hypothetical regions of the diagram and transition lines between different magnetic orders are represented vertically for simplicity. Source: Reprinted with permission from Sando et al. [54]. Copyright 2014, IOP Publishing.

moment is expected, and has been experimentally observed (sometimes obtaining orders of magnitude enhancement of the weak moment from its expected value of around 0.01 μB/Fe) [49, 63, 64]. At even larger strain state (i.e. in the T-like phase region), the magnetic order is still under debate. A pure G-type order at 300 K [48] and a mixture of C- and G-type orders [65], with different temperature dependencies, have been observed. It may imply very different magnetic properties. In the diagram, dashed lines are used for hypothetical and unexplored phase transition temperatures, and transition lines between different magnetic orders are represented vertically for simplicity. The diagram generally demonstrates the transitions of magnetic order due to temperature change and lattice misfit, while careful exploration is necessary.

6.5 Dielectric Permittivity, Electrical Conductivity, and Domain Wall Conductivity of $BiFeO_3$

As mentioned earlier, the main problem of $BiFeO_3$ is its high conductivity, which impedes the poling of $BiFeO_3$ bulk materials. The electrical properties of $BiFeO_3$, including dielectric permittivity and electrical conductivity and their relationship with the defects, are discussed in this section. In addition, a new area initialized from this material, i.e. domain wall conductivity, is also involved.

6.5.1 Dielectric Permittivity

The ~GHz dielectric permittivity (relative to the vacuum permittivity) of $BiFeO_3$ at room temperature is $\varepsilon_r \sim 30$, representing its intrinsic dielectric permittivity at radiofrequencies [66, 67]. This dielectric constant is much smaller than that of the most commercially used ferroelectrics, Pb(Zr,Ti)O_3 (which ranges from ~300 to ~3850, depending upon orientation, doping, and frequency). This small dielectric constant can be ascribed to the following two reasons [19]. Firstly, the ferroelectric Curie temperature of $BiFeO_3$ is very high, meaning that at room temperature the ferroelectric polarization is already saturated and, thus, small electric fields will barely affect it (the dielectric constant is essentially a measure of polarizability). Secondly, the ferroelectric transition is a typical first-order transition (which means that there is little phonon softening), thus the dielectric constant predicted by Lyddane–Sachs–Teller effect is expected to be very low [19]. Furthermore, the low dielectric constant is the reason why the associated piezoelectric coefficient is also small through the simplified relationship $d \sim P{\cdot}\varepsilon_r$, where P is the polarization.

Although $\varepsilon_r \sim 30$ can be regarded as the intrinsic dielectric constant of this compound at radiofrequencies, the impedance measurements in parallel-plate capacitors often yield higher values in between ~50 and ~300 depending on other factors (such as sample morphology, orientation, and measuring frequency range). This is because at the frequencies accessible by impedance analyzers (typically from 100 Hz to 1 MHz), domain-wall motion, and space-charge contributions can be significant and add to the observable permittivity. Figure 6.9 shows typical measured permittivity frequency spectra of a bulk $BiFeO_3$. It shows that at high frequency range beyond Hz frequency range, which shows the typical dielectric relaxation, the permittivity is in the range of ~100. Furthermore, the dielectric permittivity surges at low frequency range (below Hz frequency range, Figure 6.9). Room-temperature colossal permittivities ($\varepsilon_r \sim 10\,000$) were obtained in the Hz-to-MHz frequency range, in both ceramics [67] and single crystals [68]. This is due to the contribution of the electrical conductivity, which, in addition to directly affecting true DC conductivity (will be further discussed in Section 6.4.2), may contribute to both low-frequency ε' and ε'' through either the hopping or Maxwell–Wagner interfacial polarization mechanisms [69]. The Maxwell–Wagner relaxation was invoked to explain the frequency dispersion and high permittivity values of $BiFeO_3$ ceramics and single crystals [67, 68, 70].

As a function of temperature at 1 Hz (an example below MHz), the permittivity of bulk $BiFeO_3$ shows a general increasing trend, indicating a thermally activated relaxation process as shown in Figure 6.10 [71]. It is worth noting that the ε' shows a second sharp increase in the high-temperature limit, i.e. close to the maximum measured temperature of ~300 °C. The permittivity in this high-temperature limit is controlled by the macroscopic DC conductivity of the sample (estimated from the imaginary component of the permittivity, ε''). These data are in agreement with the literature, among which the sub-MHz frequency dielectric response is generally believed to be dominated by the Maxwell–Wagner mechanism at temperatures above 103 °C [67, 72].

6.5.2 Electrical Conductivity and Defects

As we have mentioned earlier, the conductivity of $BiFeO_3$ plays a significant role in the dielectric property in different frequency ranges. In a fashion similar to the

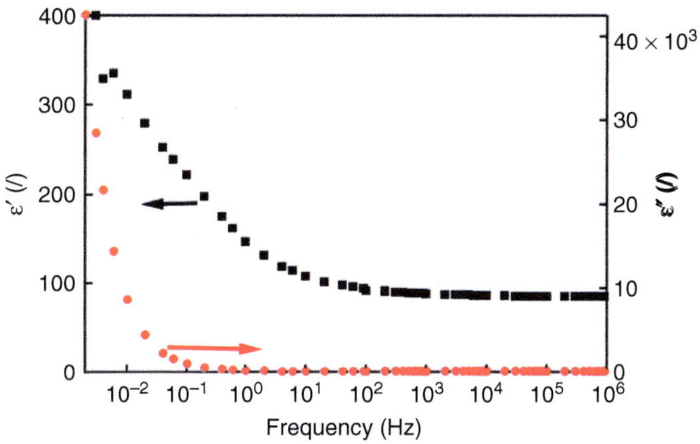

Figure 6.9 Real part and imaginary part of the dielectric permittivity as a function of frequency of a bulk $BiFeO_3$ at room temperature.

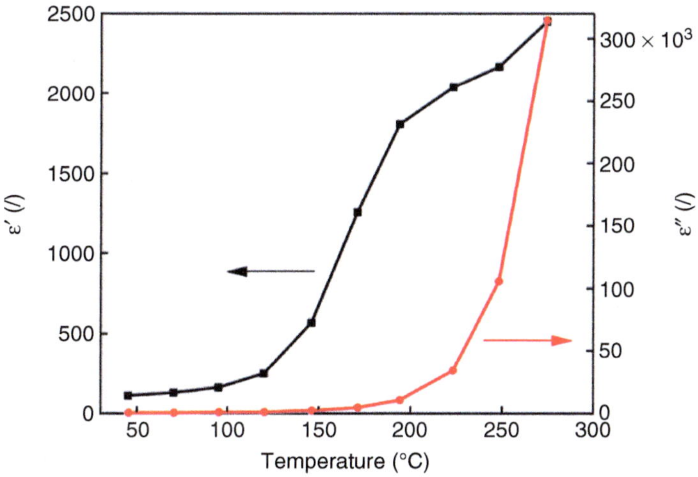

Figure 6.10 Real part and imaginary part of the dielectric permittivity of a bulk $BiFeO_3$ as a function of temperature at 1 Hz. Source: Modified from Rojac et al. [71].

low-frequency permittivity, the DC conductivity of $BiFeO_3$ ceramics varies significantly from case to case. The as-reported bulk electrical conductivity of $BiFeO_3$ ceramics and single crystals at room temperature spread over several orders of magnitude, typically between ~10^{-2} and ~10^{-10} S/m [19, 20, 68, 73]. This is generally higher than typical ferroelectrics (~10^{-12} S/m). The general high electrical conductivity, which can result in electrical leakage during polarization switch or poling, comprises one of the major issues of $BiFeO_3$ that obstacle its practical applications.

A large number of studies investigated the reasons causing the high conductivity. The initial assumption is the secondary phases due to the long-term processing difficulties of bulk $BiFeO_3$ [20, 74]. However, another cause of the high conductivity, which found a consensus over the largest number of $BiFeO_3$ studies, is commonly believed to be the presence of defects, such as reduced Fe^{3+} (Fe^{2+} sites) and/or oxygen vacancies [70, 75–81]. These defects determine the conductance type of $BiFeO_3$ as a semiconductor. The semiconducting character of $BiFeO_3$ can be attributed to the mixed Fe valance state, with its reduction/oxidation depending on the content of oxygen in the atmosphere, which can be expressed by the following defect equations:

$$2Fe_{Fe}^{\times} + V_{O}^{\cdot\cdot} + \frac{1}{2}O_2 \rightarrow 2Fe_{Fe}^{\cdot} + O_{O}^{\times} \tag{6.4}$$

$$2Fe_{Fe}^{\times} + O_{O}^{\times} \rightarrow 2Fe_{Fe}^{\prime} + V_{O}^{\cdot\cdot} + \frac{1}{2}O_2 \tag{6.5}$$

As one may see, the types of major defects and resultant conductance types are greatly influenced by the processing conditions, for example, the sintering atmosphere, changing from the p-type conductance to the n-type conductance. Under O_2-rich atmosphere (e.g. in air or O_2 atmosphere), the creation of Bi ($V_{Bi}^{\times} \rightarrow V_{Bi}^{\prime\prime\prime} + 3h^{\cdot}$) and Fe vacancies ($V_{Fe}^{\times} \rightarrow V_{Fe}^{\prime\prime\prime} + 3h^{\cdot}$) is energetically favored over oxygen vacancies ($V_{O}^{\cdot\cdot} + \frac{1}{2}O_2 + 2e\prime \rightarrow O_{O}^{\times}$). In addition, the cation vacancies (i.e. $V_{Bi}^{\prime}$ and $V_{Fe}^{\prime}$) form shallow acceptor levels within the bandgap and are thus easily ionized, releasing holes and potentially leading to p-type conduction. Theoretical studies using density-functional theory confirmed these ideas, suggesting that $BiFeO_3$ is a p-type semiconductor [82, 83], which are supported by the experimental data [84–87]. Under O_2-deficient atmosphere (e.g. in N_2 or Ar/H_2 atmosphere), calculations predicted that oxygen vacancies are the most favorable intrinsic defects in $BiFeO_3$, with the ionization transition ($V_{O}^{\times} \rightarrow V_{O}^{\cdot\cdot} + 2e\prime$) well or moderately below the conduction-band edge, which is 1.9 eV [82] or 0.6 eV [88]. Thus, in comparison with the shallow acceptor levels (cation vacancies), the donor levels associate with oxygen vacancies are rather deep, so that $BiFeO_3$ samples treated in an oxygen poor environment (leading to n-type conductance) are, in principle, not expected to have as high conductivities as those processed under ordinary oxygen-rich conditions (p-type conductance).

As an example, the comparison of conductivity of Ca-doped $BiFeO_3$ sintered under different atmospheres is shown in Figure 6.11 [84]. The material behaves as a p-type semiconductor with an activation energy (E_a) of ~0.27–0.4 eV when sintered in O_2 or air. Whereas sintered and cooled in N_2, the conductivity of Ca-doped $BiFeO_3$ could be reduced by several orders of magnitude with $E_a \sim 0.82$–1.04 eV, showing an ionic conduction mechanism.

6.5.3 Domain Wall Conductivity

Ferroic domain walls are naturally occurring nanoscale interfaces, and in ferroelectrics, these interfaces separate domains with different polarization directions. They may possess distinct properties from their parent materials [4, 21, 89]. Ferroelectric material $BiFeO_3$ has been demonstrated to possess enhanced electrical conductivity at domain walls (relative to the domain interior), found by Jan Seidel [4, 90]. This conductivity is measurable using microprobe methods as shown in Figure 6.12 (in this case, conductive-atomic force microscopy) [4]. The figure schematically shows that the current on the domain wall regions enhances

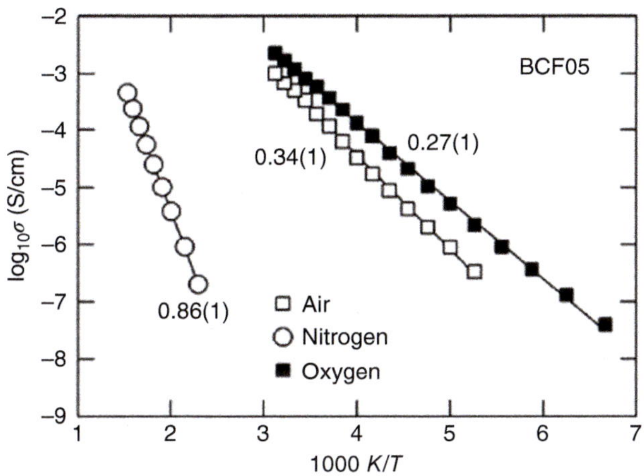

Figure 6.11 Direct current electric conductivity of 5% Ca-doped $BiFeO_3$ ceramics sintered in air, oxygen (O_2), and nitrogen (N_2), respectively. When sintered in air or O_2 the ceramics behave as a p-type semiconductor ($E_a \sim 0.3$ eV). Sintering in N_2 reduces significantly the conductivity with $E_a \sim 0.8$ eV, showing ionic conduction behavior. Source: Reprinted with permission from Masó and West [84]. Copyright 2012, American Chemical Society.

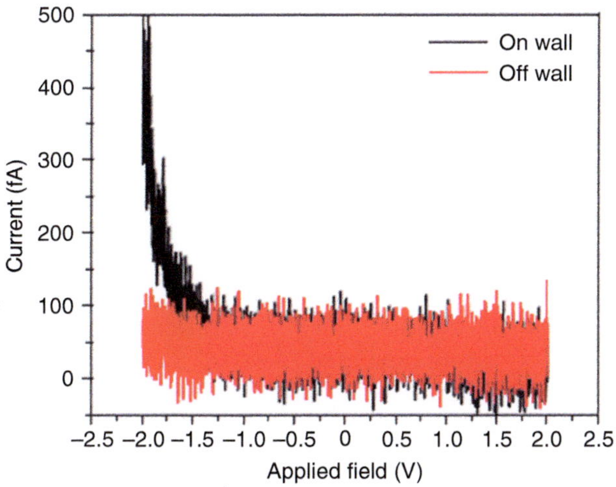

Figure 6.12 A schematic of the enhanced conductivity at domain walls compared with the domain regions. Source: Reprinted with permission from Seidel et al. [4]. Copyright 2009, Springer Nature.

significantly to hundreds of femtoampere (fA) from zero at regions off the domain walls during voltage application. The hypothesized cause of this is the structurally driven changes in both the electrostatic potential and the local electronic structure leading to a decrease in the local bandgap at the domain wall. This is always related to mobile charge effects accumulated at domain walls, which was also observed experimentally [91].

There have been a number of following studies of enhanced domain wall conductivity in $BiFeO_3$, trying to uncover its effects and mechanisms on the piezoelectric responses. For example, the domain wall conductivity would relate closely to the Maxwell–Wagner relaxation of the macroscopic piezoelectric coefficients as a function of frequency [26]. It may also decouple the domain wall motion induced strain and lattice strain (i.e. responses in different grain families) in a polycrystalline $BiFeO_3$ due to the internal field arising from the domain wall conductivity under applied electric field [92].

Since the discovery of conductive domain walls in $BiFeO_3$, a number of investigations are completing the picture. It should be mentioned that enhanced domain wall conductivity has also been observed in numerous ferroelectric materials, including $BaTiO_3$ crystals [93], $KTiOPO_4$ crystals [94], $YMnO_3$ single crystals [95, 96], $ErMnO_3$ crystals [97], $LiNbO_3$ single crystals [98], and $Pb(Zr_{0.2}Ti_{0.8})O_3$ thin films [99]. It suggests that enhanced domain wall conductivity may be a more general property than previously thought.

Due to the conductivity feature as well as the length-scale of domain walls, they have generated great interest for applications referred as domain-wall nanoelectronics [4, 21, 22, 100, 101]. Potential novel applications of conducting domain walls have been discussed and proposed in the literature [21, 22].

6.6 Ion Substitutions in $BiFeO_3$

Due to the processing and leakage problems (see also Section 6.3.1), ion substitution has been considerably investigated as a means to suppress the conductivity of $BiFeO_3$. Worth noting is that it also helps to solve other problems, i.e. high coercive field and thus the difficulty of poling [102]. In addition, it has more substantial effect to tune both the ferroelectric and magnetic properties of the parent $BiFeO_3$ in different forms (i.e. bulk, film, and nanostructures). In this section, we will briefly introduce the history of substitution in $BiFeO_3$.

At first, let us clarify our usage of "substitution" and "doping." Substitution is a crystallographic term and means the exchange of an element by another at a certain atomic position. It is used when replacing one element by another with levels reaching several percentages. By contrast, the term "doping" is used when a small amount of an element is mixed in a stoichiometric compound. This might replace some atoms existing in this phase but it does not have to. It can also sit at interstitials or in between atom layers, molecules, etc. It is often accompanied with considerable property changes, e.g. electronic properties of semiconductors by doping. Since these two terms are used in mixture under many conditions, there are no strict rules to separate their usage in general. We use them without substantial discretion in the following text.

As mentioned in Section 6.3, $BiFeO_3$ has a perovskite ABO_3 structure, with Bi^{3+} and Fe^{3+} cations on the A and B site, respectively. Surrounding the cations, O^{2-} anions form FeO_6 octahedra and BiO_{12} cuboctahedra. In principle, both cations can be substituted by other ions, provided that the radius of new ions is in a proper range as characterized by the tolerance factor. Therefore, there are three doping methods, i.e. A-site (Bi^{3+}) doping, B-site (Fe^{3+}) doping, and A/B sites co-doping. For each doping type, there are single and multiple element substitutions.

The A-site substitution refers to the case that the Bi^{3+} ions are replaced by other ions. There could be p-type, n-type, and isovalent doping. Replacing Bi^{3+} by divalent ions like Ca^{2+} and Sr^{2+} may realize novel conducting states of hole conduction (i.e. a p-type doping), unless other defects like oxygen vacancies are simultaneously produced. The latter case will be accompanied by a large amount of electrons, thus causing the doping type to change to n-type doping. On the other hand, substitution of Ce^{4+} ions is known to produce electron carriers (i.e. n-type doping). Isovalent doping, as expected, rarely alters the conduction type. Worth noting is that the A-site substitution by some ions with smaller ionic radius can induce larger buckling in the Fe—O—Fe bond angle accompanied with a smaller tolerance factor, leading to a more insulating character. This is the common reason for the reduced leakage of the materials often found with A-site doping.

The B-site substitutions are made by replacing the Fe^{3+} ions by other transition-metal ions. The B-site substitution by Cr^{3+} and Mn^{3+} can alter the magnetic property, while substitution ions such as Ti^{4+}, Ni^{3+}, and Co^{3+} reduce the leakage apart from influencing the magnetic property. Some are even found to influence the ferroelectricity through modifying its magnetic properties.

A huge number of doping studies have already been carried out for $BiFeO_3$. In fact, nowadays, dopant studies can be found to have almost spanned the entire periodic table as shown in Figure 6.13 for single-site doping. Therefore, here we do not show all references (several good reviews are [14, 103, 104] and readers can further find a number of references therein). In Figure 6.13, the experimental data of A-site and B-site doping are shown in red and blue, respectively. For those elements not yet confirmed experimentally, theoretical predictions have been made for their doping types as shown using the same color but in lighter colors. As we can see from the figure, a significant number of elements have been considered to alter the properties of pure $BiFeO_3$ in all material forms including bulk, film, and nanostructure.

In Section 6.6.1, we focus on the most influential results on the piezoelectric properties of $BiFeO_3$ through doping/substitution. Theoretical interpretations, e.g. the electronic band structure, are not included here but could be referred to in more detailed theoretical studies [104].

6.6.1 On Ferroelectricity (P_r) and Piezoelectricity (d_{33})

The aforementioned large polarization value of $BiFeO_3$ is obtained from theoretical studies and high-quality crystals or epitaxial films. However, for polycrystalline $BiFeO_3$ ceramics, saturated hysteresis loops are often difficult to be measured due to larger leakage current, and thus the polarization obtained is relatively small. While the normal remnant polarization P_r reported for pure bulk $BiFeO_3$ is approximately $20\,\mu C/cm^2$ [105], doped $BiFeO_3$ ceramics (15% Nd and 2% Mn-doped) are reported with a P_r reaching $31\,\mu C/cm^2$, owing to the phase transformation from rhombohedral $R3c$ to tetragonal $P4mm$ [106]. Rare-earth (Sm) doped $BiFeO_3$ ceramics

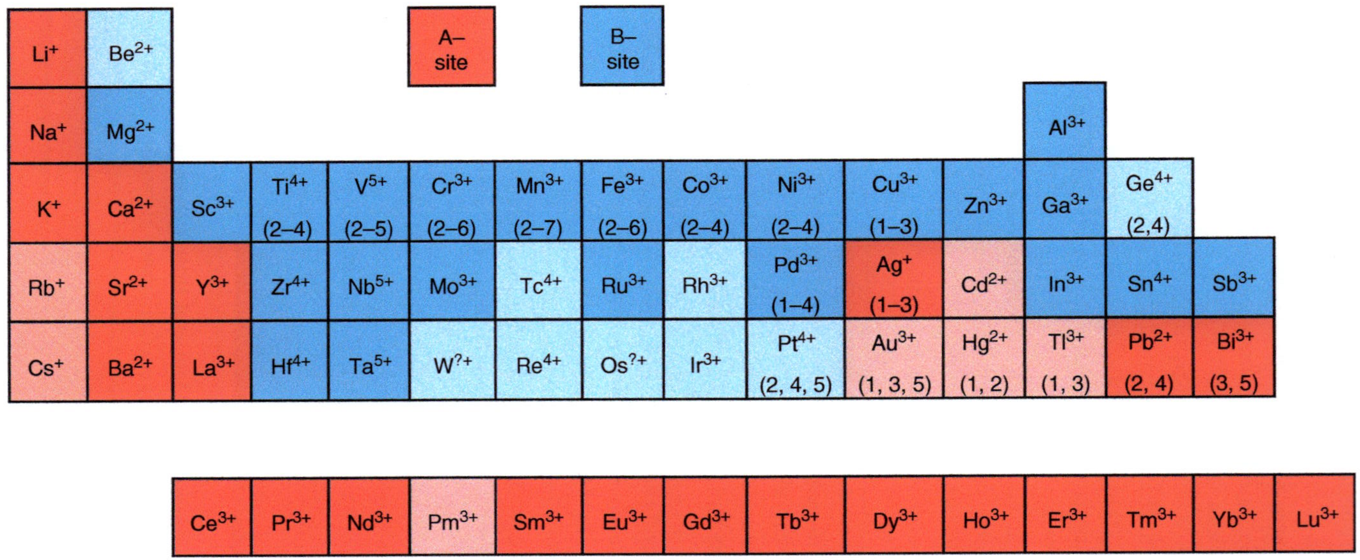

Figure 6.13 Panels of elements considered as dopants are filled with red (A-site) or blue (B-site) representing their preferred doping site. The solid colors represent experimentally achieved substitutions, while the lighter colors are only predicted by theoretical calculations. Note that these data are mainly concluded from several reviews [10, 99, 100], without listing their initial studies. Investigations that included more than one dopant were not included either.

can give a remnant polarization of 35 μC/cm^2 [107]. The values in polycrystalline films are normally larger through similar doping, due to the reduced dimension and endurability of larger electric fields. For example, in polycrystalline Sm-doped $BiFeO_3$ films, the measured P_r can reach ~90 μC/cm^2 [108]. These are only several examples out of many investigations, which are not necessarily the best results one can get [14, 103, 104]. In general, they illustrate the significant enhancement of doping on the ferroelectricity.

This is the same case for the piezoelectricity, since they are not completely independent (the piezoelectricity is related to the ferroelectricity through the simplified relationship $d \sim P{\cdot}\varepsilon_r$, where ε_r is the dielectric constant). The d_{33} values of single phase $BiFeO_3$ ceramics are typically lower than 30 pC/N, thus having room for further improvement. One example of the enhanced piezoelectric effect is obtained by La^{3+} or/and Sm^{3+} substitution on the Bi^{3+} site, achieving a d_{33} value of 50 pC/N as well as a reduction in leakage current [109, 110].

6.6.2 On Phase Transformation

Dopants provide a chemical pressure in the crystal structures and thus can alter the symmetry of the unit cell. Therefore, there is great potential for forming morphotropic phase boundary (MPB) through ion substitutions. Here we show the most established example of how the crystal structure symmetry of $BiFeO_3$ can be altered by rare-earth substitution [103].

A phase diagram was found for rare-earth (Sm, Dy, and La) doping as shown in Figure 6.14. Figure 6.14a shows the transition from a ferroelectric phase to a paraelectric phase through an antiferroelectric medium phase under strong chemical pressure with different amounts of the dopants at room temperatures. Figure 6.14b shows the same transition sequence without the antiferroelectric phase under no chemical pressure. The control parameter of the chemical pressure was the doped rare-earth ionic radius. The ionic radii of these trivalent ions with the coordination number of twelve are Bi^{3+} (1.36 Å) > La^{3+} (1.36 Å) > Sm^{3+} (1.28 Å) > Dy^{3+} (1.24 Å). Thus, for a cation with the same nominal radius as Bi^{3+}, for example, La^{3+}, there is no chemical pressure, while smaller ions, such as Sm^{3+} and Dy^{3+}, provide strong chemical pressure. Regardless of the chemical pressure status, one common feature of the phase transition with increasing temperature is that the bridging phase with a 1/4 diffraction spot disappears. Worth noting is that the scale of the composition and temperature is comparable in panels (a) and (b), i.e. the stronger chemical pressure shifts the 1/4 spot region into lower temperature and lower composition side (T_2 to T_1, x_2 to x_1, x_2' to x_1'). Such phase transformation and MPB behavior appear to be consistent in bulk $BiFeO_3$ ceramics and thin films with such doping based on many studies.

6.6.3 On Magnetic Properties

Transition-metal dopants on B-site may greatly influence the magnetic property of $BiFeO_3$ [103]. One of the interesting dopants is Cr^{3+}, which has three electrons in

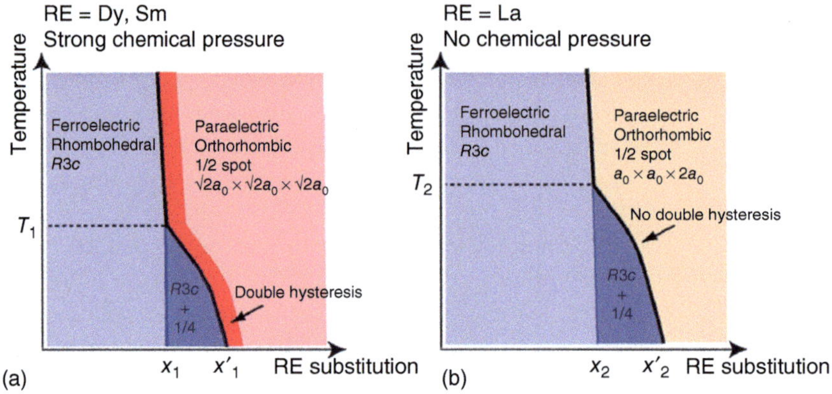

Figure 6.14 Schematic phase diagram for RE-substituted $BiFeO_3$ in the presence of (a) the strong chemical pressure due to RE substitutions (RE = Dy and Sm) and (b) no (or negligibly weak) chemical pressure (RE = La). Source: Reprinted with permission from Yang et al. [103]. Copyright 2012, Royal Society of Chemistry.

d-orbitals producing a magnetic moment of 3 μB. An ideal compound is a double perovskite of $BiFe_{0.5}Cr_{0.5}O_3$ in which Fe^{3+} and Cr^{3+} are G-type ordered making all bonds ferromagnetic type. Mn^{3+} ion with four electrons in d-orbitals shows also such promise by similar design. This shows great promise of realizing a ferromagnetic ground state but has not been realized yet, partially due to the very small amount of substitution, which is difficult to stabilize an overall ferromagnetic state. These substitutions also decrease the Néel temperature and have possibility to align the transition temperatures of magnetic order and ferroelectric order, enhancing the magnetoelectric properties at transition temperatures.

6.7 $BiFeO_3$-Based Solid Solutions

Addition of other perovskite ABO_3 compounds into $BiFeO_3$ forms a solid solution. In principle, solid solutions can reach even higher polarization values and piezoelectric coefficients because of the reduction of leakage current and more importantly, the creation of phase boundaries with other perovskite ferroelectrics with a different phase structure. Particularly, at MPBs, the piezoelectric coefficients can be enhanced significantly. These make $BiFeO_3$-based solid solutions a better objective when considering practical applications.

6.7.1 $BiFeO_3$–$BaTiO_3$

$BiFeO_3$ and $BaTiO_3$ have rhombohedral (R) and tetragonal (T) structures, respectively. $BiFeO_3$–$BaTiO_3$ solid solutions are the most investigated $BiFeO_3$-based systems ([14, 111–116]), which have recently emerged as a candidate to replace commercial $Pb(Zr,Ti)O_3$ with similar phase boundaries with increasing $BaTiO_3$ content (i.e. a general trend from the R to T transition). It is believed that $BiFeO_3$–$BaTiO_3$ forms a continuous solid solution at room temperature with crystal structures change as follows: R (100–67% $BiFeO_3$), pseudo-cubic (67–7.5% $BiFeO_3$,

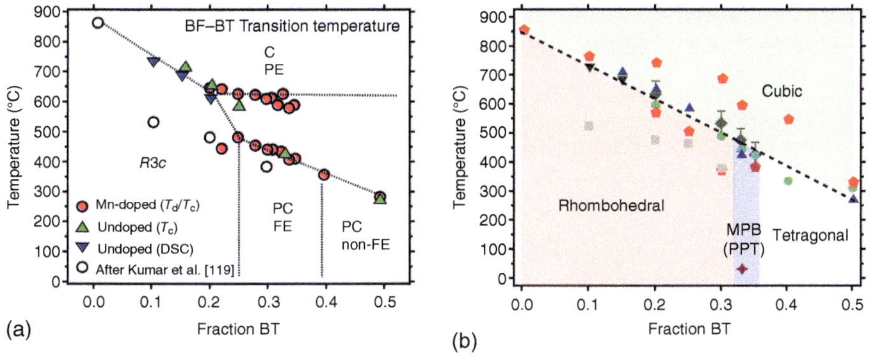

Figure 6.15 (a) Phase diagram of $BiFeO_3$–$BaTiO_3$ solid solution with an R–C boundary. (b) Phase diagram of quenched $BiFeO_3$–$BaTiO_3$ system with an R–T boundary. The abbreviations C, PC, PE, FE, and PPT denote for cubic structure, pseudo-cubic structure, paraelectric, ferroelectric, and polymorphic phase transition, respectively. Source: (a) Reprinted with permission from Leontsev and Eitel [114]. Copyright 2009, John Wiley & Sons, Inc. (b) Reproduced with permission from Lee et al. [120]. Copyright 2015, Wiley-VCH Verlag GmbH & Co. KGaA.

note that some researchers also reported R-T-like phase boundaries within this range), and T (7.5–0% $BiFeO_3$) [113, 114, 117, 118, 120–123]. It is worth noting, however, the Curie temperature is normally reduced, reaching ~400 °C for such solid solutions (this is still higher than commercial $Pb(Zr,Ti)O_3$).

In 2009, Leontsev and Eitel reported a broad, temperature-independent MPB (0.25–0.4 $BaTiO_3$) between R and C phases for furnace-cooled $BiFeO_3$–$BaTiO_3$ system, as shown in Figure 6.15a, with enhanced piezoelectricity (d_{33} = 116 pC/N) based on temperature-dependent dielectric constant and differential scanning calorimetry (DSC) measurements [114]. Enhanced piezoelectricity (d_{33} = 124–240 pC/N) can be generally observed in $(1-x)BiFeO_3$–$xBaTiO_3$ ceramics due to the involvement of phase boundaries ([113, 121–123]). Furthermore, Lee et al. reported a $Pb(Zr,Ti)O_3$-like rhombohedral–tetragonal (R–T) MPB in quenched $BiFeO_3$–$BaTiO_3$ based ceramics as shown in Figure 6.15b in 2015. A giant d_{33} value of 405 pC/N and high T_C of 454 °C was reported in $BiFeO_3$–$BaTiO_3$ ceramics by forming such phase boundary, together with the improved microstructure (i.e. dense microstructures and large grain sizes) [120].

Worth noting is that the previous system, where the phase boundary was discovered, could be modified by a third element, for instance, Mn, Cu, or Ga. Among these additional elements, some improves the microstructures (for instance, Ga, Sc [124], and Co [125]) while others form a phase boundary (such as Nd [126], Al [127], La [128], and Ga) to enhance the electromechanical responses. Similarly, the addition of a third component can show two types of effect. For example, the addition of $Bi(Mg_{0.5}Ti_{0.5})O_3$ and $Bi(Ni_{0.5}Ti_{0.5})O_3$ does not result in phase transitions, and the enhanced piezoelectricity is mainly attributed to the dense microstructure and large grain size [129–131]. Conversely, an R–T phase boundary can be formed by adding $Bi(Zn_{0.5}Ti_{0.5})O_3$ [120], $Bi(Mg_{0.5}Zr_{0.5})O_3$ [132] or $Bi_{0.5}Na_{0.5}TiO_3$ [133], and the rhombohedral–pseudocubic (R–PC) phase boundary has been reported in $BiFeO_3$–$BaTiO_3$–$Bi_{0.5}K_{0.5}TiO_3$ ceramics [134] with greatly enhanced electrical properties.

Overall, enhanced electrical properties can be mainly attributed to the phase boundary in $BiFeO_3$–$BaTiO_3$ solid solutions, while other factors (e.g. dense microstructure) are partly responsible for the improved properties in some cases. These literatures point to the potential of the $BiFeO_3$–$BaTiO_3$ solid solution as high-temperature and environmental-friendly piezoelectric devices with good performance.

6.7.2 Other Solid Solutions

$BiFeO_3$–$PbTiO_3$ is another typical solid solution for $BiFeO_3$-based ceramics, which has a very similar phase transition sequence with that of $BiFeO_3$–$BaTiO_3$ with enhanced electrical properties [135–137]. By the construction of an R–T phase boundary, the polarization of $BiFeO_3$–$PbTiO_3$ ceramics can reach ~60 μC/cm^2 [116]. The exact phase composition at the boundaries is not agreed yet within this field, with debates focusing on the intermediate phase bridging the R–T phases [136, 138, 139]. Although relatively good piezoelectricity can be achieved in $BiFeO_3$–$PbTiO_3$ ceramics, less attention has been given to this material due to the presence of the hazardous lead element.

In addition to $BaTiO_3$ and $PbTiO_3$, the phase structures and multiferroic properties of the $BiFeO_3$-based solid solutions may be strongly dependent on the type of doping compounds, for example, the addition of $SrTiO_3$ [140], $(K,Na)NbO_3$ [141], $Bi_{0.5}Na_{0.5}TiO_3$ [142], and $CoFe_2O_4$ [143] have been intensively investigated.

6.8 Application of $BiFeO_3$: Potentials and Status

Piezoelectric materials are very useful in industrial and commercial applications, covering all areas from our workplaces, homes to automobiles. Among these applications are five main categories, namely, capacitors (utilizing high dielectric constant), generators (related to the direct piezoelectric effect), motors or actuators (related to converse piezoelectric effect), sensors, and transducers [144–148]. The form of the materials desired varies from case to case as shown in Figure 6.16 [146]. Sometimes, bulk materials are more appropriate, while in other applications, films are desired. There are also cases where both bulk and film are applicable. It is noteworthy that the largest number of applications remains to be associated with bulk materials so far and bulk devices may always be an indispensable demand.

$BaTiO_3$ and $Pb(Zr,Ti)O_3$ materials are still dominating the field since their discoveries. In Section 6.8.1, the developing directions of $BiFeO_3$ utilizing its ferroelectricity, piezoelectricity, magnetoelectric coupling, and conductive domain wall properties are discussed, with extension to existing prototypical devices when necessary.

However, for such applications to ever come into fruition, important obstacles must be removed, such as (i) processing issues, (ii) the higher conductivity/leakage (and thus also dielectric losses) of $BiFeO_3$ relative to $Pb(Zr,Ti)O_3$, (iii) high coercive field (6–10 kV/mm found in prior studies) and the difficulty of poling, and (iv) its

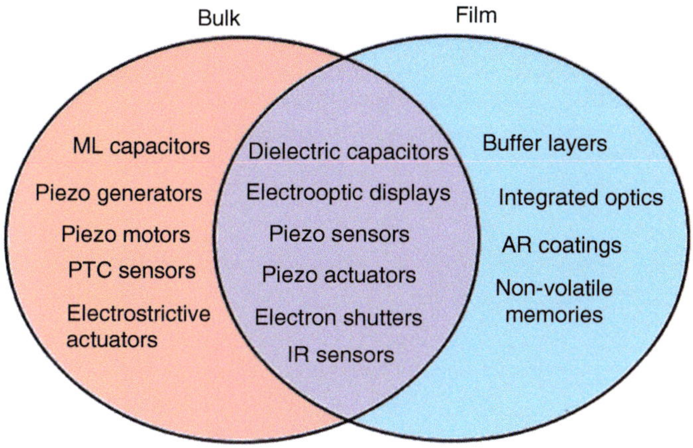

Figure 6.16 Bulk polycrystalline and thin film form applications of ferroic materials. ML, PTC, IR, and AR represent multilayer, positive temperature coefficient, infrared, and anti-reflection, respectively.

tendency to fatigue [149]. These sometimes could be suppressed by the doing and forming solid solution as have discussed earlier, but further breakthrough studies are of great demand to validate the practical applications to be discussed later.

6.8.1 Ferroelectricity and Electronics

To start with, ferroelectricity is the best and foremost advantage of $BiFeO_3$-based materials as it displays great P_r values (~100 μC/cm^2). Though there are still limitations of attaining such a value in bulk format of $BiFeO_3$ as mentioned earlier, it is normally obtainable in high-quality films, which is fortunately the desired form for integration in electronics. There is also really giant P_r value in thin films (~150 μC/cm^2), caused by symmetry distortion [150], making it more attractive.

Examples of prototypical devices based on the utilization of ferroelectric polarization of $BiFeO_3$ will be described in the following text:

One example is ferroelectric field-effect transistors (FeFETs) utilizing ferroelectric thin films and conductive oxides, which operates in a planar geometry, i.e. the current flows in a conductive channel adjacent to the ferroelectric (the schematic shown in Figure 6.17a). To achieve the largest possible modulation of the channel resistance in FeFETs, the transport response of the channel material must be very sensitive to changes in the carrier density, and the ferroelectric must allow for a large carrier density modulation [151]. Nominally, the density of surface charge accumulated or depleted by the ferroelectric is equal to its polarization. The large polarization of $BiFeO_3$ makes it an attractive gate oxide in FeFETs ($P = 100\,\mu C/cm^2$ corresponds to 6.25×10^{14} electrons/cm^2; for a typical perovskite oxide with an in-plane lattice constant of 3.8 Å this is 0.9 electrons per surface unit cell). FeFETs devices based on $BiFeO_3$ enable its transportation functionality (i.e. 0.9 electrons per surface unit cell) much better than using a typical ferroelectric, i.e. $Pb(Zr,Ti)O_3$ [152, 153].

Another operating mode of the electronics is a perpendicular geometry, such as ferroelectric tunnel junctions or ferroelectric diodes, where the current flows

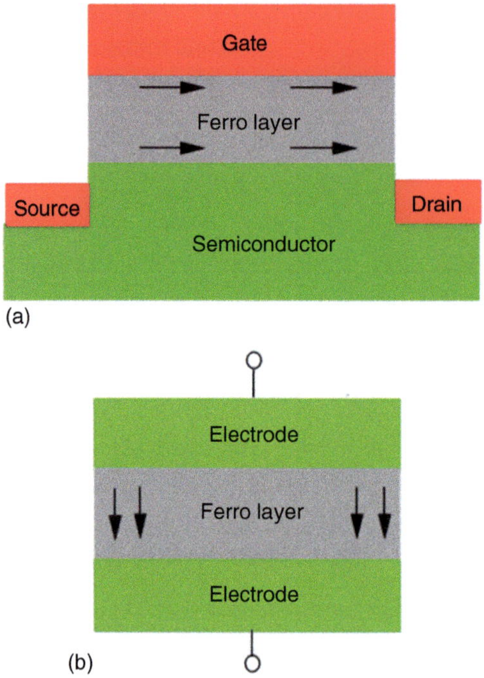

Figure 6.17 Schematic view (a) of a simplified FeFET in the planar geometry and (b) of a simplified ferroelectric diode in the perpendicular geometry.

across the ferroelectrics, as schematically shown in Figure 6.17b) [54, 154]. Ferroelectric diodes usually adopt a metal–ferroelectric–metal structure and show resistive switching behavior controlled by ferroelectric polarization. They have great potentials in non-volatile memory applications, since it allows probing of the polarization state by measuring the resistance states. This circumvents the biggest problem of destructive read-out of conventional ferroelectric memories in which device architecture the original polarization state must be reverted every time after the stored information is read [155]. Large switchable polarization and obvious hysteresis behavior of $BiFeO_3$ makes it useful for such devices. In addition, the newly found switchable diode effect in $BiFeO_3$ has made it more attractive for potential devices in the future [156–158].

6.8.2 Magnetoelectric Coupling and Spintronics

The most exciting feature of $BiFeO_3$ behind all the research is its multiferroicity at room temperature. Two critical factors to be considered for utilizing the multiferroicity of a material are: (i) the magnetoelectric coupling coefficient and (ii) control of (anti)ferromagnetism by ferroelectric switching. The magnetoelectric coupling coefficient is rather small for pure $BiFeO_3$, showing 0.067 mV/(cm Oe) [159]. The largest magnetoelectric coupling coefficient reported in the $BiFeO_3$-based material is reported in $0.3Co_{0.7}Zn_{0.3}Fe_2O_4$–$0.7Bi_{0.9}La_{0.1}FeO_3$ composite materials, showing 315 mV/(cm Oe) [160]. They are, however, generally inferior compared with other magnetoelectric materials [14]. On the other hand, although the magnetoelectric

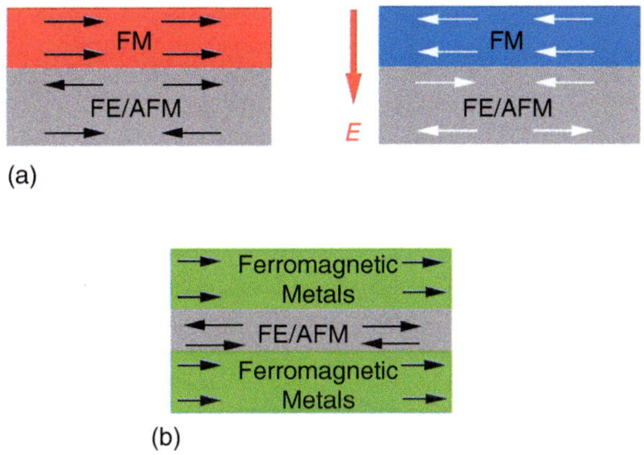

Figure 6.18 (a) Schematic of exchange bias coupling between ferromagnets (FM) and antiferromagnetic (AFM)/ferroelectric (FE) materials on application of an electric field. (b) Schematic of sandwiching $BiFeO_3$ between two ferromagnetic metals which results in a tunneling magnetoresistance.

coefficient is important, the deterministic control of (anti)ferromagnetism by ferroelectric switching is perhaps more critical to realize the practical device applications. Further understanding the control of antiferromagnetic order by electric field in $BiFeO_3$ and vice versa will promote the development and design of prototypical $BiFeO_3$-based devices.

To utilize the coupled manipulation of ferroelectricity and antiferromagnetism in memory media and spintronic devices, thin film is the required form for integration into the devices. It is worth noting that $BiFeO_3$ is by itself not technologically useful in the memory media as there is no robust magnetization as that in a ferromagnet for storing information. This is due to the operation principles of such memories, including electrically switchable and magnetically readable magnetic states. In the following we introduce prototypical devices with effective solutions for such a problem of $BiFeO_3$:

The first derives from the well-known concept of exchange bias coupling in ferromagnetic/antiferromagnetic systems, which is now extensively used in magnetic sensing and storage applications. Exchange bias is the magnetic interaction between the spins in the uppermost layer of an antiferromagnet and a thin ferromagnetic layer attached to it (as shown in Figure 6.18a). The exchange bias modifies the (usually symmetric) hysteresis loops of the ferromagnetic layer, either offsetting or widening them. The voltage-induced changes to the underlying antiferromagnetic domains will result in changes to the ferromagnetic hysteresis of the upper layer, which can then be read by conventional mechanisms. This solution was first proposed in multiferroic $YMnO_3$ [161], and subsequently implemented in $BiFeO_3$ (which is advantageous over $YMnO_3$ because it works at room temperature) with several important milestones [29, 162–164]. Devices in such geometry have been created through current attempts [165–168].

The second one utilizes ultrathin $BiFeO_3$ films as a barrier layer in spintronics. Sandwiching $BiFeO_3$ between two ferromagnetic metals (note the requirement that the $BiFeO_3$ layer should be reasonably insulated down to tunneling thicknesses) results in tunneling magnetoresistance (TMR) as shown in Figure 6.18b [10, 163]. $BiFeO_3$, which remains a robust and switchable ferroelectric down to a thickness of 2 nm [169], could be used as an electrically switchable tunnel junction. In such devices, the ferroelectric state controls the magnetic state of the thin ferromagnetic electrodes, thus modifying the TMR. A four-state memory device based on this idea was already demonstrated [170].

6.8.3 Domain Wall Based Electronics

Ferroelectric domain walls can be readily written, erased, rewritten, and displaced on demand by an applied electric field. The direct experimental demonstration of enhanced conductivity at specific domain walls in $BiFeO_3$ (see Section 6.5.3) [4, 90, 99] has led to the idea that a ferroelectric domain wall, rather than the domain itself, can store information. This is known as domain wall-based nanoelectronics as has been proposed for several years [21]. Critical advances made recently in understanding the nature of domain walls and thereby gaining the ability to precisely manipulate ferroelectric domain wall conductivity [171], have enabled the knowledge required to realize ferroelectric domain wall memory and logic devices based on these nanoscale elements [100, 172, 173]. However, since the immaturity of this field, a limited number of prototypical devices have been realized. We show two examples as follows:

One device utilizes the conductive domain walls in $BiFeO_3$ films to control the current transport between two electrodes (which are separated by a $BiFeO_3$ element) precisely through writing and erasing its domain walls [4]. Macroscopic I–V measurements reveal a step-like increase in the measured current on addition of a controlled number of conducting domain walls. The proposed functionality is of particular use in rewritable, multiconfiguration logic and memory applications, as the wall location and hence electronic conduction can be precisely controlled on the nanoscale.

Another device is a two-terminal prototype scalable (sub-100 nm) nonvolatile ferroelectric domain wall memory based on the conductive domain walls in $BiFeO_3$ thin films [22]. It is reported with high OFF/ON ratio ($\geq 10^3$), excellent retention (10^4 s), and robust endurance ($\sim 10^3$ cycles) characteristics. In particular, the specially designed in-plane electrode geometry allows for encoding and retrieval of information via moderate electric fields rather than electric currents, thus enabling low-energy operation. The low-voltage, pulse-based readout of the written states is nondestructive.

6.9 Summary

Having been discovered for more than half a century, the fundamental physics in $BiFeO_3$ as reviewed above was clearly clarified only recently with the success of processing $BiFeO_3$ materials (including large single crystals, pure-phase ceramics, and high-quality thin films) and advancement in the characterization methods. This chapter reviews the multiferrocity related aspects of $BiFeO_3$ in a sketch manner, and it is worth noting that there is a great detail of studies that could be referred to for each research direction. In addition, $BiFeO_3$ is a trendy topic that is continuing to develop at such a rapid rate. For instance, the photovoltaic effect discovered in $BiFeO_3$ is not included in this chapter, but is attracting much attention and providing another dimension for its future application. The rich physics in $BiFeO_3$ is undoubtedly attractive and draws consistent attention, while the most important thing of $BiFeO_3$ is still its potential applications, which is promising yet challenging from several aspects:

$BiFeO_3$ exhibits high Curie temperature, which spared several hundred degrees comparing with the commercial ones for applications in harsher environments. Ferroelectricity of $BiFeO_3$ is another advantage (i.e. it has one of the largest polarization among known oxide ferroelectrics) and enables great potentials in memory cells. The piezoelectricity of $BiFeO_3$ has also been doubled by doping or forming solid solutions in the past. However, the piezoelectricity is facing a bottleneck for improvement and is an order of magnitude lower than that of soft lead zirconate titanate (PZT). To enable applications based on its piezoelectricity, further improvements are required. For either advantages or shortcomings mentioned here, leakage, stability, and fatigue behaviors are the ground issues, and reduced leakage, enhanced stability, and prolonged lifespan are prerequisite to further enhance the properties of $BiFeO_3$ at whatever temperature range.

Magnetoelectric coupling of $BiFeO_3$ at room temperature and its application in spintronics is perhaps the major interests of this material, despite the low coupling coefficients that remains a real challenge. Further efforts are necessary to accomplish breakthroughs in this field. Domain wall conductivity and related electronics seems to be developing faster among all the fields. In the last decade, several orders of magnitude increase of the domain wall current in $BiFeO_3$ has been achieved and prototypical devices also developed.

We hope that the contents in this chapter are sufficient to furnish the audience with the fundamentals of $BiFeO_3$ and motivate them to accomplish breakthroughs in the field of $BiFeO_3$-based materials.

References

1 Royen, P. and Swars, K. (1957). Das system Wismutoxyd-Eisenoxyd im Bereich von 0 bis 55 Mol% Eisenoxyd. *Angewandte Chemie* 69: 779–779.

2 Teague, J.R., Gerson, R., and James, W.J. (1970). Dielectric hysteresis in single crystal $BiFeO_3$. *Solid State Communications* 8: 1073–1074.

3 Fischer, P., Polomska, M., Sosnowska, I., and Szymanski, M. (1980). Temperature dependence of the crystal and magnetic structures of $BiFeO_3$. *Journal of Physics C: Solid State Physics* 13: 1931–1940.

4 Seidel, J., Maksymovych, P., Yu, P. et al. (2009). Conduction at domain walls in oxide multiferroics. *Nature Materials* 8: 229–234.

5 Smolenski, G.A. and Chupis, I.E. (1982). Ferroelectromagnets. *Soviet Physics – Uspekhi* 25: 475–493.

6 Hill, N.A. (2000). Why are there so few magnetic ferroelectrics? *Journal of Physical Chemistry B* 104: 6694–6709.

7 Prellier, W., Singh, M., and Murugavel, P. (2005). The single-phase multiferroic oxides: from bulk to thin film. *Journal of Physics: Condensed Matter* 17: R803–R832.

8 Eerenstein, W., Mathur, N.D., and Scott, J.F. (2006). Multiferroic and magnetoelectric materials. *Nature* 442: 759–765.

9 Ramesh, R. and Spaldin, N.A. (2007). Multiferroics: progress and prospects in thin films. *Nature* 6: 21–29.

10 Bea, H., Gajek, M., Bibes, M., and Barthlemy, A. (2008). Spintronics with multiferroics. *Journal of Physics: Condensed Matter* 20: 434221.

11 Ma, J., Hu, J., Li, Z., and Nan, C.-W. (2011). Recent progress in multiferroic magnetoelectric composites: from bulk to thin films. *Advanced Materials* 23: 1062–1087.

12 Kleemann, W. (2013). Magnetoelectric spintronics. *Journal of Applied Physics* 114: 027013.

13 Ortega, N., Kumar, A., Scott, J.F., and Katiyar, R.S. (2015). Multifunctional magnetoelectric materials for device applications. *Journal of Physics: Condensed Matter* 27: 504002.

14 Wu, J.G., Fan, Z., Xiao, D.Q. et al. (2016). Multiferroic bismuth ferrite-based materials for multifunctional applications: ceramic bulks, thin films and nanostructures. *Progress in Materials Science* 84: 335–402.

15 Ascher, E., Rieder, H., Schmid, H., and Stössel, H. (1966). Some properties of ferromagnetoelectric nickel–iodine boracite, $Ni_3B_7O_{13}I$. *Journal of Applied Physics* 37: 1404.

16 Brixel, W., Rivera, J.-P., Steiner, A., and Schmid, H. (1988). Magnetic field induced magnetoelectric effects, $(ME)_H$, in the perovskites Pb_2CoWO_6 and Pb_2FeTaO_6. *Ferroelectrics* 79: 201–204.

17 Lorenz, B., Litvinchuk, A.P., Gospodinov, M.M., and Chu, C.W. (2004). Field-induced reentrant novel phase and a ferroelectric-magnetic order coupling in $HoMnO_3$. *Physical Review Letters* 92: 087204.

18 Lebeugle, D., Colson, D., Forget, A., and Viret, M. (2007). Very large spontaneous electric polarization in $BiFeO_3$ single crystals at room temperature and its evolution under cycling fields. *Applied Physics Letters* 91 (2): 022907.

19 Catalan, G. and Scott, J.F. (2009). Physics and applications of Bismuth Ferrite. *Advanced Materials* 21 (24): 2463–2485.

20 Rojac, T., Bencan, A., Malic, B. et al. (2014). $BiFeO_3$ ceramics: processing, electrical, and electromechanical properties. *Journal of the American Ceramic Society* 97 (7): 1993–2011.

21 Catalan, G., Seidel, J., Ramesh, R., and Scott, J.F. (2012). Domain wall nanoelectronics. *Reviews of Modern Physics* 84: 119–156.

22 Sharma, P., Zhang, Q., Sando, D. et al. (2017). Nonvolatile ferroelectric domain wall memory. *Science Advances* 3: e1700512.

23 Quezel, S., Tcheou, F., Rossat-Mignod, J. et al. (1977). Magnetic structure of the perovskite-like compound $TbMnO_3$. *Physica* 86–88B: 916–918.

24 Ratcliff, W., Lynn, J.W., Kiryukhin, V. et al. (2016). Magnetic structures and dynamics of multiferroic systems obtained with neutron scattering. *NPJ Quantum Materials* 1: 16003.

25 Streiffer, S.K., Parker, C.B., Romanov, A.E. et al. (1998). Domain patterns in epitaxial rhombohedral ferroelectric films. I. Geometry and experiments. *Journal of Applied Physics* 83: 2742–2753.

26 Rojac, T., Ursic, H., Bencan, A. et al. (2015). Mobile domain walls as a bridge between nanoscale conductivity and macroscopic electromechanical response. *Advanced Functional Materials* 25 (14): 2099–2108.

27 Sosnowska, I., Peterlin-Neumaier, T., and Steichele, E. (1982). Spiral magnetic ordering in bismuth ferrite. *Journal of Physics C: Solid State Physics* 15: 4835–4836.

28 Ederer, C. and Spaldin, N.A. (2005). Weak ferromagnetism and magnetoelectric coupling in bismuth ferrite. *Physical Review B* 71: 060401.

29 Chu, Y., Martin, L., Holcomb, M. et al. (2008). Electric-field control of local ferromagnetism using a magnetoelectric multiferroic. *Nature Materials* 7: 478–482.

30 Zhao, T., Scholl, A., Ederer, C. et al. (2006). Electrical control of antiferromagnetic domains inmultiferroic $BiFeO_3$. *Nature Materials* 5: 823–829.

31 Lebeugle, D., Colson, D., Forget, A. et al. (2008). Electric-field-induced spin flop in $BiFeO_3$ single crystals at room temperature. *Physical Review Letters* 100: 227602.

32 Lee, S., Ratcliff, W., Cheong, S.-W., and Kiryukhin, V. (2008). Electric field control of the magnetic state in $BiFeO_3$ single crystals. *Applied Physics Letters* 92: 192906.

33 Liu, L.S., Hinterstein, M., Rojac, T. et al. (2019). In situ study of electric-field-induced ferroelectric and antiferromagnetic domain switching in polycrystalline $BiFeO_3$. *Journal of the American Ceramic Society* 102: 1768–1775.

34 Lee, S., Choi, T., Ratcliff, W. et al. (2008). Single ferroelectric and chiral magnetic domain of single-crystalline $BiFeO_3$ in an electric field. *Physical Review B* 78: 100101.

35 Zvezdina, A.K., Kadomtseva, A.M., Krotov, S.S. et al. (2006). Magnetoelectric interaction and magnetic field control of electric polarization in multiferroics. *Journal of Magnetism and Magnetic Materials* 300: 224–228.

36 Palai, R., Katiyar, R.S., Schmid, H. et al. (2008). β phase and β–γ metal–insulator transition in multiferroic $BiFeO_3$. *Physical Review B* 77: 014110.

37 Morozov, M.I., Lomanova, N.A., and Gusarov, V.V. (2003). Specific features of $BiFeO_3$ formation in a mixture of bismuth(III) and iron(III) oxides. *Russian Journal of General Chemistry* 73: 1676–1680.

38 Shrout, T. and Zhang, S.J. (2007). Lead-free piezoelectric ceramics: alternatives for PZT? *Journal of Electroceramics* 19 (1): 113–126.

39 Takenaka, T. (2010). Lead-free piezo-ceramics. In: *Advanced Piezoelectric Materials*, Woodhead Publishing Series in Electronic and Optical Materials (ed. K. Uchino), 130–170. Woodhead Publishing Limited.

40 Kornev, I.A., Lisenkov, S., Haumont, R. et al. (2007). Finite-temperature properties of multiferroic $BiFeO_3$. *Physical Review Letters* 99: 227602.

41 Haumont, R., Kornev, I.A., Lisenkov, S. et al. (2008). Phase stability and structural temperature dependence in powdered multiferroic $BiFeO_3$. *Physical Review B* 78: 134108.

42 Amold, D.C., Knight, K.S., Morrison, F.D., and Lightfoot, P. (2009). Ferroelectric–paraelectric transition in $BiFeO_3$: crystal structure of the orthorhombic β phase. *Physical Review Letters* 102: 027602.

43 Selbach, S.M., Einarsrud, M.-A., and Grande, T. (2009). On the thermodynamic stability of $BiFeO_3$. *Chemistry of Materials* 21: 169–173.

44 Thrall, M., Freer, R., Martina, C. et al. (2008). An in situ study of the formation of multiferroic bismuth ferrite. *Journal of the European Ceramic Society* 28: 2567–2572.

45 Selbach, S.M., Tybell, T., Einarsru, M.-A., and Grande, T. (2010). Phase transitions, electrical conductivity and chemical stability of $BiFeO_3$ at high temperatures. *Journal of Solid State Chemistry* 183: 1205–1208.

46 Li, M. and MacManus-Driscoll, J.L. (2005). Phase stability, oxygen nonstoichiometry and magnetic properties. *Applied Physics Letters* 87: 252510.

47 Valant, M., Axelsson, A.-K., and Alford, N. (2007). Peculiarities of a solid-state synthesis of multiferroic polycrystalline $BiFeO_3$. *Chemistry of Materials* 19: 5431–5436.

48 Béa, H., Dupé, B., Fusil, S. et al. (2009). Evidence for room-temperature multiferroicity in a compound with a giant axial ratio. *Physical Review Letters* 102: 217603.

49 Ryu, S.W., Kim, J.-Y., Shin, Y.-H. et al. (2009). Enhanced magnetization and modulated orbital hybridization in epitaxially constrained $BiFeO_3$ thin films with rhombohedral symmetry. *Chemistry of Materials* 21: 5050–5057.

50 Yang, J.C., He, Q., Suresha, S.J. et al. (2012). Orthorhombic $BiFeO_3$. *Physical Review Letters* 109: 247606.

51 Toupet, H., Marrec, F., Lichtensteiger, C. et al. (2010). Evidence for a first-order transition from monoclinic α to monoclinic β phase in $BiFeO_3$ thin films. *Physical Review B* 81: 140101.

52 Chen, Z.H., Qi, Y.J., You, L. et al. (2013). Large tensile strain induced monoclinic MB phase in $BiFeO_3$ epitaxial thin films on $PrScO_3$ substrate. *Physical Review B* 88: 054114.

53 LiYan, Cao, H., Li, J., and Viehland, D. (2009). Triclinic phase in tilted (001) oriented $BiFeO_3$ epitaxial thin films. *Applied Physics Letters* 94: 132901.

54 Sando, D., Barthelemy, A., and Bibes, M. (2014). $BiFeO_3$ epitaxial thin films and devices: past, present and future. *Journal of Physics: Condensed Matter* 26: 473201.

55 Jia, C.L., Contreras, J.R.G., Poppe, U. et al. (2002). Lattice strain and lattice expansion of the $SrRuO_3$ layers in $SrRuO_3/PbZr_{0.52}Ti_{0.48}O_3/SrRuO_3$ multilayer thin films. *Journal of Applied Physics* 92: 101–105.

56 Sando, D., Yang, Y., Bousquet, E. et al. (2016). Large elasto-optic effect and reversible electrochromism in multiferroic $BiFeO_3$. *Nature Communications* 7: 10718.

57 Ueckera, R., Velickova, B., Klimma, D. et al. (2008). Properties of rare-earth scandate single crystals (Re = Nd–Dy). *Journal of Crystal Growth* 310: 2649–2658.

58 Johann, F., Morelli, A., Biggemann, D. et al. (2011). Epitaxial strain and electric boundary condition effects on the structural and ferroelectric properties of $BiFeO_3$ films. *Physical Review B* 84: 094105.

59 Chen, Z., You, L., Huang, C. et al. (2010). Nanoscale domains in strained epitaxial $BiFeO_3$ thin Films on $LaSrAlO_4$ substrate. *Applied Physics Letters* 96: 252903.

60 Vailionis, A., Siemons, H.B.W., Houwman, E.P. et al. (2011). Misfit strain accommodation in epitaxial ABO_3 perovskites: lattice rotations and lattice modulations. *Physical Review B* 83: 064101.

61 Guo, H.Z., Zhao, R.Q., Jin, K.J. et al. (2015). Interfacial-strain-induced structural and polarization evolutions in epitaxial multiferroic $BiFeO_3$ (001) thin films. *ACS Applied Materials & Interfaces* 7: 2944–2951.

62 Liu, H.J., Yang, P., You, L. et al. (2015). Nanoscale phase mixture in uniaxial strained $BiFeO_3$ (110) thin films. *Journal of Applied Physics* 118: 104103.

63 Wang, J., Neaton, J., Zheng, H. et al. (2003). Epitaxial $BiFeO_3$ multiferroic thin film heterostructures. *Science* 299 (5613): 1719–1722.

64 Bea, H., Bibes, M., Fusil, S. et al. (2006). Investigation on the origin of the magnetic moment of $BiFeO_3$ thin films by advanced X-ray characterizations. *Physical Review B* 74: 020101.

65 MacDougall, G.J.W., Siemons, H.M.C., Biegalski, M.D. et al. (2012). Antiferromagnetic transitions in tetragonal-like $BiFeO_3$. *Physical Review B* 85: 100406.

66 Kaczmarek, W., Pajak, Z., and Polomska, M. (1975). Differential thermal analysis of phase transitions in $(Br,La)FeO_3$ solid solution. *Solid State Communications* 17: 807–810.

67 Kamba, S., Nuzhnyy, D., Savinov, M. et al. (2007). Infrared and terahertz studies of polar phonons and magnetodielectric effect in multiferroic $BiFeO_3$ ceramics. *Physical Review B* 75: 024403.

68 Lu, J., Gunther, A., Schrettle, F. et al. (2010). On the room temperature multiferroic $BiFeO_3$: magnetic, dielectric and thermal properties. *European Physical Journal B* 75: 451–460.

69 Morozov, M.I. and Damjanovic, D. (2010). Charge migration in $Pb(Zr,Ti)O_3$ ceramics and its relation to ageing hardening, and softening. *Journal of Applied Physics* 107: 034106.

70 Hunpratub, S., Thongbai, P., Yamwong, T. et al. (2009). Dielectric relaxations and dielectric response in multiferroic $BiFeO_3$ ceramics. *Applied Physics Letters* 94: 062904.

71 Rojac, T., Makarovic, M., Walker, J. et al. (2016). Piezoelectric response of $BiFeO_3$ ceramics at elevated temperatures. *Applied Physics Letters* 109: 042904.

72 Lu, J., Günther, A., Schrettle, F. et al. (2010). On the room temperature multiferroic $BiFeO_3$: magnetic, dielectric and thermal properties. *European Physics Journal B* 75: 451–460.

73 Markiewicz, E., Hilczer, B., Baszyk, M. et al. (2011). Dielectric properties of $BiFeO_3$ ceramics obtained from mechanochemically synthesized nanopowders. *Journal of Electroceramics* 27: 154–161.

74 Yuan, G.L., Or, S.W., Wang, Y.P. et al. (2006). Preparation and multi-properties of insulated single-phase $BiFeO_3$ ceramics. *Solid State Communications* 138: 76–81.

75 Wang, Y.P., Zhou, L., Zhang, M.F. et al. (2004). Room-temperature saturated ferroelectric polarization in $BiFeO_3$ ceramics synthesized by rapid liquid phase sintering. *Applied Physics Letters* 84: 1731.

76 Pradhan, A.K., Hunter, K.Z.D., Dadson, J.B. et al. (2005). Magnetic and electrical properties of single-phase multiferroic $BiFeO_3$. *Journal of Applied Physics* 97: 093903.

77 Chen, F., Zhang, Q.F., Li, J.H. et al. (2006). Sol–gel derived multiferroic ceramics with large polarization and weak ferromagnetism. *Applied Physics Letters* 89: 092910.

78 Yuan, G.L., Baba-Kisgi, K.Z., Liu, J.-M. et al. (2006). Multiferroic properties of single phase $Bi_{0.8}5La_{0.15}FeO_3$ lead-free ceramics. *Journal of the American Ceramic Society* 89: 3136–3139.

79 Yu, B.F., Li, M.Y., Liu, J. et al. (2008). Effects of ion doping at different sites on electrical properties of multiferroic $BiFeO_3$ ceramics. *Journal of Physics D: Applied Physics* 41: 065003.

80 Yuan, G.L. and Or, S.W. (2006, 2006). Enhanced piezoelectric and pyroelectric effects in single-phase multiferroic $Bi_{1-x}Nd_xFeO_3$ (x = 0–0.15) ceramics. *Applied Physics Letters* 88: 062905.

81 Mazumder, R. and Sen, A. (2009). Effect of Pb-doping on dielectric properties of $BiFeO_3$ ceramics. *Journal of Alloys and Compounds* 475: 577–580.

82 Zhang, Z., Wu, P., Chen, L., and Wang, J.L. (2010). Density functional theory plus U study of vacancy formations in bismuth ferrite. *Applied Physics Letters* 96: 232906.

83 Paudel, T.R., Jaswal, S.S., and Tsymbal, E.Y. (2012). Intrinsic defects in multiferroic $BiFeO_3$ and their effect on magnetism. *Physical Review B* 85: 104409.

84 Masó, N. and West, A.R. (2012). Electrical properties of Ca-doped $BiFeO_3$ ceramics: from p-type semiconduction to oxide-ion conduction. *Chemistry of Materials* 24: 2127–2132.

85 Poghossian, A.S., Abovian, H.V., Avakian, S.H. et al. (1991). Bismuth ferrites: new materials for semiconductor gas sensors. *Sensors and Actuators B* 4: 545–549.

86 Wang, C., Takahashi, M., Fujino, H. et al. (2006). Leakage current of multiferroic $(Bi_{0.6}Tb_{0.3}La_{0.1})FeO_3$ thin films grown at various oxygen pressures by pulsed laser deposition and annealing effect. *Journal of Applied Physics* 99: 054104.

87 Kim, T.H., Jeon, B.C., Min, T.Y. et al. (2012). Continuous control of charge transport in Bi-deficient $BiFeO_3$ films through local ferroelectric switching. *Advanced Functional Materials* 22: 4962–4968.

88 Clark, S.J. and Robertson, J. (2009). Energy levels of oxygen vacancies in $BiFeO_3$ by screened exchange. *Applied Physics Letters* 94: 022902.

89 Salje, E.K.H. (2010). Multiferroic domain boundaries as active memory devices: trajectories towards domain boundary engineering. *ChemPhysChem* 11: 940–950.

90 Seidel, J., Maksymovych, P., Batra, Y. et al. (2010). Domain wall conductivity in La-doped $BiFeO_3$. *Physical Review Letters* 105: 197603.

91 Rojac, T., Bencan, A., Drazic, G. et al. (2017). Domain-wall conduction in ferroelectric $BiFeO_3$ controlled by accumulation of charged defects. *Nature Materials* 16: 322–327.

92 Liu, L.S., Roja, T., Damjanovic, D. et al. (2018). Frequency-dependent decoupling of domain-wall motion and lattice strain in bismuth ferrite. *Nature Communications* 9: 4928.

93 Sluka, T., Tagantsev, A.K., Bednyakov, P., and Setter, N. (2013). Free-electron gas at charged domain walls in insulating $BaTiO_3$. *Nature Communications* 4: 1808.

94 Lindgren, G. and Canalias, C. (2017). Domain wall conductivity in $KTiOPO_4$ crystals. *APL Materials* 5: 076108.

95 Du, Y., Wang, X.L., Chen, D.P. et al. (2011). Domain wall conductivity in oxygen deficient multiferroic $YMnO_3$ single crystals. *Applied Physics Letters* 99: 252107.

96 Campbell, M.P., McConville, J.P.V., McQuaid, R.G.P. et al. (2016). Hall effect in charged conducting ferroelectric domain walls. *Nature Communications* 7: 13764.

97 Meier, D., Seidel, J., Cano, A. et al. (2012). Anisotropic conductance at improper ferroelectric domain walls. *Nature Materials* 11: 284–288.

98 Schröder, M., Haußmann, A., Thiessen, A. et al. (2012). Conducting domain walls in lithium niobate single crystals. *Advanced Functional Materials* 22: 3936–3944.

99 Guyonnet, J., Gaponenko, I., Gariglio, S., and Paruch, P. (2011). Conduction at domain walls in insulating $Pb(Zr_{0.2}Ti_{0.8})O_3$ thin films. *Advanced Materials* 23: 5377–5382.

100 Whyte, J.R. and Gregg, J.M. (2015). A diode for ferroelectric domain-wall motion. *Nature Communications* 6: 7361.

101 Jiang, J., Bai, Z.L., Chen, Z.H. et al. (2017). Temporary formation of highly conducting domain walls for non-destructive read-out of ferroelectric domain-wall resistance switching memories. *Nature Materials* 17: 49–56.

102 Rojac, T., Kosec, M., and Damjanovic, D. (2011). Large electric-field induced strain in $BiFeO_3$ ceramics. *Journal of the American Ceramic Society* 94 (12): 4108–4111.

103 Yang, C.-H., Daisuke, K., Takeuchi, I. et al. (2012). Doping $BiFeO_3$: approaches and enhanced functionality. *Physical Chemistry Chemical Physics* 14: 15953–15962.

104 Gebhardt, J. and Rappe, A.M. (2018). Doping of $BiFeO_3$: a comprehensive study on substitutional doping. *Physical Review B* 98: 125202.

105 Rojac, T., Kosec, M., Budic, B. et al. (2010). Strong ferroelectric domain-wall pinning in $BiFeO_3$ ceramics. *Journal of Applied Physics* 108: 074107.

106 Hu, Z., Li, M., Yu, Y. et al. (2010). Effects of Nd and high-valence Mn co-doping on the electrical and magnetic properties of multiferroic $BiFeO_3$ ceramics. *Solid State Communications* 150: 1088–1091.

107 Yao, Y., Liu, W., Chan, Y. et al. (2011). Studies of rare-earth-doped $BiFeO_3$ ceramics. *International Journal of Applied Ceramic Technology* 8: 1246–1253.

108 Sun, W., Li, J.-F., Yu, Q., and Cheng, L.-Q. (2015). Phase transition and piezoelectricity of sol–gel processed Sm-doped $BiFeO_3$ thin films on Pt(111)/Ti/SiO_2/Si substrates. *Journal of Materials Chemistry C* 3: 2115–2122.

109 Zheng, T. and Wu, J.G. (2015). Effects of site engineering and doped element types on piezoelectric and dielectric properties of bismuth ferrite lead-free ceramics. *Journal of Materials Chemistry C* 3: 11326–11334.

110 Zheng, T. and Wu, J. (2015). Enhanced piezoelectric activity in high-temperature $Bi_{1-x-y}Sm_xLayFeO_3$ lead-free ceramics. *Journal of Materials Chemistry C* 3: 3684–3693.

111 Comyn, T.P., McBride, S.P., and Bell, A.J. (2004). Processing and electrical properties of $BiFeO_3$–$PbTiO_3$. *Materials Letters* 58: 3844–3846.

112 Comyn, T.P., Stevenson, T., and Bell, A.J. (2004). Piezoelectric properties of $BiFeO_3$–$PbTiO_3$ ceramics. In: *IEEE International Ultrasonics, Ferroelectrics, and Frequency Control Joint 50th Anniversary Conference 2004*, 122. IEEE.

113 Yang, H., Zhou, C., Liu, X. et al. (2012). Structural, microstructural and electrical properties of $BiFeO_3$–$BaTiO_3$ ceramics with high thermal stability. *Materials Research Bulletin* 47: 4233–4239.

114 Leontsev, S.O. and Eitel, R.E. (2009). Dielectric and piezoelectric properties in Mn-modified $(1-x)BiFeO_3$–$xBaTiO_3$ ceramics. *Journal of the American Ceramic Society* 92: 2957–2961.

115 Leontsev, S.O. and Eitel, R.E. (2010). Origin and magnitude of the large piezoelectric response in the lead-free $(1-x)BiFeO_3$–$xBaTiO_3$ solid solution. *Journal of Materials Research* 26: 9.

116 Amorín, H., Correas, C., Ramos, P. et al. (2012). Very high remnant polarization and phase-change electromechanical response of $BiFeO_3$–$PbTiO_3$ at the multiferroic morphotropic phase boundary. *Applied Physics Letters* 101: 172908.

117 Ismailzade, I.H., Ismailov, R.M., Alekberov, A.I., and Salaev, F.M. (1981). Investigation of the magnetoelectric $(ME)_H$: effect in solid solutions of the systems $BiFeO_3$–$BaTiO_3$ and $BiFeO_3$–$PbTiO_3$. *Physica Status Solidi A* 68.

118 Kumar, M.M., Srinivas, A., and Suryanarayana, S.V. (1999). Structure property relations in $BiFeO_3$/$BaTiO_3$ solid solutions. *Journal of Applied Physics* 87: 855.

119 Kumar, M., Srinivas, A., and Suryanarayana, S.V. (2000). Structure property relations in $BiFeO_3$–$BaTiO_3$ solid solutions. *Journal of Applied Physics* 87: 855–862.

120 Lee, M.H., Kim, D.J., Park, J.S. et al. (2015). High-performance lead-free piezoceramics with high curie temperatures. *Advanced Materials* 27: 6976–6982.

121 Yang, S.-C., Kumar, A., and Priya, S. (2013). Room-temperature magnetoelectric coupling in single-phase $BaTiO_3$–$BiFeO_3$ system. *Journal of Applied Physics* 113 (14): 144101.

122 Zheng, T., Ding, Y., and Wu, J. (2016). Bi nonstoichiometry and composition engineering in (1–*x*)$Bi_{1+y}FeO_3$–*x*$BaTiO_3$ piezoceramics. *RSC Advances* 6: 90831–90839.

123 Yang, H.B., Zhou, C.R., Liu, X.Y. et al. (2013). Piezoelectric properties and temperature stabilities of Mn- and Cu-modified $BiFeO_3$–$BaTiO_3$ high temperature ceramics. *Journal of the European Ceramic Society* 33: 1177–1183.

124 Wan, Y., Li, Y., Li, Q. et al. (2014). Microstructure, ferroelectric, piezoelectric, and ferromagnetic properties of Sc-modified $BiFeO_3$–$BaTiO_3$ multiferroic ceramics with MnO_2 addition. *Journal of the American Ceramic Society* 97: 1809–1818.

125 Zhou, C.R., Cen, Z.Y., Yang, H.B. et al. (2013). Structure, electrical properties of Bi(Fe,Co)O_3–$BaTiO_3$ piezoelectric ceramics with improved curie temperature. *Physica B: Condensed Matter* 410: 13–16.

126 Zheng, Q.J., Luo, L.L., Lam, K.H. et al. (2014). Enhanced ferroelectricity, piezoelectricity, and ferromagnetism in Nd-modified $BiFeO_3$–$BaTiO_3$ lead-free ceramics. *Journal of Applied Physics* 116: 184101.

127 Cen, Z.Y., Zhou, C.R., Yang, H.B. et al. (2013). Remarkably high-temperature stability of $Bi(Fe_{1-x}Al_x)O_3$–$BaTiO_3$ solid solution with near-zero temperature coefficient of piezoelectric properties. *Journal of the American Ceramic Society* 96: 2252–2256.

128 Wu, X.C., Tian, M.J., Guo, Y.Q. et al. (2015). Phase transition, dielectric, ferroelectric and ferromagnetic properties of La-doped $BiFeO_3$–$BaTiO_3$ multiferroic ceramics. *Journal of Materials Science: Materials in Electronics* 26: 978–984.

129 Lin, Y. and Yu, J. (2014). Ferroelectric and piezoelectric properties of high temperature perovskite-type $0.69BiFeO_3$–$0.02Bi(Mg_{1/2}Ti_{1/2})O_3$–$0.29BaTiO_3$ ceramics. *Journal of Materials Science: Materials in Electronics* 25: 5462–5466.

130 Zhou, C., Feteira, A., Shan, X. et al. (2012). Remarkably high-temperature stable piezoelectric properties of $Bi(Mg_{0.5}Ti_{0.5})O_3$ modified $BiFeO_3$–$BaTiO_3$ ceramics. *Applied Physics Letters* 101: 32901.

131 Zhou, Q., Zhou, C.R., Yang, H.B. et al. (2012). Dielectric, ferroelectric, and piezoelectric properties of $Bi(Ni_{1/2}Ti_{1/2})O_3$-modified $BiFeO_3$–$BaTiO_3$ ceramics with high curie temperature. *Journal of the American Ceramic Society* 95: 3889–3893.

132 Luo, L.L., Jiang, N., Lei, F. et al. (2014). Phase transition, ferroelectric and piezoelectric properties of $Bi(Mg_{0.5}Zr_{0.5})O_3$-modified $BiFeO_{3\_}BaTiO_3$ lead-free ceramics. *Journal of Materials Science: Materials in Electronics* 25: 1736–1744.

133 Li, Y., Guo, Y.Q., Zheng, Q.J. et al. (2015). Enhancement in multiferroic and piezoelectric properties of $BiFeO_3$–$BaTiO_3$–$Bi0.5Na_{0.5}TiO_3$ lead-free ceramics with MnO_2 addition by optimizing sintering temperature and dwell time. *Materials Research Bulletin* 68: 92–99.

134 Lin, D.M., Zheng, Q.J., Li, Y. et al. (2013). Microstructure, ferroelectric and piezoelectric properties of $Bi_{0.5}K_{0.5}TiO_3$-modified $BiFeO_3$–$BaTiO_3$ lead-free ceramics with high curie temperature. *Journal of the European Ceramic Society* 33: 3023–3036.

135 Bhattacharjee, S. and Pandey, D. (2010). Stability of the various crystallographic phases of the multiferroic (1−*x*)$BiFeO_3$–*x*$PbTiO_3$ system as a function of composition and temperature. *Journal of Applied Physics* 107: 124112.

136 Bhattacharjee, S., Tripathi, S., and Pandey, D. (2007). Morphotropic phase boundary in (1−*x*)$BiFeO_3$–*x*$PbTiO_3$: phase coexistence region and unusually large tetragonality. *Applied Physics Letters* 91: 042903.

137 Fedulov, S.A., Ladyzhinskii, P.B., Pyatigorskaya, I.L., and Venevtsev, Y.N. (1964). Complete phase diagram of the $PbTiO_3$–$BiFeO_3$ system. *Soviet Physics-Solid State* 6: 375–378.

138 Woodward, D.I., Reaney, I.M., Eitel, R.E., and Randall, C.A. (2003). Crystal and domain structure of the $BiFeO_3$–$PbTiO_3$ solid solution. *Journal of Applied Physics* 94: 3313.

139 Zhu, W.-M., Guo, H.-Y., and Ye, Z.-G. (2008). Structural and magnetic characterization of multiferroic $(BiFeO_3)_{1-x}(PbTiO_3)_x$ solid solutions. *Physical Review B* 78: 014401.

140 Ma, Z.Z., Tian, Z.M., Li, J.Q. et al. (2011). Enhanced polarization and magnetization in multiferroic (1−*x*)$BiFeO_3$–*x*$SrTiO_3$ solid solution. *Solid State Sciences* 13: 2196–2200.

141 Huo, S.X., Yuan, S.L., Qiu, Y. et al. (2012). Crystal structure and multiferroic properties of $BiFeO_3$–$Na_{0.5}K_{0.5}NbO_3$ solid solution ceramics prepared by Pechini method. *Materials Letters* 68: 8–10.

142 Tian, Z.M., Zhang, Y.S., Yuan, S.L. et al. (2012). Enhanced multiferroic properties and tunable magnetic behavior in multiferroic $BiFeO_3$–$Bi_{0.5}Na_{0.5}TiO_3$ solid solutions. *Materials Science Engineering B: Advanced Functional Solid-State Materials* 177: 74–78.

143 Mo, H.L., Jiang, D.-M., Wang, C.-M. et al. (2013). Magnetic, dielectric and magnetoelectric properties of $CoFe_2O_4$–$Bi_{0.85}La_{0.15}FeO_3$ multiferroic composites. *Journal of Alloys and Compounds* 579: 187–191.

144 Uchino, K. (1994). Piezoelectric actuators/ultrasonic motors – their developments and markets. In: *Proceedings of the Ninth IEEE International Symposium on Applications of Ferroelectrics*, 319–324. IEEE.

145 Alcock, C. (1992). Solid-state sensors and process control. *Solid State Ionics* 53–56: 3–17.

146 Haertling, G.H. (1999). Ferroelectric ceramics: history and technology. *Journal of the American Ceramic Society* 82: 797–818.

147 Zhao, L.L., Steinhart, M., Yosef, M. et al. (2005). Long scale template-assisted growth of $LiNbO_3$ one-dimensional nanostructures for nano-sensors. *Sensors and Actuators B* 109: 86–90.

148 Uchino, K. (1998). High electromechanical coupling piezoelectrics: relaxor and normal ferroelectric solid solutions. *Solid State Ionics* 108: 43–52.

149 Jang, H.W., Baek, S.H., Ortiz, D. et al. (2008). Epitaxial (001) $BiFeO_3$ membranes with substantially reduced fatigue and leakage. *Applied Physics Letters* 92: 062910.

150 Yun, K.Y., Ricinshi, D., Kanashima, T. et al. (2004). Giant ferroelectric polarization beyond 150 μC/cm^2 in $BiFeO_3$ thin film. *Japanese Journal of Applied Physic* 43: L647–L648.

151 Ahn, C.H., Bhattacharya, A., Ventra, D. et al. (2006). Electrostatic modification of novel materials. *Reviews of Modern Physics* 78: 1185–1212.

152 Yamada, H., Marinova, M., Altuntas, P. et al. (2013). Ferroelectric control of a Mott insulator. *Scientific Reports* 3: 2834.

153 Crassous, A., Bernard, R., Fusil, S. et al. (2011). Nanoscale electrostatic manipulation of magnetic flux quanta in ferroelectric/superconductor $BiFeO_3$–YBCO heterostructures. *Physical Review Letters* 107: 247002.

154 Ge, C., Wang, C., Jin, K.-J. et al. (2013). Recent progress in ferroelectric diodes: explorations in switchable diode effect. *Nano-Micro Letters* 5: 81–87.

155 Triscone, P. and Triscone, J.-M. (2009). A leak of information. *Nature* 460: 45–46.

156 Wang, C., Jin, K.-J., Xu, Z.-T. et al. (2011). Switchable diode effect and ferroelectric resistive switching in epitaxial $BiFeO_3$ thin films. *Applied Physics Letters* 98: 192901–192903.

157 Lee, D., Baek, S.H., Kim, T.H. et al. (2011). Polarity control of carrier injection at ferroelectric/metal interfaces for electrically switchable diode and photovoltaic effects. *Physical Review B* 2011 (84): 125305–125313.

158 Hong, S., Choi, T., Jeon, J.H. et al. (2013). Large resistive switching in ferroelectric $BiFeO_3$ nano-island based switchable diodes. *Advanced Materials* 25: 2339–2343.

159 Suryanarayana, S.V. (1994). Magnetoelectric interaction phenomena in materials. *Bulletin of Materials Science* 17: 1259–1270.

160 Kumar, A. and Yadav, K.L. (2013). Enhanced magnetoelectric sensitivity in $Co_{0.7}Zn_{0.3}Fe_2O_4$–$Bi_{0.9}La_{0.1}FeO_3$ nanocomposites. *Materials Research Bulletin* 48: 1312–1315.

161 Laukhin, V., Skumryev, V., Marti, X. et al. (2006). Electric-field control of exchange bias in multiferroic epitaxial heterostructures. *Physical Review Letters* 97: 227201.

162 Dho, J., Qi, X., Kim, H. et al. (2006). Large electric polarization and exchange bias in multiferroic $BiFeO_3$. *Advanced Materials* 18: 1445–1448.

163 Béa, H., Bibes, M., Cherifi, S. et al. (2006). Tunnel magnetoresistance and robust room temperature exchange bias with multiferroic $BiFeO_3$ epitaxial thin films. *Applied Physics Letters* 89: 242114.

164 Bea, H., Bibes, M., Ott, F. et al. (2008). Mechanisms of exchange bias with multiferroic $BiFeO_3$ epitaxial thin films. *Physical Review Letters* 100: 017204.

165 Bibes, M. and Barthélémy, A. (2008). Multiferroics towards a magnetoelectric memory. *Nature Materials* 7: 425.

166 Zhang, X., Wang, Y.H., Zhang, D.L. et al. (2011). Electric-field-induced change of the magnetoresistance in the multiferroic spin-valve based on $BiFeO_3$ film. *IEEE Transactions on Magnetics* 47: 3139–3142.

167 Allibe, J., Infante, I.C., Fusil, S. et al. (2009). Coengineering of ferroelectric and exchange bias properties in $BiFeO_3$ based heterostructures. *Applied Physics Letters* 95: 182503.

168 Allibe, J., Fusil, S., Bouzehouane, K. et al. (2012). Room temperature electrical manipulation of giant magnetoresistance in spin valves exchange-biased with $BiFeO_3$. *Nano Letters* 12: 1141–1145.

169 Bea, H., Fusil, S., Bouzehouane, K. et al. (2006). Ferroelectricity down to at least 2 nm in multiferroic $BiFeO_3$ epitaxial thin films. *Japanese Journal of Applied Physics* 45: L187–L189.

170 Gajek, M., Bibes, M., Fusil, S. et al. (2007). Tunnel junctions with multiferroic barriers. *Nature Materials* 6: 296–302.

171 Vasudevan, R.K., Morozovska, A.N., Eliseev, E.A. et al. (2012). Domain wall geometry controls conduction in ferroelectrics. *Nano Letters* 12: 5524–5531.

172 Whyte, J.R., McQuaid, R.G.P., Sharma, P. et al. (2014). Ferroelectric domain wall injection. *Advanced Materials* 26: 293–298.

173 Li, L.Z., Britson, J., Jokisaari, J.R. et al. (2016). Giant resistive switching via control of ferroelectric charged domain walls. *Advanced Materials* 28: 6574–6580.

7
Applications

7.1 Introduction

Piezoelectric materials have been widely used in the fields of biomedical, information technology, consumer electronics, automotive, aerospace, etc. because of the functionality of interconverting electrical and mechanical energies. After years of research, lead-free piezoelectric materials have been recently applied in real applications. Although their average performances are still slightly inferior to lead zirconate titanate (PZT)-based materials, some potential applications have been reported. This chapter covers some noteworthy lead-free prototype devices, such as multilayer actuators (MLAs), inkjet printing elements, knocking sensors, ultrasonic transducers (e.g. cleaning, nondestructive testing, and medical imaging), energy harvesting devices, and high-temperature applications.

Piezoelectric materials can be utilized for sensing and actuating applications, including accelerators, viscosity sensors, Langevin actuators, multilayer bimorph, ultrasonic transducers, ultrasonic motors, transformers, acoustic devices, generators, etc., as shown in Figure 7.1a [1–5]. The global market for piezoelectric devices is estimated to be valued at $25.1 billion in 2019 and projected to expand at a compounded annual growth rate (CAGR) of 6.2% through the next five years. Emerging applications, increased funding, and rising demand for energy-harvesting techniques are expected to boost market growth [6]. Considering the lead toxicity to both the environment and human beings, as well as RoHS (Restriction of the Use of Certain Hazardous Substances in Electrical and Electronic equipment) legislation, tremendous effort has been pour into the study of lead-free piezoelectric materials [7, 8]. In 2004, Saito et al. reported a high-performance $(K,Na)NbO_3$ (KNN)-based lead-free piezoceramic with properties comparable to PZT, stimulating widespread interest in developing lead-free alternatives [9]. The past decade has witnessed rapid development in the research of application on lead-free piezoelectrics. Yearly patents involving (lead-free) piezoelectric application since 2004 are shown in Figure 7.1b. It is observed that thousands of patents referring to piezoelectric applications were granted every year. Until 2018, the number of granted patents has grown nearly double compared with 2004, which shows the industries' intention of developing revolutionary technologies and possible market growth in the future. Exploiting lead-free piezoelectrics in applications has significantly increased on

Lead-Free Piezoelectric Materials, First Edition. Jing-Feng Li.

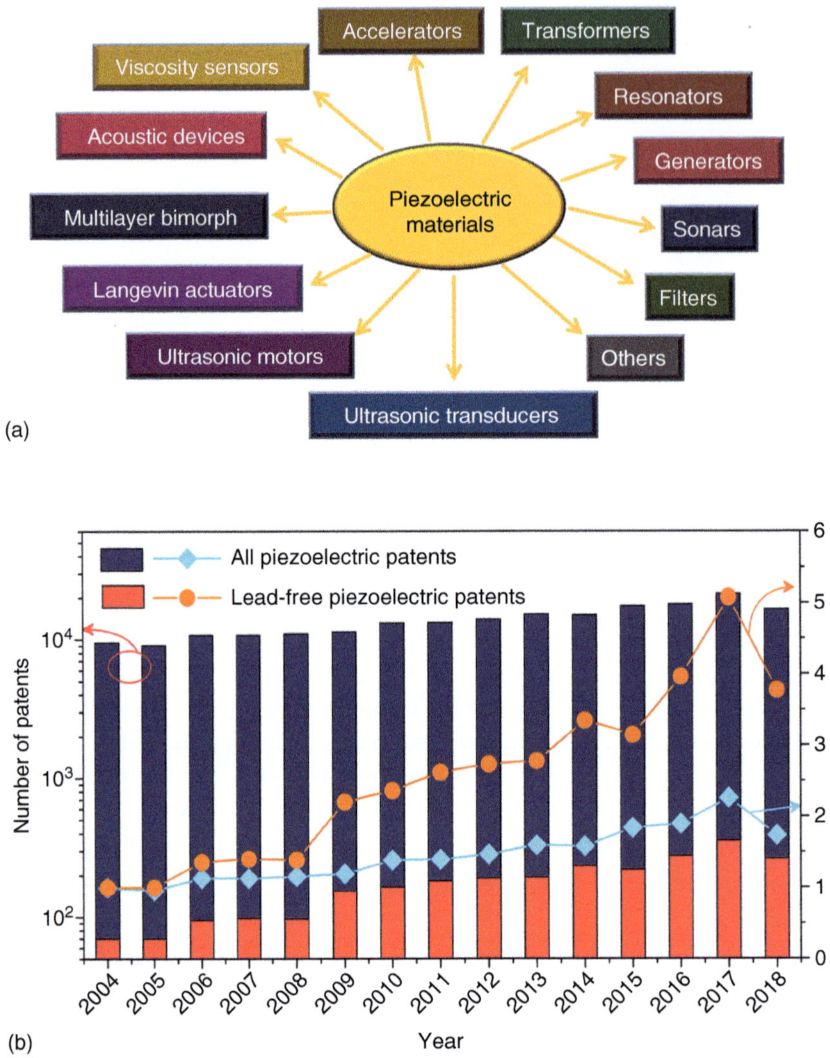

Figure 7.1 (a) Representative applications of piezoelectric materials [1–5]. (b) Yearly growth of patents involving overall piezoelectric applications and lead-free piezoelectric applications since 2004. This graph is compiled by searching in Google patents with following keywords: "piezoelectric," and "lead-free and piezoelectric." "Lead-free" mentioned herein covers not only inorganic perovskite and non-perovskite materials but also organic piezoelectrics.

both the academic and industrial aspects. Approximately 300 patents on lead-free piezoelectric applications were granted yearly in the past few years. Nevertheless, the proportion of lead-free patents is still low so far, which is only around 1–1.5% of the overall patents, in spite of the possible delayed transfer from academic research to patent development. As a result, intensive research is still in need of realizing lead-free piezoelectric applications. Certainly, this process may be accelerated because of the incentives, like stricter regulation on lead-based products, increased funding, cost reduction on manufacturing, and rising demand for revolutionary techniques [2, 4, 6, 8].

7.2 Representative Applications of Lead-Free Piezoelectric Ceramics

7.2.1 Piezoelectric Multilayer Actuators

Piezoelectric actuators are a state-of-the-art technique for precise positioning systems with small displacement and high force over a short time scale. Applications of piezoelectric actuators can be classified into three major categories: positioners, motors, and vibration suppressors [1, 3, 10, 11]. Nowadays, piezoelectric actuators are at a vital stage of their development into a large component market with over 35% contribution on whole piezoelectric markets [2]. Performance of a piezoelectric actuator is mainly evaluated by the electric-field-induced strain, blocking force (force produced in the fully clamped state), and Curie temperature [1, 12]. The conventional single-layer disk has very limited performance because of the large thickness and high driving voltage. Therefore, a modified multilayer structure is proposed. An MLA of the same size can obtain equivalent strain under a relatively small driving voltage or create a larger displacement under the same driving voltage. Similarly, the blocking force can be enhanced when the multilayer structure is adopted [12, 13]. The structure of an MLA is displayed in Figure 7.2a, where piezoelectric ceramic layers are co-fired with thin electrodes. Theoretically, the piezoelectric displacement (ΔL) is proportional to the number of stacked layers (n), converse piezoelectric strain coefficient (d_{33}^*), and driving electric field (E) [13]. Therefore, by increasing the number of stacked layers, the performance of the MLAs can be improved, as shown in Figure 7.2b.

The choice of electrode materials is of utmost importance to MLAs because it determines the overall device cost and the manufacturing method. Common metal electrode materials include base metal, e.g. Ni and Cu, and noble metal, e.g. Ag, Pd, Pt, and their alloys. It has been demonstrated that KNN-based lead-free ceramics can be co-fired with base metal Ni or Cu electrodes, while $(Bi_{1/2}Na_{1/2})TiO_3$ (BNT)-based ceramics are thermodynamically compatible with Cu electrodes in a low oxygen partial pressure atmosphere, i.e. a reducing atmosphere [13, 15–17]. However, in PZT-based MLAs, the oxygen partial pressure required to retain the metallic Ni and Cu electrodes can easily reduce PbO. Additionally, Ni is not compatible with PZT because a secondary phase $NiTiO_3$ will form, which can drastically destroy the structure and properties of piezoelectric actuators [13, 15, 18]. Therefore, expensive noble metal materials are irreplaceable and used in PZT-based MLA, instead of cheap Ni or Cu electrodes. From Figure 7.2c, the cost of Pd or Pt is extremely expensive, which is at about the order of 10^2–10^4 times higher than the base metals. From the cost perspective, base-metal electrode should be used in commercialized piezoelectric actuators. However, co-firing PZT with the cheaper Ni/Cu electrodes is limited because of the thermodynamic incompatibility between them. It is known that, in many cases,

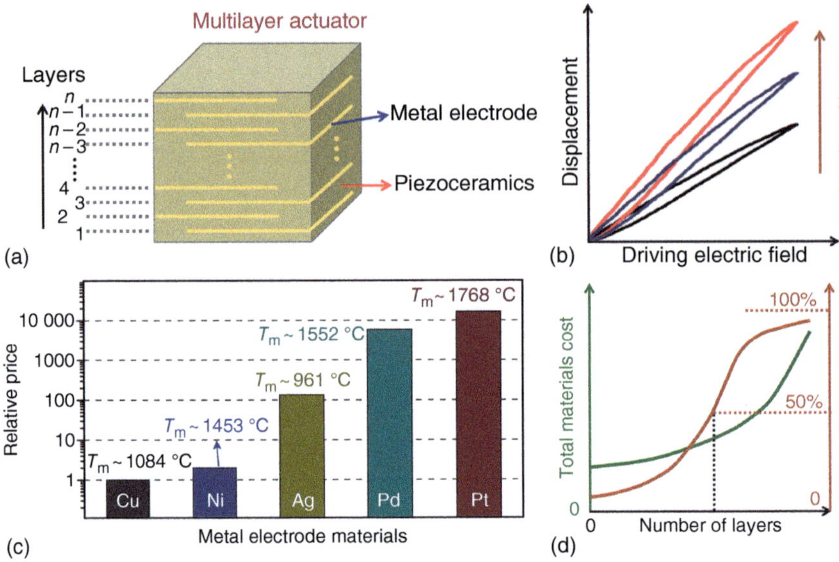

Figure 7.2 (a) Schematic of multilayer actuators. (b) Electric-field-induced displacement of multilayer actuators as a function of the number of layers. (c) The relative price and melting temperature (T_m) of common metal electrode materials. (d) Total materials cost of the MLAs and the proportion of electrode cost as a function of the number of layers. Source: (c) Modified from Gao et al. [13]; (d) Modified from Randall et al. [14].

the cost of electrodes determines a very high proportion of the production cost of the MLAs, especially when the number of stacked layers is large, as plotted in Figure 7.2d [14]. Therefore, the piezoelectric industries that manufacture PZT-based MLAs have been continually suffering cost pressure from using precious electrode materials. Fortunately, KNN-based and BNT-based ceramics possess good compatibility with base metals, which stimulates the development of lead-free MLAs.

KNN-based piezoceramics have long been regarded as one of the most promising candidates to replace PZT counterparts because of their comparable piezoelectric properties (200–600 pC/N or pm/V), high Curie temperature ($T_C \sim 200$–$400\,°C$), relatively small density, and good mechanical performance [19–21]. BNT-based piezoceramics also draw attention because of the large electric-field-induced strain ($d_{33}^* \sim 300$–1000 pm/V) in addition to moderate direct piezoelectric coefficient ($d_{33} \sim 100$–230 pC/N), as well as better processing controllability. Especially BNT-based piezoceramics show feasibility for actuators in high power applications [22, 23]. In the past few years, a series of KNN-based and BNT-based prototype MLAs have been developed by research institutes and industry companies such as *Murata Manufacturing Co., Ltd.* (Japan) [16, 24], *Taiyo Yuden Co., Ltd.* (Japan) [25], *PI Ceramic GmbH* (Germany) [26], and *Noliac A/S* (Denmark) [27], as shown in Figure 7.3. These prototype MLAs are manufactured by tape casting and consist of 10–30 active layers, with the layer thickness of 30–120 μm and several-microns-thickness inner electrodes. Properties of some KNN-based and BNT-based lead-free MLAs are presented in Table 7.1. KNN-based MLAs co-fired

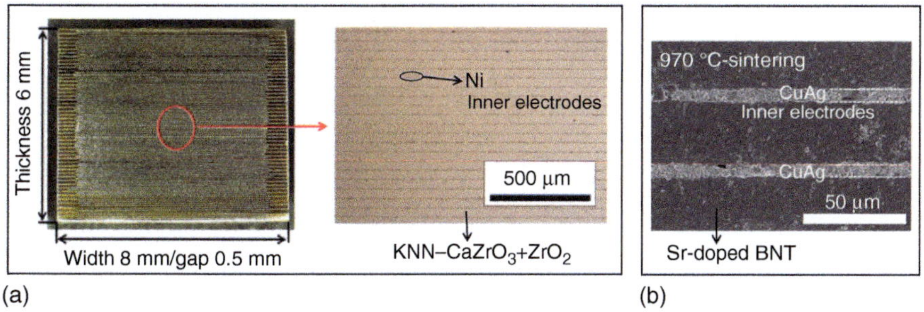

Figure 7.3 Cross-section microscope images of (a) KNN–$CaZrO_3$ + ZrO_2 and (b) Sr-doped BNT multilayer actuators. Source: (a) Kawada et al. [16]; (b) Koo et al. [28].

Table 7.1 Electrical properties of KNN-based and BNT-based lead-free MLAs with base metal Ni or Cu electrode.

Material system	Inner electrode	$\varepsilon_{33}{}^{T}/\varepsilon_0$	Tan δ	d^*_{33} (pm/V)	E (kV/mm)	References
KNN–0.04$CaZrO_3$ + 0.03ZrO_2	Ni	1180		360	2	[16]
KNN$Ta_{0.2}$–0.02NaF–0.04Nb_2O_5	Ni	900	~0.07	385	5	[25]
Li, Ta modified KNN	Cu	830	0.036	220	2	[26]
KNN–0.05LiF	Ni	415	0.033	210	2	[29]
BNT–0.26$SrTiO_3$	Cu/Ag			~860	3	[28]

with Ni/Cu electrodes can produce a normalized strain d^*_{33} of 200–360 pm/V at a low driving electric field of 2 kV/mm, while BNT-based MLAs co-fired with Pt or Ag/Pd electrodes can achieve a giant normalized strain d^*_{33}, more than 1000 pm/V [30]. Besides, Sr-doped BNT-based MLAs co-fired with Cu electrodes can also produce a large d^*_{33} of 860 pm/V [28]. BNT-based MLAs do not only produce comparable strain performance to PZT-based ones but also show excellent fatigue resistance and thermal stability. For example, BNT-8BT MLA can withstand the unipolar electric field for more than 10^8 cycles and deliver 0.07% strain at 6 kV/mm and a frequency of 50 Hz [27]. Li and Nd co-doped $(Bi_{0.50}Na_{0.335}K_{0.125}Li_{0.04})TiO_3$ (BNKLT) MLA shows a maximum strain of 0.19% at 7 kV/mm with a slight variation of only 10% between 25 and 150 °C [31].

7.2.2 KNN-Based Actuation Structure in Inkjet Printhead

Piezoelectric (multilayer) actuators are largely used as an important component in automotive industry, information technology, etc., including applications such as fuel injector, hard disk drive, haptic system, and feedback control system [3, 10, 11, 14, 17, 32]. Among these applications, the shear-mode inkjet printhead is the most recognized application. Inkjet printheads that utilize KNN-based piezoelectric actuators have been developed by many companies, e.g. *Konica Minolta Inc.*, *Toshiba Tec Corporation*, and *TDK Corporation* from Japan [33–35]. The actuation structure in

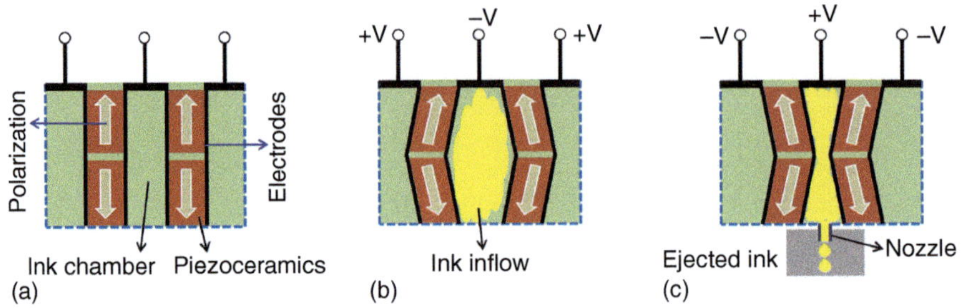

Figure 7.4 (a) Cross-sectional view of actuation structure in a shear-mode inkjet printhead. (b) Ink flows from the cartridge to the chamber when the chamber is enlarged. (c) Ink is ejected from the chamber when the chamber is squeezed by the piezoceramics. Source: Reproduced with permission from Shibata et al. [33]. © 2018, Cambridge University Press.

the inkjet printhead is shown in Figure 7.4a. Two poled piezo-actuator sheets are conglutinated along inverse polarization direction and then combined with another set of piezoceramic plates to assemble a sealed ink chamber. The conductive electrodes are painted on the internal surface of the ink chamber to offer AC driving electric field. When a positive electric field is applied, initiation of shear-mode actuation of piezoceramic will lead to an expansion of the chamber, allowing ink inflows form the storage, as presented in Figure 7.4b. When a negative signal is applied, the inverse shear motion of actuators will cause the chamber shrinkage, ejecting ink out of the nozzle, as displayed in Figure 7.4c. The KNN-based piezo-actuators can provide a rapid response speed with easy starting and high efficiency. For example, the KNN-based inkjet printhead developed by *Toshiba Tec Corporation* consists of 60 nozzles and the density of nozzles is 150 dots/in., with the minimum ink droplet as small as 6 pL.

7.2.3 Ultrasonic Transducers

Ultrasonic transducers can convert electrical signals into acoustic waves as an actuator, and do the reverse as a sensor. Ultrasonic transducers have been widely used in ultrasonic cleaner, ultrasonic motor, sonar, anemometer, humidifier, burglar alarm, nondestructive testing, medical ultrasonography, wireless charging, etc. [1, 3, 36–38]. When a piezoelectric element is operated near its resonant frequency, which is normally within its ultrasound frequency range, the electromechanical response can be highly enhanced and the conversion between electrical and mechanical energy can be most efficient. Therefore, ultrasonic transducers are always operated at the resonant frequency to mitigate energy loss and overheating. Additionally, a high performance transducer needs a broad bandwidth (BW) and low insertion loss (IL), which help further decrease the power consumption and heat generation. A piezoelectric element with high electromechanical coupling coefficient k, large mechanical quality factor Q_m, high dielectric properties, and small acoustic impedance Z, is necessary for obtaining a broad BW and small IL, and facilitating the output power and ultrasonic intensity. In recent years, with the advantages of environmentally friendly and bio-compatibility, much attention has been paid to the feasible adoption of lead-free piezoelectric materials in ultrasonic transducers [39, 40]. The BW of lead-free transducers is in the range of 45–76% with IL in the range of 18–29%, which is comparable with the PZT-based transducers [5].

An ultrasonic cleaner is a common high-power actuator, which transmits electrical energy to ultrasound wave for cleaning purpose. As shown in Figure 7.5a, the first commercialized lead-free bolt-clamped Langevin ultrasonic cleaner using $(Bi_{1/2}Na_{1/2})TiO_3$–$BaTiO_3$–$(Bi_{1/2}Na_{1/2})(Mn_{1/3}Nb_{2/3})O_3$ piezoceramics is developed by the company of *Honda Electronics Co., Ltd.* in Japan [43]. Under high-speed vibration, the Q_m of this ceramic is higher than that of hard PZT. Under a fixed input power, a higher Q_m will allow one material to vibrate more effectively. As a result, the vibration velocity of this BNT-based transducer is 30–40% larger than that of hard PZT at the same input power, leading to a ~20% higher ultrasonic intensity than PZT-based transducer, which provides a better cleaning effect for the lead-free ultrasonic cleaner product.

An ultrasonic motor can provide high torque and quick response; therefore it is widely used in miniature devices, such as smartphones, microrobots, and auto-focusing modules. A ring-type rotary ultrasonic motor, which uses a $0.95(K_{0.49}Na_{0.49}Li_{0.02})(Nb_{0.8}Ta_{0.2})O_3$–$0.05CaZrO_3$ (CZ5) piezoceramic as the active material, is used in an auto-focusing module of a digital camera [44]. As presented in Figure 7.5b, the vibration of the CZ5 stator is observed at 45.1 kHz. Although this ultrasonic motor has a smaller revolution speed of 5–7 rpm at the operating frequency of 45.1 kHz than the PZT motor with a revolution speed of 45–55 rpm (at the operating frequency of 31 kHz), this motor is found to have a lower power loss and enhanced electric power efficiency in operation. The results suggest that KNN-based piezoceramics have comparable performance in the ultrasonic motor application.

Ultrasonic transducers can be used in structural health monitoring applications, that is, ultrasonic nondestructive testing. Based on the propagation of ultrasonic waves induced by transducers in the object or material, this technique can detect the feedback with embedded sensors, without causing any damage to the tested structures. Since the ultrasonic wave can propagate for a very long distance, this technique possesses good sensitivity to detect defects, such as cracks and thickness [45, 46]. As displayed in Figure 7.5c, a thermal-sprayed $(K_{0.44}Na_{0.52}Li_{0.04})(Nb_{0.84}Ta_{0.10}Sb_{0.06})O_3$ piezoceramic coated on a stainless substrate is fabricated to study the capability of generation and detection of ultrasonic waves from the actuator and sensor [42]. When the actuator is excited by 50 V at 416 kHz for five cycles to generate ultrasonic waves, the strong signal could be detected by the piezoelectric sensor in the range 350–450 kHz with a good signal-to-noise ratio. The voltage amplitude of the received signal is around 100 mV, which is 2 orders of magnitude higher than similar testing with poly(vinylidene difluoride-co-trifluoroethylene) (P(VDF-TrFE)) polymer. The results demonstrate that KNN-based materials are also feasible for structural health monitoring.

Lead-free piezoelectric materials show tremendous prospect in disposable patient monitors, heart monitors, catheters, and ultrasonic transducers for noninvasive therapy and medical imaging, due to their less toxicity and good biocompatibility

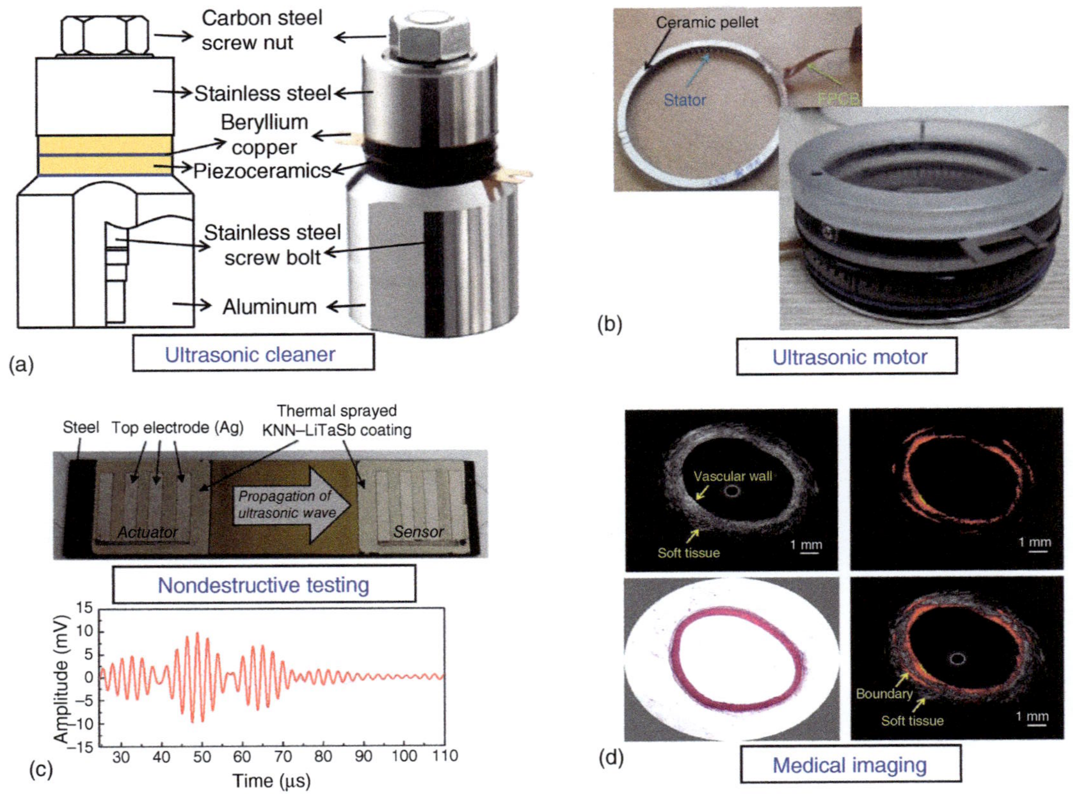

Figure 7.5 (a) Schematic diagram and picture of commercial BNT-based ultrasonic cleaner. (b) A ring-type rotary ultrasonic motor using KNN-based piezoceramics. (c) Ultrasonic nondestructive testing including the actuator and sensor made of the thermal-sprayed KNN-based piezoceramics. (d) Medical imaging of intravascular ultrasound and photoacoustic images in a healthy rabbit aorta using an ultrasonic transducer utilizing KNN-based single crystal thick film [41]. Source: (a) Shibata et al. [33]; (c) Reprinted with permission from Chen et al. [42]. © 2018, John Wiley & Sons, Inc.

[39, 40, 47]. As shown in Figure 7.5d, using the $(K_{0.45}Na_{0.55})_{0.96}Li_{0.04}NbO_3$ single crystal thick films with high d_{33} (490 pm/V) and k_t (0.55), a side-looking miniature transducer is fabricated with a central frequency of 82 MHz and a bandwidth of 57.3% at −6 dB, and shows the sensitivity of pulse-echo signal with 420 mV [41]. Employing this transducer, the vascular wall and its outer boundary as well as soft tissue of a rabbit aorta are clearly observed through the monitor. This research indicates that the KNN-based material is a good lead-free candidate for high-frequency ultrasound medical imaging. Until now, lead-free transducers with a wide range of frequency (~3–80 MHz) and good axial resolution (~20–60 μm) and lateral resolution (~90–160 μm) for variety of imaging applications have been researched [40, 47], such as a 30 MHz needle type transducer for intravascular imaging using $0.5Ba(Ti_{0.8}Zr_{0.2})O_3$–$0.5(Ba0_{.7}Ca_{0.3})TiO_3$ ceramics [48], a ~20 MHz lens-focused transducer with high electrical matching and lateral resolution using $0.88Bi_{0.5}Na_{0.5}TiO_3$–$0.08Bi_{0.5}K_{0.5}TiO_3$–$0.04Bi_{0.5}Li_{0.5}TiO_3$ ceramics [49], a 5.5–25 MHz single element $(K_{0.44}Na_{0.52}Li_{0.04})(Nb_{0.84}Ta_{0.10}Sb_{0.06})O_3$ transducer with small IL acoustic performance [50], etc. These lead-free transducers have even superior resolution compared with transducers using the PZT-5H [50] and PZT–0.26PT piezoceramics [51].

7.2.4 KNN-Based Knocking Sensors

Piezoelectric sensors can detect external mechanical signals or acoustic impulses, and subsequently convert them into electrical signals [10, 32]. The performance indicator of a piezoelectric sensor is the practical sensitivity (S_P), which is prominently influenced by the piezoelectric charge/voltage coefficient of the piezoelectric elements [52]. A piezoelectric knocking sensor is a representative sensing device in automotive vehicles. With high sensitivity, it can responsively detect the knocking of a car engine, monitor the combustion process, and adjust the ignition timing of automotive engines timely, as shown in Figure 7.6a. Figure 7.6b illustrates the common structure of a piezoelectric element used for knocking sensors. An example of knocking sensors using high-performance CuO-added

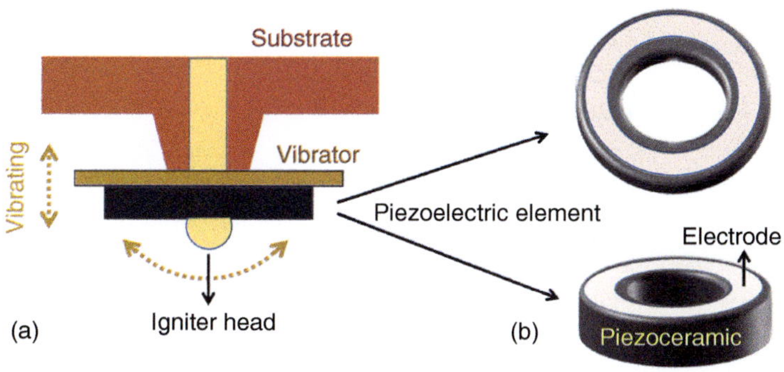

Figure 7.6 Schematics of (a) resonating-type knocking sensors and (b) common piezoelectric elements. Source: Reprinted with permission from Kim et al. [52]. © 2017, John Wiley & Sons, Inc.

$0.95(K_{0.5}Na_{0.5})(Nb_{0.95}Sb_{0.05})O_3$–$0.05CaTiO_3$ piezoceramics (d_{33} = 245pC/N, g_{33} = 25.7 Vm/N, k_p = 0.46) was fabricated by universities and *Lattron Co., Ltd.* in Korea [52]. This knocking sensor exhibited a higher S_P of 119 mV/g at the resonance frequency than a commercial PZT knocking sensor with S_P = 112 mV/g, suggesting a possibility to replace the PZT-based commercial knocking sensors. Nowadays, there are some commercial lead-free knocking sensors, such as KNN-$KTiNbO_5$ (KNN-KTN) knocking sensors, which exhibit the same level of output voltage and superior durability compared with PZT, developed by *NGK Spark Plug Co., Ltd.* (Japan) [53], and lead-free knocking sensors and combustion pressure sensors supplied by *Kyungwon Industry Co., Ltd.* (Korea) [33], etc.

7.3 Other Potential Applications

7.3.1 Energy Harvesting

Piezoelectric energy harvesting, with a simple structure, easily minimization, and integration, enables harvest and conversion of available energy from the surroundings into electrical energy, such as wind, solar, ocean wave, etc. [54–56]. Two major types of cantilever configurations for piezoelectric vibration energy harvesting are developed. As shown in Figure 7.7a, the piezoelectric materials are embedded into two metal electrode layers, and the device operates in the d_{31} mode when the cantilever is vibrated. If the metal electrodes are printed on one side of the piezoelectric material with interdigitated structure displayed in Figure 7.7b, the piezoelectric cantilevers can exploit the longitudinal d_{33} piezoelectric strain constants. Additionally, a proof mass is usually added to the top end of the cantilever to match the available environmental vibration sources and decrease the fundamental resonance of the device.

In recent years, piezoelectric energy harvesting has become an interesting research direction for its promising potential applications in self-sustainable power system and portable electronics [61, 62], which has also paid attention on the lead-free importance. For example, a cantilever-type energy harvester using $(1-x)Ba(Zr_{0.1985}Cu_{0.0015}Ti0.8)O_{3-\delta}$–$x(Ba_{0.7}Ca_{0.3})TiO_3$ (BZT-BCT) piezoceramics shown in Figure 7.7c is fabricated [57]. A high output power of 70–700 μW and an output voltage of 8–25 V is obtained at the resonance frequency of 90 Hz and acceleration of 10–50 m/s^2, due to the high figure of merit ($d_{33} \times g_{33} = 4200 \times 10^{-15}$ m^2/N). Another practical energy harvesting device with mask-patterned d_{31} mode MEMS cantilevers is developed utilizing MnO_2-doped $(K_{0.5}Na_{0.5})NbO_3$ thin films [58]. The MEMS energy harvesting element is shown in Figure 7.7d. The maximum power density and the corresponding peak voltage are 1800 μW/cm^3 and 520 mV at the resonance frequency of 132 Hz, respectively, which is comparable device performance with PZT-based MEMS devices, showing the strong feasibility for miniaturized self-powered electronic device applications.

Flexible energy harvesters, e.g. nanogenerators, composed of piezoelectric ceramics, piezoelectric polymers, or their composites, can be used for

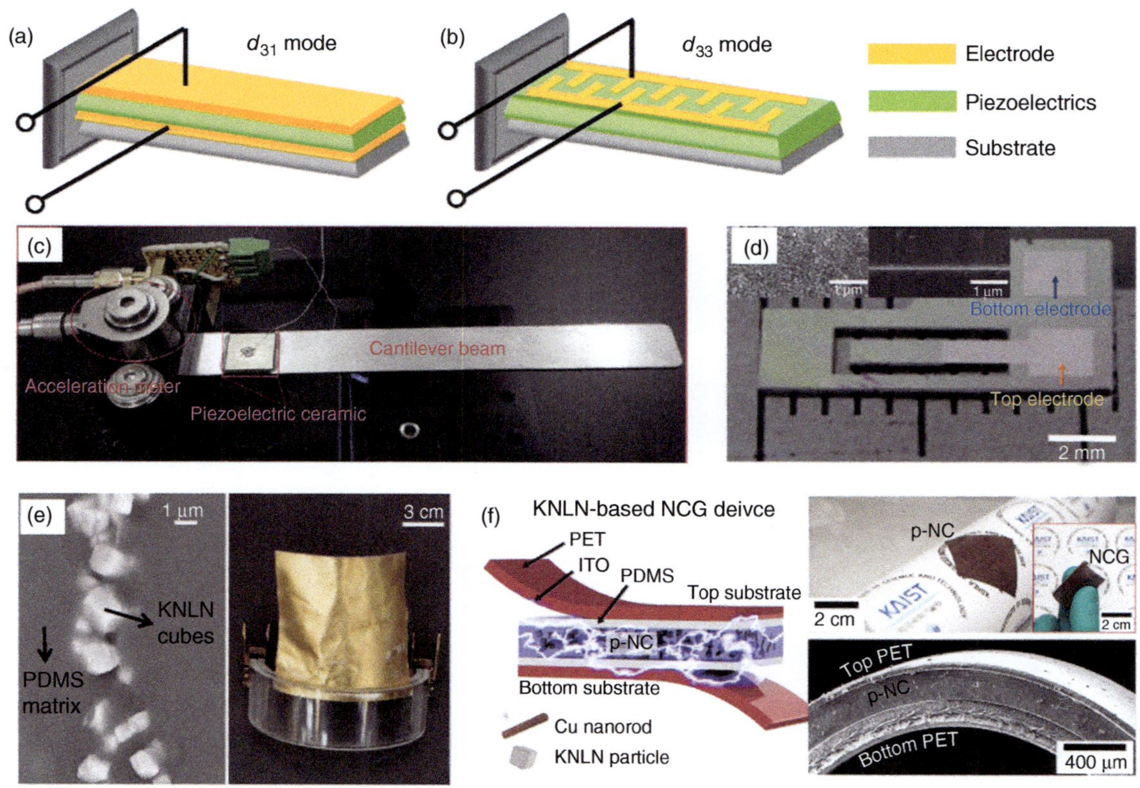

Figure 7.7 Schematics of common cantilever configurations with (a) d_{31} mode and (b) d_{33} mode. (c) A cantilever-type energy harvester using the CuO-doped BZT–BCT piezoceramics [57]. (d) A fabricated MEMS energy harvester using the Mn-doped KNN thin film [58]. (e) A flexible piezoelectric touch sensor using the KNLN particles aligned in a PDMS matrix [59]. (f) A flexible piezoelectric energy harvester using the $LiNbO_3$-doped KNN thin film. Source: (f) Reprinted with permission from Jeong et al. [60]. © 2014, John Wiley & Sons, Inc.

realizing self-powered flexible electronics for bio-implantable and medical applications. The structure of these generators includes thin films, nano- or micro-particles/wires/fibers/cubes/tubes, etc. The polymer covers poly-dimethylsiloxane (PDMS), polyvinylidene difluoride (PVDF), and P(VDF-TrFE) [58–60, 63–65]. As shown in Figure 7.7e, a high-performance flexible composite film is achieved by using the $K_{0.485}Na_{0.485}Li_{0.03}NbO_3$ (KNLN) powders aligned in a PDMS matrix [59]. This composite film exhibits a high and temperature stable piezoelectric voltage coefficient of ~220 mV•m/N. Then the composite films were integrated into a flexible touch sensor, which shows a high sensitivity under a controlled dynamical change of applied force via measuring the open circuit voltage. This work demonstrates a new flexible sensor system for temperature, infrared, impact, and torque measurements, which can be used in medical diagnostics and wireless sensors networks. Another example, as shown in Figure 7.7f, is a flexible nanocomposite generator made by a simple spin-casting method. Lead-free $0.942(K_{0.480}Na_{0.535})NbO_3$–$0.058LiNbO_3$ particles synthesized using a solid-state method and the Cu nanorods are distributed in a PDMS matrix [60]. A large-area generator device (30 cm × 30 cm) fabricated by the bar-coating technique is able to generate ~mW power (≈140 V and ≈8 μA) from biomechanical energy. This generator technology demonstrates a simple, cost-effective, scalable, and high-throughput approach for practical flexible electronics.

7.3.2 High-Frequency Medical Imaging Transducers Using 1–3 Composites

Piezoelectric 1–3 composites consist of a parallel array of piezoelectric rods embedded in a polymer matrix with 1–3 connectivity. Compared with the monolithic single crystals or isotropic bulk ceramics, they can provide superior electromechanical coupling property in the longitudinal length mode, and lower acoustic impedance as well as mechanical/electric loss due to the incorporation of a passive polymer phase. In recent years, lead-free piezoelectric 1–3 composites have received much attention in ultrasonic transducer applications.

KNN-based ceramics have been developed as the most promising lead-free piezoceramics due to their high electromechanical performances, which are extremely favorable for medical imaging transducers. For example, fine-grained $(Na_{0.535}K_{0.485})_{0.95}Li_{0.05}(Nb_{0.8}Ta_{0.2})O_3$ (NKLNT) piezoceramics prepared by spark plasma sintering (SPS) is selected as the piezoelectric element to fabricate NKLNT/epoxy 1–3 composite utilizing a modified dice-fill method [66]. NKLNT ceramic rods could be miniaturized to a lateral width of 50 μm because of enhanced strength and hence good machinability achieved in the SPS-processed fine-grained materials (Figure 7.8a). Because of the thin piezoelectric layer showing a high electromechanical coupling coefficient (k_t = 0.66) and yielding a high-frequency resonance, the fabricated NKLNT/epoxy 1–3 composite shows low acoustic impedance (Z = 6.6 Mrayls). Finally, as shown in Figure 7.8b,c, the fabricated composite transducer exhibits a broad bandwidth nearly 90% at 6 dB and a center

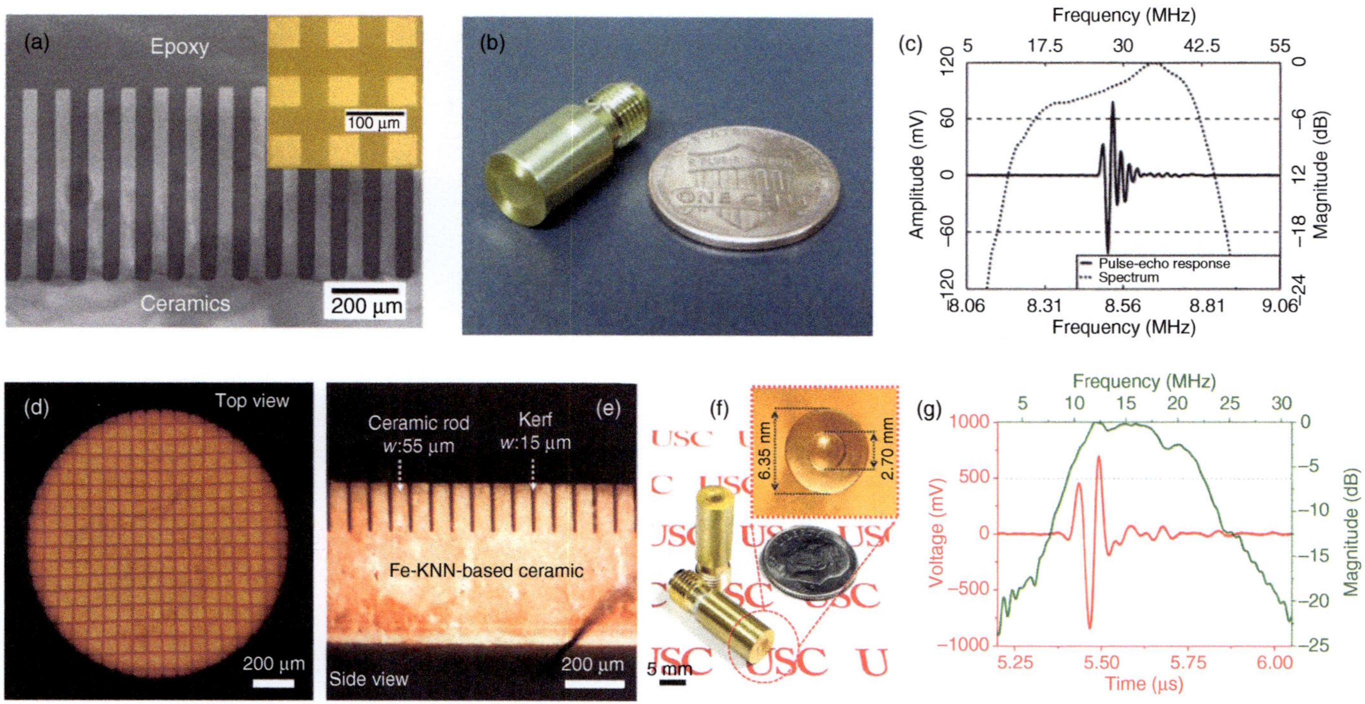

Figure 7.8 (a) Optical images of the end surface and the cross-section of NKLNT rods array after dicing. (b) The photograph of 1–3 NKLNT/epoxy composite transducer. (c) Pulse-echo response and spectrum of the 1–3 composite for the transducer in (b). (d, e) Optical images of the KNN–BNZ–BS ceramic rods array after dicing. (f) The photograph of KNN–BNZ–BS-based 1–3 composite transducers. (g) Pulse-echo waveform (red) and its frequency spectrum (green) of the KNN–BNZ–BS-based 1–3 composite transducer. Source: (c) Reprinted with permission from Shen et al. [66]. © 2011, John Wiley & Sons, Inc.; (g) Reprinted from, Jiang et al. 2019 [67]. Used with permission of AIP Publishing.

frequency of 29 MHz, which is very promising as an alternative to lead-based piezoceramics in numerous high-frequency ultrasonic transducer applications.

Besides, a Fe_2O_3-added (K,Na)NbO_3–(Bi,Na)ZrO_3–$BiScO_3$ (KNN–BNZ–BS) piezoceramic with high piezoelectric property is used to fabricate the 1–3 piezoelectric composites and applied to high-frequency imaging transducers [67]. As shown in Figure 7.8d,e the ceramic pillars were miniaturized to a width of 55 μm with a kerf of 15 μm in the epoxy matrix using the modified dice-and-fill technique. Because of its high k_t of ~0.55 and thin ceramic pillars, the fabricated 1–3 composite exhibits a high center frequency (16 MHz), a broad bandwidth (83%), and a very low IL (9.8 dB) (Figure 7.8f,g). Finally, evaluation of imaging capability indicates this 1–3 composite transducer show high sensitivity, making it suitable for high-frequency biomedical imaging applications.

7.3.3 High-Temperature Piezoelectrics and Applications

Although the widely used PZT-based piezoceramics possess high piezoelectric coefficient, temperature stability, and mechanical strength, their practical operating temperature is less than one half of their Curie temperature (T_C), in order to ensure stable electromechanical properties. T_C around 200–400 °C in PZT-based ceramics prevents them from high-temperature applications, e.g. aerospace and nuclear reactors. The high-temperature piezoelectrics are feasible to stably operate in a high-temperature environment of $1/2T_C$ ~200–1000 °C [68–70]. The most used high-temperature lead-free piezoelectrics include bismuth layer-structured ferroelectric oxides and $LiNbO_3$ single crystal. The former can be defined as $(Bi_2O_2)^{2+}(A_{m-1}B_mO_{3m+1})^{2-}$, accompanied with high T_C of ~500–900 °C, giant Q_m of ~2000–8000, and usable d_{33} of ~15–30 pC/N. $LiNbO_3$ single crystal (T_C~1150 °C, k_p~0.6, d_{15}~75 pC/N) is another important high-temperature piezoelectric material. They are usually applied in piezoelectric filters, surface acoustic wave devices, microwave acoustic devices, transducers, and optical devices [68, 70]. Recently, high-performance lead-free $BiFeO_3$-based piezoceramics with high T_C of ~460 °C and high d_{33} of 300–400 pC/N are fabricated by hot-pressing sintering combined with water-quenching process, displaying a broad application prospect [71]. Nowadays, high-temperature piezoelectrics are desperately needed for the specific devices and applications like fuel nozzles of an internal combustion engine, sensing in aerospace and energy fields, energy harvesting and transform in a high-temperature vibrational environment, nonmagnetic piezoelectric motors and nanogenerators in varies of instruments operated in a high-temperature environment, etc.

7.4 Summary and Outlooks

Lead-free piezoelectric ceramics have entered the market, which is estimated to be about $US 172 million in 2019 [6]. Despite such a small value, its annual growth rate from 2019 to 2024 is projected to be 20.8%, being much higher than that for the whole market for piezoelectric devices. Because of the well-balanced high piezoelectricity and Curie temperature as well as the ability to be co-fired with nickel electrodes, KNN-based piezoelectric ceramics represent the largest share of the market.

After reading this chapter, readers should have understood the status of lead-free piezoelectric applications. First, viewing from the granted patent number, the applications based on lead-free piezoelectric materials have been getting more in the past few years and the number will keep growing fast in the future. The reason for the rapid growth might be several, including the extrinsic pressure from government regulations and the intrinsic advantages in lead-free materials, such as thermodynamic compatibility between materials and electrodes, biocompatibility of materials for in vivo applications, and high-temperature stability. Second, many prototype devices were found to have comparable performances with PZT-based piezoelectric devices. These prototypes can boost investors' confidence to a certain degree and subsequently lead to more investment into the development of lead-free applications, forming a positive developing cycle. Finally, while all these developments are meant to replace PZT-based piezoelectric ceramics with the lead-free ones, many new and interesting applications based on lead-free materials have been proposed. Of course, there may be still a long way ahead us to push lead-free piezoelectrics to large scale applications, but from now we must and can gradually extend the applications of lead-free piezoelectric materials, which are equally important to the development of lead-free piezoelectric materials.

References

1 Uchino, K. (2009). *Ferroelectric Devices*, 2e. CRC Press.
2 Hong, C.H., Kim, H.P., Choi, B.Y. et al. (2016). Lead-free piezoceramics – where to move on? *Journal of Materiomics* 2: 1–24.
3 Uchino, K. (2017). *Advanced Piezoelectric Materials: Science and Technology*. Woodhead Publishing.
4 Priya, S. and Nahm, S. (2012). *Lead-Free Piezoelectrics*. New York, NY: Springer.
5 Wu, J. (2018). *Advances in Lead-Free Piezoelectric Materials*. Springer.
6 Gagliardi, M. (2019). Lead-free piezoelectric ceramics market projected to grow at much faster pace through 2024. *American Ceramic Society Bulletin* 99:7.
7 Rödel, J. and Li, J.-F. (2018). Lead-free piezoceramics: status and perspectives. *MRS Bulletin* 43: 576–580.
8 Bell, A.J. and Deubzer, O. (2018). Lead-free piezoelectrics – the environmental and regulatory issues. *MRS Bulletin* 43: 581–587.
9 Saito, Y., Takao, H., Tani, T. et al. (2004). Lead-free piezoceramics. *Nature* 432: 84–87.
10 Aksel, E. and Jones, J.L. (2010). Advances in lead-free piezoelectric materials for sensors and actuators. *Sensors* 10: 1935–1954.
11 Gao, J., Xue, D., Liu, W. et al. (2017). Recent progress on $BaTiO_3$-based piezoelectric ceramics for actuator applications. *Actuators* 6 (24).
12 Rödel, J., Webber, K.G., Dittmer, R. et al. (2015). Transferring lead-free piezoelectric ceramics into application. *Journal of the European Ceramic Society* 35: 1659–1681.

13 Gao, L., Guo, H., Zhang, S.J., and Randall, C.A. (2016). Base metal co-fired multilayer piezoelectrics. *Actuators* 5: 8.

14 Randall, C., Kelnberger, A., Yang, G.Y. et al. (2005). High strain piezoelectric multilayer actuators – a material science and engineering challenge. *Journal of Electroceramics* 14: 177–191.

15 Koruza, J., Bell, A.J., Frömling, T. et al. (2018). Requirements for the transfer of lead-free piezoceramics into application. *Journal of Materiomics* 4: 13–26.

16 Kawada, S., Hayashi, H., Ishii, H. et al. (2015). Potassium sodium niobate-based lead-free piezoelectric multilayer ceramics co-fired with nickel electrodes. *Materials* 8: 7423–7438.

17 Reichmann, K., Feteira, A., and Li, M. (2015). Bismuth sodium titanate based materials for piezoelectric actuators. *Materials* 8: 8467–8495.

18 Jacob, K. and Saji, V. (2006). Interaction between Ni/NiO and $PbTiO_3$: phase reversal with redox switching. *Journal of Phase Equiliburia and Diffusion* 27: 456–461.

19 Wang, K., Malič, B., and Wu, J. (2018). Shifting the phase boundary: potassium sodium niobate derivates. *MRS Bulletin* 43: 607–611.

20 Wu, J., Xiao, D., and Zhu, J. (2015). Potassium-sodium niobate lead-free piezoelectric materials: past, present, and future of phase boundaries. *Chemical Reviews* 115: 2559–2595.

21 Liu, Q., Gao, J., Zhou, Z. et al. (2018). High-performance lead-free piezoelectrics with local structural heterogeneity. *Energy & Environmental Science* 11: 3531–3539.

22 Jo, W., Dittmer, R., Acosta, M. et al. (2012). Giant electric-field-induced strains in lead-free ceramics for actuator applications: status and perspective. *Journal of Electroceramics* 29: 71–93.

23 Hao, J., Li, W., and Zhai, J. (2019). Progress in high-strain perovskite piezoelectric ceramics. *Materials Science & Engineering R: Reports* 135: 1–57.

24 Kawada, S., Kimura, M., Higuchi, Y., and Takagi, H. (2009). $(K,Na)NbO_3$-based multilayer piezoelectric ceramics with nickel inner electrodes. *Applied Physics Express* 2: 111401.

25 Liu, C., Liu, P., Kobayashi, K., and Randall, C.A. (2014). Base metal co-fired $(Na,K)NbO_3$ structures with enhanced piezoelectric performance. *Journal of Electroceramics* 32: 301–306.

26 Gao, L., Ko, S.W., Guo, H. et al. (2016). Demonstration of copper co-fired $(Na,K)NbO_3$ multilayer structures for piezoelectric applications. *Journal of the American Ceramic Society* 99: 2017–2023.

27 Sapper, E., Gassmann, A., Gjødvad, L. et al. (2014). Cycling stability of lead-free BNT–8BT and BNT–6BT–3KNN multilayer actuators and bulk ceramics. *Journal of the European Ceramic Society* 34: 653–661.

28 Koo, B.-K., Saleem, M., Lim, D.-H. et al. (2018). Fabrication of borosilicate-glass-coated CuAg inner electrode for multilayer ceramic actuator. *Sensors and Actuators A: Physical* 277: 8–17.

29 Kobayashi, K., Randall, C.A., Ryu, M. et al. (2012). New opportunity in alkali niobate ceramics processed in low oxygen partial pressure. In: *Proceedings of ISAF-ECAPD-PFM 2012*, 1–4. Piscataway, NJ, USA: IEEE.

30 Choy, S., Jiang, X., Kwok, K., and Chan, H.L.W. (2010). Piezoelectric and dielectric characteristics of lead-free BNKLBT ceramic thick film and multilayered piezoelectric actuators. *Ceramics International* 36: 2345–2350.

31 Krauss, W., Schütz, D., Naderer, M. et al. (2011). BNT-based multilayer device with large and temperature independent strain made by a water-based preparation process. *Journal of the European Ceramic Society* 31: 1857–1860.

32 Turner, R., Fuierer, P.A., Newnham, R., and Shrout, T.R. (1994). Materials for high temperature acoustic and vibration sensors: a review. *Applied Acoustics* 41: 299–324.

33 Shibata, K., Wang, R., Tou, T., and Koruza, J. (2018). Applications of lead-free piezoelectric materials. *MRS Bulletion* 43: 612–616.

34 Aida, Y., Tanaka, Y., Chihara, H., et al. (2015). Piezoelectric element, piezoelectric actuator, piezoelectric sensor, hard disk drive, and ink-jet printer device. US Patent No. 9 022 532, filed 05 May 2015 and issued 21 October 2013.

35 Sakuma, H. (2016). Piezoelectric thin film element, piezoelectric actuator, piezoelectric sensor, hard-disk drive and ink jet printer device. US Patent Application No. 15/006,395, filed 26 January 2016 and issued 20 June 2017.

36 Gururaja, T., Schulze, W., Cross, L., and Newnham, R.E. (1985). Piezoelectric composite materials for ultrasonic transducer applications. Part II: Evaluation of ultrasonic medical applications. *IEEE Transactions of Sonics and Ultrasonics* 32: 499–513.

37 Safari, A. and Akdogan, E.K. (2008). *Piezoelectric and Acoustic Materials for Transducer Applications*. Springer Science & Business Media.

38 Wang, Y., Xu, Y., and Huang, Q. (2017). Progress on ultrasonic guided waves de-icing techniques in improving aviation energy efficiency. *Renewable & Sustainable Energy Reviews* 79: 638–645.

39 Lam, K.H., Ji, H.F., Zheng, F. et al. (2013). Development of lead-free single-element ultrahigh frequency (170–320 MHz) ultrasonic transducers. *Ultrasonics* 53: 1033–1038.

40 Taghaddos, E., Hejazi, M., and Safari, A. (2015). Lead-free piezoelectric materials and ultrasonic transducers for medical imaging. *Journal of Advanced Dielectrics* 5: 1530002.

41 Zhang, T., Ou-Yang, J., Yang, X. et al. (2019). High performance KNN-based single crystal thick film for ultrasound application. *Electronic Materials Letters* 15: 1–6.

42 Chen, S., Tan, C.K.I., Tan, S.Y. et al. (2018). Potassium sodium niobate (KNN)-based lead-free piezoelectric ceramic coatings on steel structure by thermal spray method. *Journal of the American Ceramic Society* 101: 5524–5533.

43 Tou, T., Hamaguti, Y., Maida, Y. et al. (2009). Properties of $(Bi_{0.5}Na_{0.5})TiO_3$–$BaTiO_3$–$(Bi_{0.5}Na_{0.5})(Mn_{1/3}Nb_{2/3})O_3$ lead-free piezoelectric ceramics and its application to ultrasonic cleaner. *Japaness Journal of Applied Physics* 48: 07GM3.

44 Hong, C.-H., Han, H.-S., Lee, J.-S. et al. (2015). Ring-type rotary ultrasonic motor using lead-free ceramics. *Journal of Sensors Science and Technology* 24: 228–231.

45 Silk, M.G. (1984). *Ultrasonic Transducers for Nondestructive Testing*. Elsevier Science.

46 Chen, C.-H. (2007). *Ultrasonic and Advanced Methods for Nondestructive Testing and Material Characterization*. World Scientific.

47 Zhang, Z., Chen, R., Wang, B. et al. (2018). Development of a KNN ceramic-based lead-free linear array ultrasonic transducer. *IEEE Transactions of Ultrasonics Ferroelectrics and Frequency Control* 65: 2113–2120.

48 Yan, X., Lam, K.H., Li, X. et al. (2013). Correspondence: lead-free intravascular ultrasound transducer using BZT−50BCT ceramics. *IEEE Transactions of Ultrasonics Ferroelectrics and Frequency Control* 60: 1272–1276.

49 Hejazi, M.M., Jadidian, B., and Safari, A. (2012). Fabrication and evaluation of a single-element $Bi_{0.5}Na_{0.5}TiO_3$-based ultrasonic transducer. *IEEE Transactions of Ultrasonics Ferroelectrics and Frequency Control* 59: 1840–1847.

50 Hagh, N.M., Jadidian, B., Ashbahian, E., and Safari, A. (2008). Lead-free piezoelectric ceramic transducer in the donor-doped $K_{1/2}Na_{1/2}NbO_3$ solid solution system. *IEEE Transactions of Ultrasonics Ferroelectrics and Frequency Control* 55: 214–224.

51 Chen, Y., Qiu, W., Lam, K. et al. (2015). Focused intravascular ultrasonic probe using dimpled transducer elements. *Ultrasonics* 56: 227–231.

52 Kim, D.H., Lee, T.G., Cho, S.H. et al. (2017). Piezoelectric properties of (Na1−xKx)NbO3-based lead-free piezoelectric ceramics and their application in knocking sensor. *Journal of the American Ceramic Society* 100: 5367–5373.

53 Matsuoka, T., Kozuka, H., Kitamura, K. et al. (2014). KNN-NTK composite lead-free piezoelectric ceramic. *Journal of Applied Physics* 116: 154104.

54 Li, L., Jie, X., Liu, J., and Gao, F. (2018). Recent progress on piezoelectric energy harvesting: structures and materials. *Advanced Composites and Hybrid Materials* 1: 478–505.

55 Todaro, M.T., Guido, F., Mastronardi, V. et al. (2017). Piezoelectric MEMS vibrational energy harvesters: advances and outlook. *Microelectronic Engineering* 183: 23–36.

56 Priya, S. and Inman, D.J. (2009). *Energy Harvesting Technologies*. Springer.

57 Yan, X., Zheng, M., Hou, Y.D., and Zhu, M.K. (2017). Composition-driven phase boundary and its energy harvesting performance of BCZT lead-free piezoelectric ceramic. *Journal of the European Ceramic Society* 37: 2583–2589.

58 Won, S.S., Lee, J., Venugopal, V. et al. (2016). Lead-free Mn-doped $(K_{0.5},Na_{0.5})NbO_3$ piezoelectric thin films for MEMS-based vibrational energy harvester applications. *Applied Physics Letters* 108: 232908.

59 Deutz, D.B., Mascarenhas, N.T., Schelen, J.B.J. et al. (2017). Flexible piezoelectric touch sensor by alignment of lead-free alkaline niobate microcubes in PDMS. *Advanced Functional Materials* 27: 1700728.

60 Jeong, C.K., Park, K.I., Ryu, J. et al. (2014). Large-area and flexible lead-free nanocomposite generator using alkaline niobate particles and metal nanorod filler. *Advanced Functional Materials* 24: 2620–2629.

61 Maurya, D., Peddigari, M., Kang, M.-G. et al. (2018). Lead-free piezoelectric materials and composites for high power density energy harvesting. *Journal of Materials Research* 33: 2235–2263.

62 Machado, S.P., Febbo, M., Rubio-Marcos, F. et al. (2015). Evaluation of the performance of a lead-free piezoelectric material for energy harvesting. *Smart Materials and Structures* 24: 115011.

63 Gao, M., Li, L., Li, W. et al. (2016). Direct writing of patterned, lead-free nanowire aligned flexible piezoelectric device. *Advanced Science* 3: 1600120.

64 Kim, B.Y., Lee, W.H., Hwang, H.G. et al. (2016). Resistive switching memory integrated with nanogenerator for self-powered bioimplantable devices. *Advanced Functional Materials* 26: 5211–5221.

65 Baek, C., Yun, J.H., Wang, J.E. et al. (2016). A flexible energy harvester based on a lead-free and piezoelectric BCTZ nanoparticle–polymer composite. *Nanoscale* 8: 17632–17638.

66 Shen, Z.-Y., Li, J.-F., Chen, R. et al. (2011). Microscale 1–3-type (Na,K) NbO_3-based Pb-free piezocomposites for high-frequency ultrasonic transducer applications. *Journal of the American Ceramic Society* 94: 1346–1349.

67 Jiang, L., Chen, R., Xing, J. et al. (2019). Fabrication of a (K,Na)NbO_3-based lead-free 1–3 piezocomposite for high-sensitivity ultrasonic transducers application. *Journal of Applied Physics* 125: 214501.

68 Wu, J.E., Gao, X.Y., Chen, J.G. et al. (2018). Review of high temperature piezoelectric materials, devices, and applications. *Acta Physica Sinica* 67: 207701.

69 Damjanovic, D. (1998). Materials for high temperature piezoelectric transducers. *Current Opinion in Solid State & Materials Science* 3: 469–473.

70 Zhang, S., Xia, R., Lebrun, L. et al. (2005). Piezoelectric materials for high power, high temperature applications. *Materials Letters* 59: 3471–3475.

71 Lee, M.H., Kim, D.J., Park, J.S. et al. (2015). High-performance lead-free piezoceramics with high curie temperatures. *Advanced Materials* 27: 6976–6982.

Index

Lead-Free Piezoelectric Materials, First Edition. Jing-Feng Li.

p

q